5GtoB From Theory to Practice

Liang Song · Mingsong Chen · Zhibing Xu

5GtoB From Theory to Practice

中国工信出版集团
電子工業出版社
PUBLISHING HOUSE OF ELECTRONICS INDUSTRY
http://www.phei.com.cn
Springer

Liang Song
Fudan University
Shanghai, China

Mingsong Chen
East China Normal University
Shanghai, China

Zhibing Xu
Huawei
Shanghai, China

ISBN 978-981-97-8452-3 ISBN 978-981-97-8453-0 (eBook)
https://doi.org/10.1007/978-981-97-8453-0

Jointly published with Publishing House of Electronics Industry
The print edition is not for sale in China (Mainland). Customers from China (Mainland) please order the print book from: Publishing House of Electronics Industry.
ISBN of the Co-Publisher's Edition: 978-712-14-3748-9.

This Springer imprint is published by the registered company Springer Nature Singapore Pte Ltd.
The registered company address is: 152 Beach Road, #21-01/04 Gateway East, Singapore 189721, Singapore

Foreword by Hequan Wu

The year 2022 marks the third year of 5G commercialization. In April, the number of 5G users in China reached 413 million, accounting for 24.83% of the total number of mobile communication users. 5G significantly enhances the broadband download rate of mobile communication. By the end of 2021, the average download rate for mobile broadband users in China reached 59.34 Mbps, which was close to the average download rate of fixed-line users (66.55 Mbps) during the same period. The download rate for 5G users reached 146.34 Mbps, approximately four times that of 4G users (36.33 Mbps) during the same period. However, these figures do not fully reflect the true potential of 5G in its ultra-wideband, wide-coverage, low-latency, and large-scale connectivity. On the one hand, in terms of consumer Internet applications, it is challenging to showcase the difference from 4G due to the limited availability of ultra-high-definition video content. Although 5G supports VR/AR and can consume ultra-wideband, it faces difficulties in gaining traction due to the scarcity of corresponding programs and the absence of mature and convenient VR/AR headsets. On the other hand, the industrial Internet is considered the "second half" of the Internet. The key performance indicators (KPIs) of 5G are designed with industrial Internet applications in mind. However, the fragmented nature of industrial control system standards and the lack of openness among original industrial enterprises have raised the entry barrier for 5G. It is challenging to directly transfer the 5G network architecture, terminal functions, service applications, and business models that are effective in the consumer Internet to the industrial Internet. Although 5G finds some applications in areas like machine vision and remote control, its implementation in the industrial sector has not yet formed a systematic approach and has not played a significant role in the core link of the industrial chain.

From 5GtoC to 5GtoB, the expansion of application fields requires improvement in capabilities. While consumer Internet users have common applications, industrial Internet applications are personalized and often have high-performance requirements. 5GtoC primarily focuses on ultra-wideband downlink bandwidth, whereas 5GtoB applications require large uplink bandwidth reaching Gbps. Moreover, while 5GtoC aims for low air interface latency, 5GtoB applications demand low end-to-end delay and even deterministic delay. In terms of positioning capabilities, 5GtoC can achieve high-precision indoor and outdoor positioning using multiple base stations. However, 5GtoB applications aspire to achieve high-precision positioning in a single base station environment with rapid response. While 5GtoC can provide reliability through multi-base station cooperation, certain 5GtoB scenarios may rely only on the signal from a single base station, necessitating ultra-high reliability guarantees of up to 99.9999%. 5GtoB also has higher network and information security requirements compared to 5GtoC, with enterprises being particularly sensitive to the protection of internal data. Additionally, 5GtoB allows for better meeting of the needs for large-scale connectivity in expanding the application of the IoT through technologies like integrated access and backhaul (IAB) and direct communication between terminals (D2D). Despite the strict requirements of 5GtoB, its application scenarios possess specific characteristics that can be leveraged to improve network architecture. For example, the core network of 5GtoB can be sunk, with some core network functions integrated into base stations, enabling a flattened network architecture and facilitating low latency. In contrast to 5GtoC terminals, 5GtoB terminals do not mandate power consumption or multi-frequency and multi-mode universality. Instead, they prioritize the number and coverage of connected devices, computing, storage and programming capabilities, embedded edge computing, smart network functions, and stricter working environment and intrinsic safety requirements. Furthermore, if 5GtoB can integrate the functions of industrial control equipment such as PLC, and function as a new industrial control gateway, it could realize the integration of IT and OT, solving the issue of incompatible standards and simplifying enterprise network hierarchies. While 5GtoC applications focus on switching, roaming, channel benefits, and billing management, 5GtoB prioritizes reliability over economic channel benefits. It can establish redundant, concurrent, and always-online channels to ensure safety and reliability. In summary, 5GtoB represents a profound transformation in the combination of existing 5G and enterprise network technologies, catering to the characteristics and requirements of industrial applications. It opens up new opportunities for integrating 5G with IPv6, cloud computing, artificial intelligence, and other technologies, signifying a technological breakthrough in the "second half" of the Internet.

This book introduces both the theory and practice of 5GtoB, focusing on the innovation chain, industrial chain, and industrial applications. Based on Huawei's experience in promoting the application of 5G in industries, this book contains rich case studies and presents the content in a novel and simple manner, making it practical and valuable as a unique teaching and engineering reference book. Although this book aims for promoting 5G innovation and industrial applications, since 5GtoB is still in its infancy, different enterprises have varying volumes and digital foundations, leading to diverse requirements ranging from digitalization and networking to intelligence and low carbonization. Digital transformation necessitates not only technological innovation but also management innovation and process reengineering. Considering the wide-ranging and complex applications, it is difficult for this book to comprehensively cover all aspects. All in all, it is hoped that this book will attract more scientific and technological workers to contribute to the research, development, and application promotion of 5GtoB.

Beijing, China

Hequan Wu
Academician of the Chinese Academy of Engineering

Foreword by Chongqi Huang

In the development of mobile communication system standards from 1G to 5G, transitioning from analog to digital systems and from narrowband voice services to broadband mobile interconnection services, 5G stands out as the first mobile communication system that is truly oriented toward industrial applications. If we consider that from 1G to 4G, the global telecommunication industry was focused on building an information superhighway, improving the bandwidth and coverage of mobile communications, and reducing costs, then the development of the 5G standard, starting from 3GPP Rel-15, placed significant emphasis on meeting the comprehensive service needs of numerous industrial applications. Whether it is the initial enhanced mobile broadband (eMBB), massive machine-type communications (mMTC), ultra-reliable low-latency communications (URLLC), or the additions made in 5.5G such as uplink ultra-broadband (UCBC), real-time broadband interaction (RTBC), and communication perception convergence (HCS), all of these components embody the development trend of the new generation of mobile communication systems for industrial applications, which can be flexibly customized.

5G to Business (5GtoB) not only means that mobile communication systems should be flexibly customized according to the needs of industrial applications but also bring great opportunities and challenges to the development of each industrial application itself. Whether it is autonomous driving or unmanned factories, telemedicine, future education, power cables, or rail transportation, the innovative applications brought by 5G to the industry signify the empowerment and transformation of the entire industry. The synergistic optimization of the mobile communication system and industrial application system integrates network and computing, representing an unprecedented event for both the telecom industry itself and the empowerment of numerous industries. This new round of native driving force is bringing technological innovation and industrial upgrading to the entire electronic information world.

5GtoB will have a profound impact on the existing telecom service and operation models. The IT departments of traditional enterprises need to transform toward the ICT direction. This means that some industrial enterprises will gradually require their own employees to be responsible for maintaining their 5G network, which poses a new challenge to talent cultivation and management within the enterprise. In light of this new era, this book provides an in-depth summary of the current network architecture and key technologies, intelligent terminals and service capabilities, major industrial application scenarios, and successful cases involved in 5GtoB. To promote the development of the 5GtoB industry ecosystem, this book focuses on training ICT talents within enterprises with innovative 5G applications, which can play a positive and critical role. Additionally, this book is recommended to scientific and technological workers and researchers across various related industries.

The great thing about this book is that its content covers theory and practice, and it is published at the right time. Engineers and technicians in the manufacturing industry require new knowledge, and enterprises require new technology for upgrading and transformation. Therefore, it is hoped that relevant individuals can read this book as soon as possible, and that the dissemination of 5GtoB-related theories and application technologies will continue to nourish the growth of all things empowered by 5G.

Beijing, China

Chongqi Huang
Academician of the Chinese Academy of Engineering

Foreword by Guodong Wang

With the development and application of new-generation information technologies such as industrial Internet, cloud computing, big data, and artificial intelligence gradually becoming mature, human society has entered the digital era. In the digital era, data has become the most important element of the production, and the data mining method of "data + machine information" provided by data science has become a new tool for human beings to understand the world. As a result, the digital economy is becoming a key force in restructuring global factor resources, reshaping the global economic structure, and changing the pattern of global competition.

Against this backdrop, digital transformation is imperative. We must promote the integrated development of the digital economy and the real economy, grasp the direction of digitization, networking, and intelligence, and promote the digitization of manufacturing, services, agriculture, and other industries to achieve the optimal allocation of enterprise resources. In this way, the right data can be delivered to the right people and machines at the right time and in the right way, thus achieving data flow automation. From the perspective of data flow, digitalization solves the problem of "having data," networking solves the problem of "being able to flow," and intelligence solves the problem of "automatic flow."

Networking can interconnect units within the industrial Internet and other network systems. When applied to industrial production scenarios, there are special requirements for network performance, such as network connection delay, reliability, network flexibility, and power consumption. In addition, the challenges of heterogeneous network convergence, service support efficiency, and intelligence must be solved. Each device, module, unit, enterprise, and other entities that constitute the industrial Internet should have ubiquitous connectivity and achieve cross-network, cross-industry, heterogeneous multi-technology convergence, and synergy to guarantee the free flow of data within the system. Ubiquitous connectivity provides ubiquitous optimized decision-making and intelligent services through real-time acquisition and transmission of the physical world state and real-time feedback of control instructions from the information world.

The iron and steel industry involves a large and complex process, where the whole process is a "black box," so real-time information is extremely scarce; the iron and steel production process is characterized by multi-variable, strong coupling, nonlinearity, and large hysteresis; and each unit adopts an isolated control strategy, so there is not yet a seamless and precise interface between the units. These serious uncertainties are the major challenges facing iron and steel production. However, the iron and steel industry has rich application scenario resources and data resources. The iron and steel industry has a good informatization foundation, well-developed data acquisition systems, automated control systems and R&D facilities, which can generate a large amount of experimental data (historical, real-world, and production big data from laboratory experiments, pilot experiments, and production line experiments). These massive data contain all the laws of the enterprise's production process, and they are the most valuable resources and key production factors. Therefore, the iron and steel industry has great potential in fully and rationally utilizing the rich data resources and realizing digital industrialization.

5G is an advanced information and communication technology with three characteristics; enhanced mobile broadband (eMBB) is suitable for coping with the explosive growth of Internet traffic; ultra-high reliable low-latency communication (URLLC) is suitable for the application needs of vertical industries with extremely high requirements for latency and reliability; and massive machine-type communication (mMTC) is geared toward the needs of applications that are targeted at sensing and data collection. These features are particularly suited to the specific needs of the iron and steel industry.

As a typical process industry, each process in the iron and steel production process should be handled as a vertical industry, from the bottom layer to the edge, and then to the upper layer of the cloud platform. In order to handle massive data, massive machine class communication is an important hard requirement. In the vertical direction, the massive data generated from the bottom layer of the production line can be processed using a big data/machine learning platform and a high-fidelity digital twin model can be built to replace the traditional machine model. High-precision setting calculations are performed on the edge side to complete high-precision real-time control with high certainty and low latency. The highly reliable, low-latency communication down to a few milliseconds provides strong support for the real-time control of each edge and production line. At the platform layer of the infrastructure cloud, extremely complex enterprise resource deployment and management is carried out using massive data information from the bottom of the workshop, such as production planning, scheduling management, logistics, energy, equipment, safety, and so on, as a way to ensure stable, efficient, safe, and high-quality operation of the enterprise as a whole.

5G is an enhanced mobile Internet with a peak transmission rate of up to 20 Gbps, which will be relied upon for the transmission and processing of huge amounts of data between manufacturing units in the process direction, the transmission and processing of unstructured data, and the transmission and processing of virtual reality and augmented reality. The harsh environment, high temperature, and hazardous, repetitive on-site operating positions in the iron and steel production process urgently need remote and automated operation and maintenance. Unstructured data captured through mobile phones/inspectors and other audio/video can be used for real-time analysis, computing, monitoring, and management of equipment operating status. In remote assembly scenarios, technical experts rely on AR for real-time annotation, audio and video communication, desktop sharing, and other technologies to remotely guide production line assembly work. The application of 5G not only promotes the development of detection, processing, transmission, and control technologies for unstructured data, such as images, sound, and video but also provides a powerful development impetus for the analysis, decision-making, and control of complex working conditions.

In recent years, the 5G slicing scheme that has been proposed above innovation, guarantees unified infrastructure to adapt to differentiated business needs, which is the key to entering the vertical industry. 5G, with its multi-access edge computing (MEC) function, deploys multiple forms of access functions, content, applications, etc. to the edge of the network close to the access side, sinks the core network user interface and applications to a location closer to the user, and reduces the latency to milliseconds, thus ensuring low-latency real-time interaction between the edge and the underlying layer in the vertical direction. The new intelligent edge cloud, combined with 5G, can empower new applications and services, and network performance, security, and privacy protection can be improved. 5G low-latency big data transmission, sliced network architecture, multi-access edge computing, and intelligent edge cloud provide a powerful driving force for the application and development of the Internet in the iron and steel industry.

Huawei has summarized its engineering practice of "5G + industrial Internet" in industrial enterprises and organized experts and scholars in related fields to write 5GtoB From Theory to Practice. This book systematically introduces the latest 5G technology theories and industrial application achievements, which are of great value for further exploiting the advantages and value of 5G, continuously deepening and expanding industrial applications, broadening the knowledge horizons of related professional students in higher education institutions, and enhancing self-confidence in self-owned technologies.

Beijing, China

Guodong Wang
Academician of the Chinese Academy
of Engineering

Preface

With the vigorous development of technologies such as digital twinning, smart cities, and industrial Internet, the Fifth Generation Mobile Communication System (5G), as a leading technology independently controllable by China, is playing an increasingly important role in digital transformation. As a significant component of the construction of new infrastructure in the era of smart economy, abbreviated as "New Infrastructure" (NI), 5G is characterized by high speed, low latency, and high connectivity. It has been applied in enhancing mobile broadband, ultra-reliable low-latency communications, massive machine-type communications, and other business scenarios, gradually achieving ubiquitous intelligent interconnection integrating "cyber-physical-human systems." By the end of 2021, China has constructed 1.425 million active 5G base stations. Accelerating the construction of new infrastructures like 5G and actively enriching 5G technology application scenarios has become an overarching trend.

Currently, 5G technology has achieved large-scale commercial deployment across various sectors, covering multiple industrial stages, such as research and development, production, operation management, and product services. While greatly reducing labor costs, it has enhanced production efficiency and commercial value. Broadly speaking, 5G has commenced an extensive integration with the industrial economy nationwide, empowering the digitization, networking, and intelligence of China's industries and other sectors. At present, China has witnessed the emergence of a multitude of innovative fusion applications like 5G+ smart factories, 5G+ smart grids, and 5G+ intelligent ports, accelerating the digital transformation of numerous industries. Recognizing the significance of 5G for the economy, the Ministry of Industry and Information Technology, along with ten other ministries, jointly issued the *"Set Sail" Action Plan for 5G Applications (2021–2023)*, providing policy support to strengthen the pivotal role of enterprises in 5G application development. These favorable signals will further unleash the potential demand for 5G applications in consumer markets, vertical industries, and social livelihoods, among other aspects, stimulating innovation vitality in 5G applications. It can be seen in the future that 5G technology will empower new modes of production and operation in sectors such as industry, transportation, energy, education, healthcare, cultural tourism, information consumption, finance, agriculture, and public security, among others, facilitating the profound integration of information technology (IT), communication technology (CT), and operational technology (OT), thus robustly propelling industry transformation and upgrading.

With the rapid development of communication and computing infrastructure, such as the construction of new infrastructure in China, the integration of 5G and artificial intelligence (AI) will emerge as a significant national requirement for enhancing productivity in new infrastructure services. In the future, communication networks will be seamlessly integrated with AI, establishing the groundbreaking ecosystem over Networking Systems of AI (NSAI), i.e., "Networks as AI Systems, and AI as Networking Systems." Diverse intelligent terminals connecting "man-machine-thing" will be unified into a cohesive entity, empowering the development of human society. Ultimately, an interconnected society will come into being where the physical world, cyberspace, and human society blend harmoniously—a pervasive fusion system of artificial intelligence networks and human intelligence. In alignment with industrial development trends and crucial national demands, the theory and practice of 5GtoB represent a solid initial step toward a future interconnected society.

Most of the 5G-related books available in the market mainly focus on introducing basic 5G technologies, making it difficult for readers to gain a profound understanding of the characteristics of 5G. Most readers have difficulties independently deploying differentiated business scenarios after the study. This book is a textbook for application-oriented undergraduate education or as training material for 5G-related certifications. It targets ordinary readers interested in engineering and technology, striving to minimize redundancy with traditional computer network knowledge. The book concentrates on the novel service-oriented architecture and design methodologies of 5G, integrating scenario-based design solutions and industry cases. This approach enables readers to grasp the potential of IT technologies comprehensively and 5G network communications. Besides, it can be used as a reference for R&D and industry professionals engaged in the development of 5G network applications within fields like networking and communication, computer science, software engineering, electronic engineering, and related domains. Readers are only required to possess fundamental mathematical and computer network knowledge without requiring prerequisites from other courses.

The book consists of eight chapters, primarily covering two main sections. The first section (the first five chapters) emphasizes fundamental knowledge related to 5G. It mainly expounds on the 5G end-to-end network architecture, 5GtoB software and hardware technologies, and basic service capabilities. This section encompasses the 5G applications in emerging fields such as Internet of Things (IoT), cloud computing, and artificial intelligence. The second section (the last three chapters), in conjunction with the previous concepts, highlights the scaled applications of 5G technology in various sectors. By introducing core theories, technologies, and exemplary cases of 5G, this book aims to provide readers with a comprehensive understanding of 5G technology from theory to practice. It endeavors to assist readers inefficiently and logically constructing 5G application scenarios suitable for their respective domains.

Finally, we extend our heartfelt gratitude to the teams from Fudan University, East China Normal University, Huawei Technologies Co., Ltd., and others for their significant contributions to the creation of this book. We would like to express our appreciation to Zhejiang Huawei Communication Technology Co., Ltd. for providing resources and support throughout the writing process. We are deeply grateful to Academician Hequan Wu, Academician Chongqi Huang, and Academician Guodong Wang for graciously reviewing the manuscript amidst their busy schedules and providing forewords for this book. Special thanks to Yun Ding, President of Carrier BG at Huawei Technologies Co., Ltd., Weidong Qian, Scientist at China Baowu Steel Group Corporation, and Ping Zhong, General Manager of China Telecom Tianyi IoT Technology Co., Ltd., for reviewing the manuscript and offering recommendations for this book. We acknowledge the diligent efforts of the editorial committee colleagues whose dedicated work was pivotal in the timely completion of this book. Without their contributions, this book would have been challenging to complete promptly. Additionally, we appreciate the substantial support received from teams such as the Publishing House of Electronics Industry and Zanwei Information Technology—without their diligent efforts, this book could not have been published so efficiently. We would also like to thank numerous behind-the-scenes friends who provided invaluable assistance in the writing process, although they cannot all be mentioned here. Our heartfelt gratitude goes out to them collectively!

Special note: Readers requiring the teaching materials and exercise answers for this book can download them from the website of Huaxin Educational Resources (https://www.hxedu.com.cn).

Due to the limited time available for compilation, this book may inevitably contain deficiencies and oversights. Your constructive criticism and feedback are sincerely welcomed and appreciated.

Shanghai China
June 2024

Liang Song
Mingsong Chen
Zhibing Xu

Acknowledgements

This is to acknowledge the work that the translating team has achieved in the composition of the English version of this book. In this team, those who have proofread the English version are Jingjing Huang from China Electronics Standardization Institute who is the third reviewer, Jun Huang from Anhui University of Technology and Xiaolan Li from Taizhou Institute of Science and Technology, NUST, who are the second reviewers, Jingjing Xu, Xue Han, Guixiang Xu, Ming Wang, Mengyao Liu, Yueting Tang, Zhaodong Wang, and Xinran Wang from Taizhou Institute of Science and Technology, NUST, are the first reviewers. Those who have translated the first draft are Chenchen Liu, Jing Wu, Yuge Chen, Shiyi Fang, Xinchen Zhang, and Yifei Wang from Anhui University of Technology. Ruilan Cao, who is in charge of the final review, would also like to thank Dongyang Li, the project manager of the translation, and all the authors of this book who have contributed their time in proofreading the English version to ensure its accuracy.

Translators

Chief Senior Translator

Ruilan Cao, Anhui University of Technology

Associate Senior Translators

Jingjing Huang, China Electronics Standardization Institute
Xiaolan Li, Taizhou Institute of Science and Technology, NUST
Jun Huang, Anhui University of Technology
Jingjing Xu, Taizhou Institute of Science and Technology, NUST

Xue Han, Taizhou Institute of Science and Technology, NUST
Guixiang Xu, Taizhou Institute of Science and Technology, NUST
Ming Wang, Taizhou Institute of Science and Technology, NUST
Mengyao Liu, Taizhou Institute of Science and Technology, NUST
Yueting Tang, Taizhou Institute of Science and Technology, NUST
Zhaodong Wang, Taizhou Institute of Science and Technology, NUST
Xinran Wang, Taizhou Institute of Science and Technology, NUST

Assistant Translators

Chenchen Liu, Anhui University of Technology
Jing Wu, Anhui University of Technology
Yuge Chen, Anhui University of Technology
Shiyi Fang, Anhui University of Technology
Xinchen Zhang, Anhui University of Technology
Yifei Wang, Anhui University of Technology

Contents

Chapter 1
Introduction

The mobile communication system has evolved into its fifth generation. Each generation has demonstrated the progress of technology and brought better experience in mobile communication services to mankind. It is also a booster of the rapid development of multiple industries in the global arena. Compared with the previous generations, the Fifth Generation Mobile Communication System (5G) bears strong network capabilities in different dimensions, such as ultra-high rate, ultra-low latency and hyperscale connectivity, all industry-oriented.

This chapter mainly introduces the progress of each mobile communication system, the driving force in 5G development, the process of 5G standardization, the development of 5G industrial chain and its commercial use worldwide.

Learning Objectives

- To know the development of each generation of mobile communication systems.
- To understand the background of 5G development and its main application scenarios.
- To master the process of 5G standardization.
- To know the development of 5G industrial chain and its commercial use worldwide.

1.1 Evolvement of Mobile Communication Systems

The mobile communication technology has been developing rapidly since the 1980s. Meanwhile, people's demand for such services has also increased. As is shown in Fig. 1.1, the mobile communication system has experienced a revolution from its first to fifth generations to meet people's ever-growing demands for mobile communication services.

L. Song et al., *5GtoB From Theory to Practice*,
https://doi.org/10.1007/978-981-97-8453-0_1

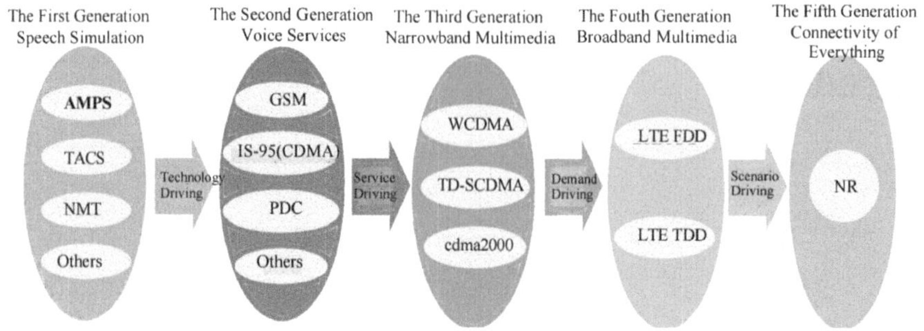

Fig. 1.1 Evolvement of mobile communication systems

The First Generation Mobile Communication System (1G) has been used commercially since the 1980s, with the help of analog telecommunication technology based on wireless radio waves. The system achieved analog voice call services through portable terminals. In this way, people were finally freed from the cables that had been fixed. They could communicate with each other even when they were on the move. The first generation is also called “1G”. Although it has helped people step into the era of mobile communication, “1G” was characterized by poor connection quality, lower security, and costly expenses due to the defects and restrictions of analog technology.

With the development of digital communication technology, the Second-Generation Mobile Communication System (2G), based on digital communication, was available commercially by the 1990s. Compared with analog communication, digital communication is a lot safer and has higher quality. Mobile terminals then have more powerful functions. They can be used to make digital voice phone calls and send short messages. In addition, compared with the first portable phones, they are smaller in size and easier to carry. People call them “cell phones”.

In the following ten years, the demand for mobile data services is mounting with the development of mobile Internet. In the early twenty-first century, the Third Generation Mobile Communication System (3G) was born. 3G can support not only digital voice services but high-speed data services, such as web browsing, online video, etc. With the ever-improving technology and widespread 3G network, mobile terminals were transformed into “smart phones”. Various applications can be installed on smart phones, such as games, maps, videos and entertainment, news, etc. These applications have made smart phones very powerful.

3G has brought great convenience to people’s daily lives and diversified online experience. At the same time, there has been an explosive increase in the number of mobile users and the demand for high-speed data services. In the early twenty-first century, the Fourth Generation Mobile Communication System (4G) was put into commercial use. This system is equipped with a smarter terminal and greater network bandwidth, making it possible for users to watch videos online and enjoy “HD digital voice” service, a concurrence of internet and voice service.

While personal users have an increasing demand for network and service, industries crave mobile communication under different scenarios, and communication of ultra-high rate and ultra-low latency. The Fifth Generation Mobile Communication System (5G) came into being to match these demands. The communication scenarios have been transformed from traditional personal communication to a world in which everything can be interconnected with each other. Mobile terminals have been extended from cell phones to smart terminals such as smart equipment, Virtual Reality (VR) glasses, and unmanned vehicles.

1.1.1 The First-Generation Mobile Communication System

1G was first born in the 1940s. It was first applied to the vehicular wireless radio system used by polices in Detroit, US. The system mainly adopted the simulation technology of large-area systems. By the end of 1978, the Advanced Mobile Phone System (AMPS) was developed successfully in Bell Labs and then the cellular mobile communication network was built, which represents the first truly cellular mobile communication system with a large capacity for instant communication. In 1983, AMPS was put into commercial use for the first time in Chicago. Later, it has been promoted rapidly and extensively. By 1985, AMPS applications have been promoted to 47 regions.

At the same time, other countries have developed their cellular mobile communication networks in succession. In 1985, UK developed Total Access Communication System (TACS) which adopts 900 MHz frequency band. Canada launched 450 MHz Mobile Telephone System (MTS). In 1980, the four Nordic countries, including Sweden, developed a Nordic Mobile Telephone (NMT) mobile communication network that adopts a 450 MHz frequency band. On 18 November 1987, right before the Sixth National Games in Guangdong, China's 1G system, which adopts the TACS system, was initiated and officially used in business. From November 1987, when China Telecom began to operate analog mobile phone services, to the end of December 2001, when China Mobile closed its analog mobile communication network, 1G system has been applied in China for 14 years with a peak number of its users hitting 6.6 million. The terminal used in the1G era (Fig. 1.2) has become history.

Since 1G system is based on analog telecommunication technology, it has many weaknesses, such as low spectrum utilization, poor confidentiality of system security, difficulties in carrying out data bearer services, high cost of equipment, large volume, high expenses, etc. The most critical problem lies in the low system capacity which cannot meet the increasing requirements of mobile users. Hence Second-Generation, the Second-Generation Mobile Communication System (2G) has been able to solve these problems.

Fig. 1.2 1G terminal

1.1.2 The Second-Generation Mobile Communication System

In the middle of the 1980s, Europe took the lead in launching Global System for Mobile Communications (GSM), a digital communication system. Afterwards, US and Japan also established their digital communication systems, respectively.

2G includes GSM, IS-95 Code Division Multiple Access (CDMA), Personal Digital Cellular (PDC), and Digital Advanced Mobile Phone System (DAMPS). The GSM system, in particular, has become the main public land mobile network public land mobile network for its open system, mature technology, and extensive application.

The GSM system using a 900 MHz frequency band was called GSM900, while the one using an 1800 MHz frequency band was called DCS1800. It is designed according to the Time Division Multiple Access (TDMA) global digital cellular communication standard. GSM supports low-speed data service and can be interconnected with the Integrated Services Digital Network (ISDN). GSM adopts Frequency Division Duplex (FDD) and Time Division Multiple Access (TDMA) modes. Each frequency band supports eight channels with 200 kHz bandwidth. With the introduction of General Packet Radio System (GPRS) and Enhanced Data Rate for GSM Evolution (EDGE), GSM network has been enhanced. The network was at its initial stage of supporting multi-media services, which can send pictures and e-mails.

IS-95 is the standard of North America digital cellular network, which uses 800 MHz frequency band or 1.9 GHz frequency band, using Code Division Multiple Access (CDMA) with CDMA One, the brand name. CDMA 2000, the standard of wireless communication, has evolved from IS-95. It can be divided into two phases: IS-95A and IS-95B.

PDC is the standard proposed by Japan, namely Personal Handy-phone System (PHS) in China afterward, which is commonly known as "Personal Handphone System (PHS)" network. However, PHS network has already been shut down due to outdated technology and demands for the follow-up development of mobile communication.

2G system in China is mainly GSM network. For example, China Mobile and China Unicom both deployed GSM networks. In 2001, China Unicom began to deploy the IS-95 CDMA network (C network for short). In May 2008, China Telecom took over C network from China Unicom and made it a priority for future development.

Voice service is the main service of 2G system, which mainly provides digital voice service and low-speed data service. Compared with the first-generation analog communication system, a digital communication system is endowed with high spectrum effectiveness, large capacity, various categories of service, high security, high voice quality, and strong capabilities in network management. Meanwhile, the application of digital technology has made 2G terminals smaller. Examples of 2G terminals are shown in Fig. 1.3, which shows that 2G terminal is easier for users to carry around.

Fig. 1.3 Examples of 2G terminals

Although 2G has been improved in its development, people still have a higher demand for mobile data service. They hope to enjoy the rates similar to those of fixed broadbands, even when they are on the move. Therefore, a new generation of mobile communication technology is needed to provide high-speed aerial bearing to provide colorful high-speed data services, such as movie on-demand, profile downloading, video calls, online games, etc. Hence, the third-generation of mobile communication system met these new service demands.

1.1.3 The Third-Generation Mobile Communication System

The Third-Generation Mobile Communication System (3G) was officially named IMT-2000 by International Telecommunication Union (ITU) in 1996, which refers to the international mobile communication system that was put into commercial use around the year 2000, operating on the 2000 MHz frequency band with 2000 kbps downlink rate. The standardization of IMT-2000 began in 1985. The specifications of 3G standards were under the charge of the 3rd Generation Partnership Project (3GPP) and the 3rd Generation Partnership Project 2 (3GPP2).

Initially, there are three main standards for 3G systems, namely Wideband Code Division Multiple Access (WCDMA) proposed by Europe and Japan, Code Division Multiple Access 2000 (CDMA2000) by the US, and Time Division-Synchronous Code Division Multiple Access (TD-SCDMA) by China. Among them, the 3G WCDMA standard of 3GPP had been set from Release 99 and the TD-SCDMA standard from Release 4. Subsequent versions (Release 5, Release 6, Release 7, etc.) with enhancements further promoted the performance of uplink and downlink rates. 3GPP2 put forward the evolving strategy of IS-95: (CDMA) (2G) → CDMA2000 1x → CDMA2000 3x (3G).

The 3G system adopts CDMA and packet switching instead of TDMA and circuit switching often used by 2G systems. Compared with 2G terminals, the functions of 3G terminals are more powerful with the appearance of "smart phones" (Fig. 1.4). In terms of service and performance, 3G is capable of transmitting voices and data and providing high-quality multi-media services, such as the data of variable rates, mobile videos, and HD images. In doing so, it can integrate information, thus providing a fast and convenient wireless application.

Although the 3G system is endowed with low cost, quality service, high confidentiality, and good safety performance, it still has weaknesses:

First, 3G standard has three branches: WCDMA, CDMA2000 and TD-SCDMA. However, these systems are incompatible with each other.

Second, the low spectrum utilization rate of 3G systems makes it impossible to fully utilize valuable spectrum resources.

Third, 3G speed is not fast enough.

These shortcomings are far from meeting the needs of the development of mobile communication, so a new system is needed to embrace the future better.

Fig. 1.4 An example of 3G terminals

1.1.4 The Fourth-Generation Mobile Communication System

Since the international standard of 3G system was set in 2000, ITU has started relevant work on the Fourth-Generation Mobile Communication System (4G). In 2008, ITU began to collect 4G standards publicly. Three programs became the alternatives, including Long Term Evolution (LTE) of 3GPP, Ultra Mobile Broadband (UMB) of 3GPP2 and WiMAX (IEEE 802.16 m, also known as Wireless MAN-Advanced or WiMax2) by the Institute of Electrical and Electronics Engineers (IEEE), among which LTE is the most favored by the industrial community. Although LTE, UMB, and mobile WiMAX are different, they still have something in common: they all employ Orthogonal Frequency Division Multiplexing (OFDM) and Multiple-Input Multiple Output (MIMO) technologies to offer higher spectrum utilization rates. Therein, the formulation of LTE standardization was started from Release 8 of 3GPP, and the following versions (Release 9, Release 10, etc.) were enhanced and improved in their performances.

In 2012, LTE-Advanced was officially established as the international standard of IMT-Advanced (also known as 4G). Meanwhile, TD-LTE-Advanced, mainly formulated by China, became the international standard of IMT-Advanced. LTE includes LTE TDD (also known as TD-LTE, namely Time Division Duplex) system and LTE FDD (Frequency Division Duplex) system. China took the lead in developing TD-LTE, which has inherited and expanded key technologies and proprietary intellectual property rights of TD-SCDMA in areas, such as smart antenna and systematic design. In addition, its system capability is equivalent to that of LTE FDD. In October 2015, new LTE standard was officially named as LTE-Advanced Pro by 3GPP on the 35th Session of Project Coordination Group (PCG). This is the official name of the 4.5G standard.

Compared with the 3G system, 4G employs many new technologies, such as OFDM and MIMO. The larger carrier bandwidth of 4G has greatly improved 4G peak rates than those of 3G. That explains why 4G is capable of supporting media services, such as high-speed Internet access and online video playing. In addition, the 4G network also supports "HD digital voice" (carrying voice by LTE network, i.e., Voice over LTE, VoLTE) service. It can achieve the concurrence of data service and voice service. Examples of 4G terminals are shown in Fig. 1.5. With the ever-improving technologies in mobile terminals, 4G terminals have been more diversified and intelligent. They have greatly enriched people's lives.

Fig. 1.5 4G communication terminals

1.1.5 The Fifth-Generation Mobile Communication System

The Radio communication Assembly 2015 was held in Geneva, Switzerland, from 26 to 30 October 2015. At the assembly, the Radio communication Department of the International Telecommunication Union (ITU-R) officially ratified three resolutions that help to promote the research progress of 5G in the future and officially determined the legal name of 5G as "IMT-2020".

ITU has defined eight targets for 5G capabilities (shown in Fig. 1.6) to meet the requirements of different service applications for network capabilities in the future. 5G is required to meet the following standards: downlink peak rate at 20 Gbps, downlink user-experienced data rate at 100 Mbps, three times the spectrum efficiency of IMT-A, 500 km/h mobility, 1 ms latency, 1 million devices per square kilometer for connection density, 100 times the network power usage effectiveness of IMT-A, and area traffic capacity at 10 Mbps/m^2.

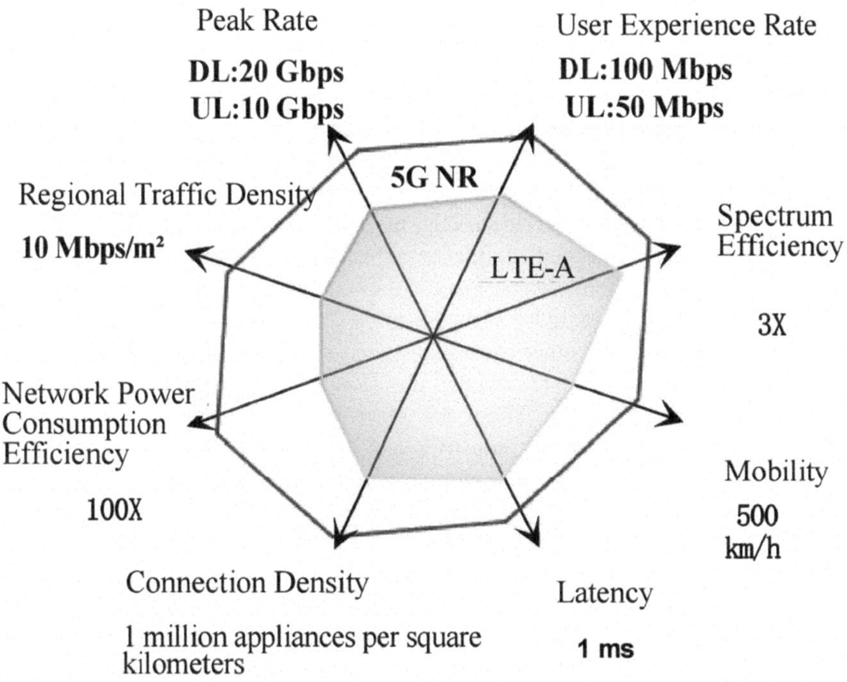

Fig. 1.6 Eight targets for 5G capabilities

1.2 Driving Forces for 5G Development

The continuous progress of human society is reflected in all aspects, such as higher productivity, safer production mode, and more diversified and convenient daily life, which cannot be separated from the non-stop development of science and technology. Meanwhile, technological progress has facilitated the continuous renewal of mobile networks, bringing more applications with powerful functions and greater user experiences, all of which can better meet personal and family needs for daily life and recreation. More importantly, these applications have an extensive impact on various fields, such as industry, agriculture, finance, energy, transportation, and even the military. Numerous industries have achieved rapid development with the support of mobile communication technology. It is a fact that the progress of mobile communication technology is the epitome of the ever-progressing human society.

1.2.1 Background of 5G Development

5G cannot be simply interprcted as a result of the updating of mobile communication systems every 10 years. Instead, its development background needs to be analyzed from multiple perspectives, such as service needs, economic development, and scientific and technological progress.

From the perspective of service needs, mobile communication services in the 4G era are mainly high-speed mobile broadband services. However, many new services will appear in mobile scenarios, and they cannot be supported by the current network. These services include: Virtual Reality (VR), Augmented Reality (AR), Mixed Reality (MR), and other ultra-high-speed services, ultra-large connection services such as large-scale IoT, as well as ultra-low latency services, such as remote control and unmanned driving. These new services will be of great importance to digital transformation in various industries and will also bring ultimate experience to people's production and life. Therefore, it will definitely gain more popularity in the future. However, the capacity of the 4G network can no longer meet the needs for these services. So, mobile communication networks must be developed to a higher level to accommodate the needs for these future services.

From the perspective of economic development, the world has been in economic stagnation for a long time, and the innovation of applications brought by the renewal of mobile communication systems will give rise to a large number of new products in many industries, which will boost the social economy and help the world step out of the economic downturn, especially in industrial production. Therefore, most countries around the world are now attaching great importance to 5G construction, with several countries (such as Germany, the US, the UK, Japan, South Korea, and other traditional industrial powers) even setting 5G construction and application as a national strategic goal. All those countries are in the hope of catalyzing a new generation of industrial revolution through the integration of information and communication technologies, such as 5G and manufacturing technologies. The Chinese government also expects to use advanced network and information technologies such as 5G to drive changes in production organization and manufacturing models, thus achieving intelligent production, networked collaboration, customization, service-oriented transformation and completing the transformation and upgrading of industrial development. In the Central Economic Work Conference held from 19 to 21 December 2018, seven new types of infrastructure construction (new infrastructure) were defined, with 5G topping the list.

From the perspective of scientific progress, on the one hand, the ever-improving mobile communication and other related technologies, such as more sophisticated modulation technology, more advanced coding technology and increasingly powerful antenna technology, ensure the unprecedentedly powerful air interface of 5G. Compared with the previous networks, the performance of the 5G network has achieved a leap in terms of rate, latency, linking number, reliability, etc.; On the other hand, the booming new technologies such as artificial intelligence, big data, and cloud computing can be integrated perfectly with 5G network because of the adoption of the Network Functions Virtualization (NFV) technology. Therefore, 5G was placed with high hopes by the international community and was seen as a good opportunity for industries to participate in the digital revolution.

All competitions in traditional communication networks are confined to the communication field, such as competitions among major network equipment manufacturers and operators. However, 5G competition is no longer a competition in the communication field; it is seen as industrial and economic competition among countries and regions. At the technical level of 5G, China is leading the world. Take 5G patents as an example (Table 1.1). According to the survey on 5G standard patents released by IPlytics, a German patent statistics company, by 1 January 2020, the number of 5G patent applications worldwide totaled 21,571 with Huawei ranking first with 3,147 applications, and two Chinese companies were among the TOP 10. In the past, China has been lagging and following the US, Europe, Japan, and Korea in mobile communication technology, but now China has completed the "overtaking" in 5G technology and is ahead of other communication powerhouses.

1.2.2 Three Major 5G Application Scenarios

As mentioned above, the demand for ultra-high speed, ultra-low latency, and ultra-large linking number service is one of the driving forces for 5G development. Accordingly, ITU has defined three major application scenarios for 5G, as shown in Fig. 1.7. They are enhanced Mobile Broadband (eMBB), Ultra Reliable and Low Latency Communications (URLLC) and massive Machine Type Communications (mMTC).

Typical services of eMBB scenarios include VR, AR, MR, etc. Such services bring users the ultimate immersive experience with ultra-high resolution, 360-degree

Table 1.1 Ranking of the number of 5G essential patents

Enterprise	Number/piece of 5G patents publishes	Enterprise	Number/piece of 5G patents published
Huawei	3147	Qualcomm	1293
Sumsung	2795	Intel	570
ZTE	2561	Sharp	747
LG Electronics	2300	NTT Docomo	721
Nokia	2149	OPPO	647
Ericsson	1494	China Academy of Information and Communications Research (CAICR)	570

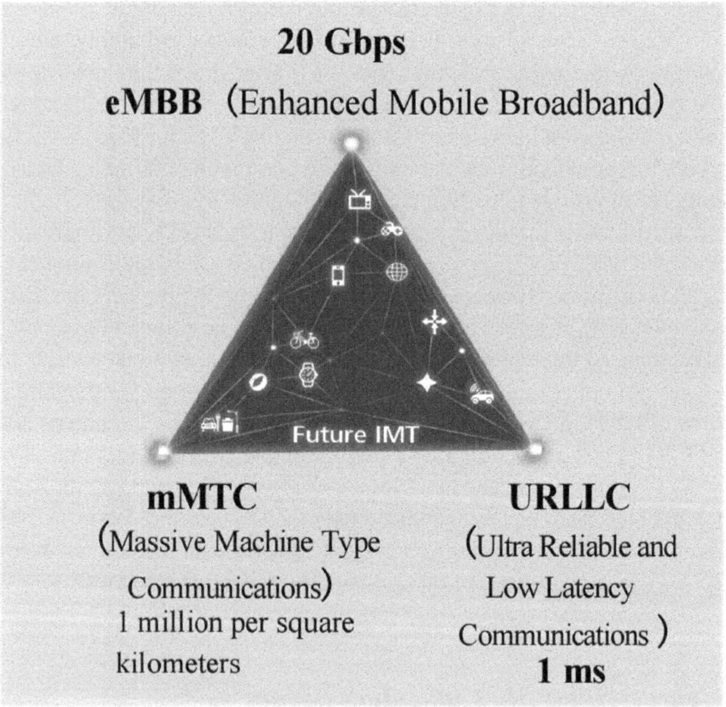

Fig. 1.7 Three major application scenarios in 5G

panoramic view and real-time interactivity. Take VR service as an example (Fig. 1.8). Professionals in this field believe that VR development will be dominated by Cloud VR, which uses the Cloud to process images so that the terminals worn by users only need to display the screen and perform simple coding/decoding. Cloud VR is characterized by light weight, convenient mobility, and low cost. In mobile scenarios,

Fig. 1.8 VR service scenario

VR requires about 50 Mbps air interface user rate (AIUR) to achieve 4 K resolution and about 1.8 Gbps AIUR to achieve a good carrier-class service experience. 5G can meet the service needs of Cloud VR since it has a peak rate of 20 Gbps.

Typical services for URLLC scenarios include unmanned driving, remote control, etc. Such services require ultra-low end-to-end service latency as well as ultra-high reliability. Take unmanned driving as an example (Fig. 1.9), and the service allows a maximum latency of 5 to 10 ms in terms of the two-way communication between a vehicle and its cloud control center. However, the end-to-end communication latency in traditional mobile communication networks is generally over 50 ms, which cannot meet the needs of this service. With the help of layered deployment of the core network and Multi-access Edge Computing (MEC) technology, 5G is capable of combining and deploying the core network and service control center to the edge of the network, which meets the latency requirements of the service by greatly reducing the distance from the vehicle to the cloud. In addition, 5G network slicing technology can achieve resource isolation for individual services in wireless networks, transport networks and core networks, guaranteeing the service quality and reliability of special services.

Typical services of mMTC scenarios include smart cities, industrial sensors interconnection, etc. Such services are characterized by a huge demand for linking numbers on the network side. As shown in Fig. 1.10, the world will move towards the "Connectivity of Everything" trend. The number and density of various networked terminals will increase geometrically. At the same time, 4G networks cannot meet the demand for connectivity of everything because they can only connect about 10,000 terminals per km^2. Nevertheless, characterized by ultra-large connectivity, 5G can perfectly match the requirements of connectivity of everything with 1 million linking numbers per km^2.

Fig. 1.9 Unmanned driving needs ultra-low latency

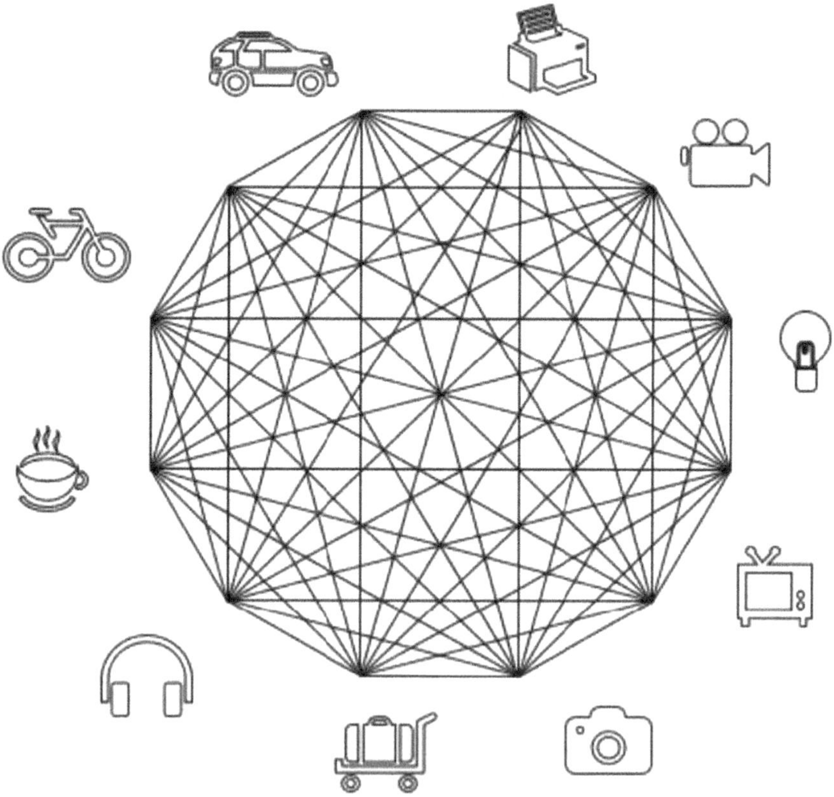

Fig. 1.10 The future world will achieve "connectivity of everything"

1.3 Process in 5G Standardization

The 3GPP organization has led in developing 5G technical standards for communication technology. On 27 June 2016, 3GPP announced that they agreed on a detailed work plan for the first version of 5G standards—Release-15 (Rel-15) on the 72nd Plenary Session of 3GPP Technical Specifications Groups (TSG). The plan confirmed the coordinated items and inspection priorities for each working group and specified that 5G-related specifications in Rel-15 would be finalized in June 2018. As shown in Fig. 1.11, the major 5G versions include Rel-15 and Release 16 (Rel-16). According to the original plan, Rel-15 was to be frozen in June 2018, and Rel-16 was to be frozen in December 2019. Afterwards, due to the competitive pressure from the Open Trial Specification Alliance (OTSA) established by US, Japan, and Korea, 3GPP launched the acceleration of 5G standards in February 2017 to split Rel-15 into Non-Standalone (NSA) and Stand-alone (SA). The freeze time of NSA was also moved forward to December 2017 to meet the needs of a few operators expecting rapid 5G commercialization. The entire Rel-15 focuses on the specification of standards for eMBB application scenarios. Rel-16, which was postponed to freeze in July 2020 for various reasons, has mainly enhanced technologies related to eMBB scenario to improve network performance. Related technical standards of URLLC scenario have also been defined. As for the mMTC scenario, Rel-16 has not frozen related technical standards of mMTC because the application scale of machine communication and the density of network connections has not reached the magnitude defined by 5G, and the technical standards for this scenario are expected to be frozen in Release 17 (Rel-17).

In terms of 3GPP TSG RAN, regarding the scope of 5G New Radio (NR) for Rel-15, the technical specification group has agreed unanimously to provide support for both Stand-alone (SA) and Non-Standalone (NSA) architectures. 4G base station and 4G core network are needed for 5G NSA networking. At the initial stage, 4G will be used as the control plane's anchor to meet the operators' demands to rapidly deploy 5G NR with existing LTE network resources. As a transitional solution, NSA mainly aims to enhance the bandwidth of hotspot areas. With no independent signaling plane, NSA relies on a 4G base and core network. Thus, the corresponding standards have

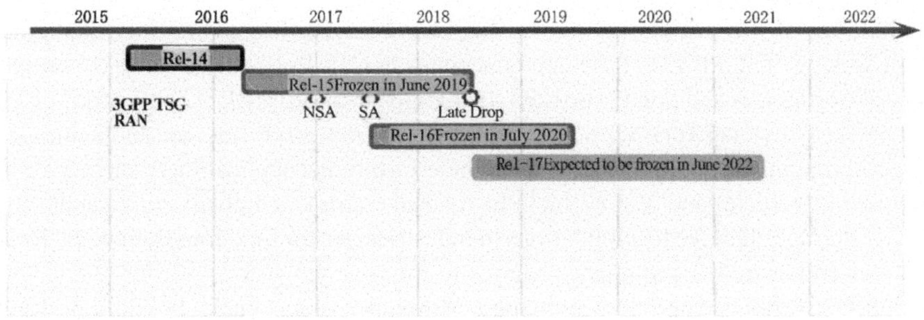

Fig. 1.11 Evolution of 5G protocol versions

been improved quickly. The existing 4G network needs to be upgraded to achieve 5G NSA, and this process will have a certain influence on the performance and smooth operation of existing networks, which needs operators' attention. Meanwhile, Rel-15 has also identified target use cases (eMBB, URLLC and mMTC) and target frequency bands, which include ranges below or above 6 GHz. In addition, the 72nd Plenary Session of 3GPP TSG RAN emphasized that 5G standards should have upward compatibility in wireless and protocol designs. Besides, importing functions in stages is the key point for implementing each use case.

On 21 December 2017, the 5G NSA standard, the world's first commercially deployable 5G standard, was frozen at the 78th Plenary Session of the International Telecommunications Standards Organization 3GPP RAN. Completing the NSA program of the 5G standard is an important milestone in the process of 5G standardization, marking the acceleration of the 5G standard and industrial process. The freeze of standards is of great significance to the communication industry, which means the core standard has been determined. Even if some minor adjustments might be made to the formal standard in the future, it will not affect the previous product development. 5G commercial use is counting down.

On 14 June 2018, the 80th Plenary Session of 3GPP TSG RAN ratified the 5G SA standard freeze. The freeze of SA standard not only enables 5G NR to deploy independently but brings a brand-new end-to-end architecture. Enterprise users and vertical industries are empowered with smart development. Operators and industrial partners are embracing new business models. A new era of full connectivity has begun. However, as Rel-15 came too fast, many details were not fully considered. Meanwhile, the operators have already started to deploy networks, facing many demands. So, the 3GPP TSG RAN Plenary Session finally decided to add Late Drop to Rel-15 and introduce networking architecture such as Option4 and Option7 to supplement NSA networking. Besides, two scenarios are added: (1) 5G base stations as the primary and 4G base stations as the auxiliary; (2) 4G base stations as the primary and 5G base stations as the auxiliary. With the Core Network evolving into a 5G Core Network (5GC), operators can deploy 5G networks more efficiently. Rel 15 was originally scheduled to freeze Late Drop in December 2018, but it wasn't completed until June 2019 for certain reasons. Since then, Rel-15 has had three versions in total, and 5G has completed the first stage of standardization.

As an enhanced version of Rel-15, the formulation of Rel-16 is affected by Rel-15 Late Drop, with the freeze time having been delayed twice. On 3 July 2020, the 88th Session of 3GPP TSG RAN announced the 5G NR Rel-16 freeze. Rel-16 brings new performances and functions to 5G, including enhanced URLLC, industrial-class latency-sensitive network service capabilities (for industrial Internet applications), direct communication capabilities (for vehicle-to-Internet applications), such as V2V (vehicle-to-vehicle) and V2I (vehicle-to-roadside units), and positioning capability of air interface adaptive to applications in industries. As for functions defined by Rel-15, Rel-16 has added some new features to improve network spectrum efficiency, handoff performance and power-saving performance of terminals. In addition, Rel-16 also mentioned automatic road tests and network operation and maintenance technologies to reduce the cost of network operation and maintenance, such as the remote interference management for base stations targeting atmospheric ducts, and other technologies.

On the 86th Plenary Session of 3GPP TSG RAN held in Sitges, Spain, 5G Rel-17 was planned and a total of 23 standard items were approved, which mainly includes data acquisition and application enhancement for intelligent network operation and maintenance, NR multicast and broadcast services, precision positioning technology for vertical industries, NR V2X enhancement, NR Light technology (applied to industrial IoT sensors, surveillance cameras, wearable devices and other scenarios) and URLLC enhancement. 5G Rel-17 will continue to promote 5G standards and sustainable industry development.

1.4 Process of 5G Industrial Chain

As a super pipeline, 5G achieves its functions by integrating multiple components. Regarding overall structure, 5G can be divided into two parts: terminal and network.

5G terminals can be classified as consumer and industrial terminals according to different clients. Consumer terminals are mainly devices facing personal users, such as cell phones and smart watches. 5G cell phones contain many components, such as a camera and screen. However, chips are the core components that determine the performance of cell phones. Compared with the consumer Internet, a more important mission for 5G is to sustain the industrial Internet. In other words, 5G must be applied in every trade, such as industry, transportation, and energy. Terminals vary in different industries and application scenarios. Take 5G vehicle networking as an example. A vehicle is a terminal. 5G module installation is required if a vehicle needs to connect to 5G network. Other industries, such as unmanned aerial vehicles, do as well. The key of the modules is still a chip.

5G network can be divided into three parts: radio access, transport, and core network. The radio access network comprises base stations, and 5G base stations mainly consists of Base Band Unit (BBU) and Remote Radio Unit (RRU). Compared with 4G base stations, 5G BBU can be separated into Centralized Unit (CU) and Distributed Unit (DU) while RRU mainly includes Active Antenna Unit (AAU). Transport networks still use optical fiber communication, mainly including optical modules, optical fibers, and cables as primary devices for optical communication. As for the core network, in order to meet differentiated network demands for all walks of life in the 5G era, the 5G core network can develop a powerful carrier-class capability based on common infrastructure, with the help of stateless architecture. At the same time, introducing some key technologies, such as edge calculation and network slicing, guarantees SLA (Service Level Agreement) for 5G networks.

Since the whole 5G network involves all sorts of hardware and software, the 5G industrial chain consists of multiple interrelated industrial chains rather than a single one. In terms of 5G industrial chain development, this chapter mainly introduces relevant processes of 5G chips, 5G terminals, and 5G network devices.

1.4.1 Process of 5G Chips

The key to 5G chips lies in 5G baseband chips and 5G System on Chip (SoC). The powerful functions of the 5G network lead to extremely high requirements for chips. Chip production includes three parts: design, processing and manufacturing, and package test. At present, not many manufacturers are capable of designing 5G chips in the world. As shown in Fig. 1.12, companies, such as Qualcomm, Huawei, MediaTek, Samsung, and UNISOC can design.

In October 2016, Qualcomm rolled out the first 5G baseband chip-Snapdragon X50, which merely supports NSA networking mode and 5G system. 2G, 3G, and 4G systems are incompatible with Snapdragon X50. As the first 5G baseband chip for commercial use globally, it was officially put into use in December 2018. In February 2020, Qualcomm launched a Snapdragon X60 baseband chip with a 5 nm process. This chip supports NSA or SA networking and 2G, 3G, 4G, and 5G. Compared with Snapdragon X50, the performance of Snapdragon X60 is more powerful.

Huawei is the first Chinese company to release 5G baseband chips. In January 2019, Huawei released the first generation of a multi-mode terminal baseband chip—Balong 5000. Balong 5000 uses a 5G module with multi-chip modules. It supports multiple networks, including 2G, 3G, 4G and 5G. Meanwhile, the chipset is also the first to support NSA or SA network mode worldwide. Balong 5000 supports FDD and TDD and can be used across all frequency bands. In September 2020, Huawei released Kirin 9000, a mobile System-on-Chip (SoC) based on a 5 nm process. Compared with plug-in baseband chips, SoC is smaller in size and lower in power consumption.

In China, MediaTek and UNISOC also have their places in the industrial chain of the design of 5G chips. At the end of 2021, MediaTek mass-produced Dimensity 9000. In February 2020, UNISOC released Huben T7520, the second-generation

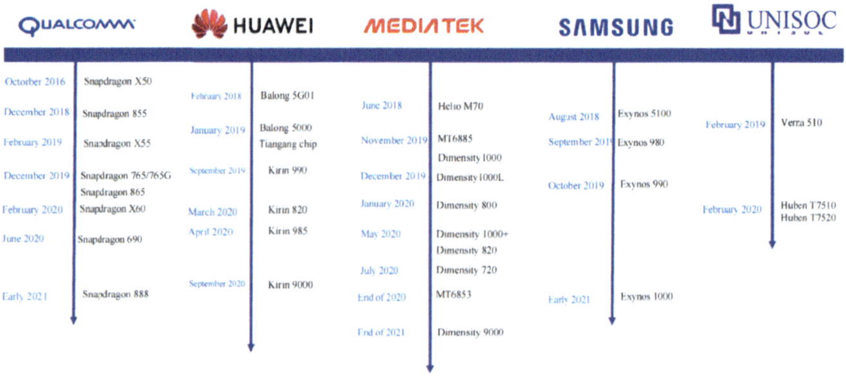

Fig. 1.12 Timeline of 5G chip releases

of 5G chips by UNISOC, which adopts a 6 nm EUV process and supports Sub-6 GHz frequency band, NSA/SA dual-mode networks, and various network modes including 2G, 3G, 4G and 5G.

In the manufacturing of 5G chips, current high-end chips all adopt the 7 nm process and below. The main manufacturers capable of producing such chips are TSMC and Samsung. Regarding the 5G chip package test, the three main manufacturers in packaging and testing from the Chinese mainland (JCET, H-T Technology and Tongfu Microelectronics) account for more than 20% market share.

1.4.2 Process of 5G Terminals

Terminals are crucial in 5G industrial chains. As shown in Fig. 1.13, 5G terminals mainly include smartphones, mobile hotspot devices, Helmet-Mounted Displays (HMD) and modules, etc. In the smartphone area, the main manufacturers worldwide are Huawei, Samsung, Xiaomi, VIVO, OPPO, One Plus, etc. To bring fresh 5G experiences to users, big screens, foldable screens, and multiple screens will be the mainstream in developing 5G terminals.

In 2020, 70% of the smart phones delivered worldwide support the 5G system. 5G cell phones with prices lower than 300 US dollars have sprung up. In China, there are more than 200 types of listed 5G smart phones, among which the sales of medium and high-end 5G cell phones (priced above 2,000 RMB yuan) have accounted for over 90%.

The core object of 5G service is all industries. In the 5G industrial chain division, the upstream is about chips while the downstream is about applications in a variety of vertical industries. Modules, as the intermediate link, are important to the development of 5G industries. 3GPP announced that it has frozen Rel-16 on 3 July 2020. It means that 5G has a mature technical standard which will accelerate the development of industrial applications on downstream of the 5G industrial chain and promote

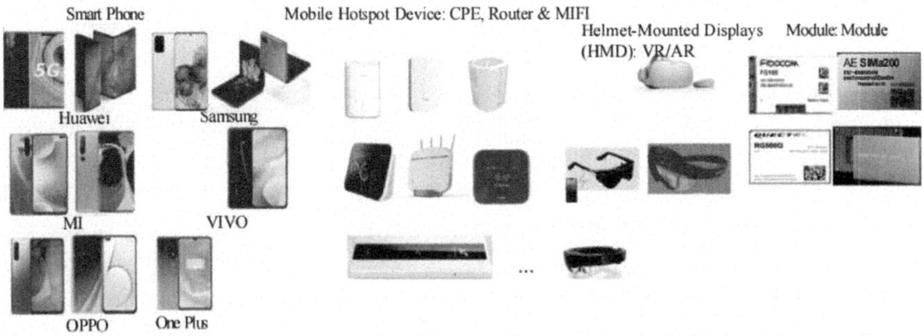

Fig. 1.13 5G terminals

Table 1.2 Main manufacturers and products of 5G modules in China

Manufacturer	Module type	Chip used
Huawei	MI-15000	Ba long 5000
Quectel Wireless Solutions Co., Ltd	RG500Q/RG510Q	Snapdragon X55
	RG800H	HI-SILICON 5G module middleware
ShenzhenGuanghe Tong	FG150/FB101	Snapdragon X55
SiMCom	SIMS200EA	Snapdragon X55
China Mobile	CMCC M5	HI-SILICON 5G module middleware
MeiG Smart Technology Co., Ltd	SRMS15 SRM 825	Snapdragon X55
Wingtech Group	WM518	Snapdragon X55
Sichuan Ai-Link Technology Co., Ltd	AI-NR10	HI-SILICON 5G module middleware
LongSung Technology (Shanghai) Co., Ltd	EX510	Snapdragon X55
Chongqing CMO IOT	OneMO FO2X1OneM 0 F03X	Snapdragon X55
GosuncnWel ink Technology Co., Ltd	GM8001GMS01	SnapdragonX55

the maturity of the technologies of 5G modules. At present, many manufacturers in China produce 5G module products (Table 1.2).

1.4.3 Process of 5G Network Devices

5G network devices are developing fast, with well-developed products in radio access networks, bearer networks and core networks.

The main devices of 5G radio access network are the Base Band Unit (BBU) and Remote Radio Unit (RRU). At present, Huawei, ZTE, Ericsson, Nokia and other main manufacturers worldwide have all released new generation of 5G BBU products. Compared with BBU3900/3910 for 4G, Huawei's BBU5900 for 5G has significantly improved performance. The frame and size of such devices remain unchanged to ensure its backward compatibility with 4G devices, which is beneficial for constructing the co-site of 4G and 5G. As for RRU, the type used in the 4G era is no longer the main type for the 5G era because 5G uses a higher frequency band. 5G mainly uses AAU supporting Massive MIMO technology to ensure network coverage in outdoor scenarios. 5G AAU supporting Massive MIMO technology can achieve 3D beamforming, which can substantially improve the coverage quality for cell-edge users due to its great coverage on tall buildings and high altitudes. In addition, AAU utilizes enhanced space division multiplexing (SDM) technology with its capability of 5 to 8 times larger than 4G, which perfectly matches the demand of 5G for air interface capacity and coverage quality. As for the types of base stations, in addition to the outdoor macro station and digital indoor distribution system, all kinds of small stations and micro stations are deployed in the current network to meet the requirements for the coverage and deployment of various scenes, such as high-density urban areas, areas with high telephone traffic, inner part of large buildings, and subways.

The Front-haul Network (RRU to DU) of the 5G transport network can use Centralized Radio Access Network (CRAN) and Distributed Radio Access Network (DRAN). When CRAN is applied, AAU is deployed far away, and BBU from multiple sites is deployed in one server room. The optical transmission of 5G Front-haul Network usually adopts Wavelength Division Multiplexing (WDM), active WDM, or Optical Transport Network (OTN). For DRAN, BBU for each site is deployed respectively in its server room, with AAU being closer to BBU. 5G Front-haul Network mainly uses optical fiber direct connection for optical transmission. Compared with 4G, a middlehual network (the network between DU and CU) is added to 5G. The middlehaul and backhaul networks both adopt an IP transport network. Compared with 4G, there is no substantial updating in optical modules and optical fibers/cables used in the bottom layer of the 5G bearer network. New equipment is needed for 5G bearer network deployment because the 4G transport network cannot support new technologies needed for network element devices, such as Slicing Packet Network (SPN) or Metro-optimized OTN (M-OTN) targeting the optimization of the mobile bearer. Currently, the main providers of transport network equipment and solutions in China are Huawei, ZTE, FiberHome, H3C, etc.

A 5G core network is very different from a 4G core network. It adopts network function virtualization instead of proprietary hardware to provide network element functions. The underlying hardware of the network is a universal server, and network element functions are made possible in the form of software. In addition, the control plane and user plane of the 5G core network are completely separated. The user plane can be deployed flexibly in layers. Combined with MEC technology, it is able to flexibly control the latency of end-to-end service. In the universal server industry, manufacturers with well-developed products include IBM, Huawei, Inspur, etc. In the field of network element software, all providers of network equipment and solutions, such as Huawei, ZTE, Ericsson, Nokia, etc., have their developed products.

Figure 1.14 shows some 5G end-to-end network devices. Available network equipment and terminals that can be mass-produced and commercialized are gaining momentum with unprecedented maturity. The 5G standard freezing and commercial use of end-to-end network devices have been developing rapidly at an unprecedented speed. That is unparalleled in the history of mobile communication development.

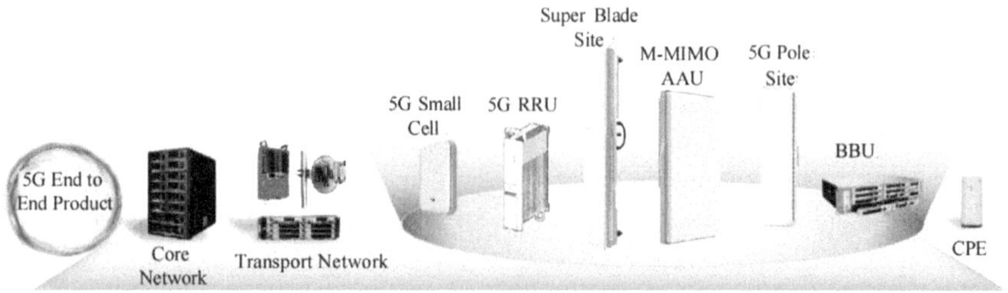

Fig. 1.14 5G end-to-end network devices

1.5 Process of Global Commercial 5G

Commercial 5G in the globe has been developing rapidly under the influence of factors such as demands for industrial application, current global economic status, different national strategies, and the development of mobile communication technologies. At the same time, more and more enterprises have realized the importance of digital transformation due to the COVID-19 pandemic. During the pandemic, many services like telecommuting, teleconferences, and remote education have been operated online thanks to the mobile communication networks. The extensive application of digital technology and big data in pandemic prevention and control has further accelerated 5G development.

1.5.1 Overview of Commercial 5G Worldwide

According to statistics by the Global mobile Suppliers Association (GSA), by the end of May 2021, 5G network coverage can be accessed in every continent in the world, among which 169 operators in 70 countries and regions have rolled out commercial 5G network, 443 operators in 133 countries and regions have invested in 5G. Since 2015, 34 countries have finished the allotment or auction of C-band spectrum, and operators in 28 countries have gained 5G mm Wave licenses.

In the field of terminals, more than 500 commercial 5G terminals have hit the global market. 340 million 5G connections cover 15% of the global population. It is estimated that by 2025, the investment in 5G network construction will reach over 700 billion US dollars, accounting for 80% of operators' total network construction investment. In addition, the investment in Asia, Europe, and North America will be significantly higher than that in other regions.

1.5.2 Commercial 5G Stages

2019 marks the first year of commercial 5G worldwide. China, the US, and South Korea had all completed the first release of commercial 5G licenses in 2019. In general, commercial 5G focuses on the eMBB scenario first, with URLLC scenario under continuous cultivation. Furthermore, the mMTC scenario will follow the IoT connectivity of 4G.

The first stage of commercial 5G focuses on the eMBB scenario, mainly for personal users and Wireless To The x (WTTx) service. The main reasons for prioritizing eMBB service are: (1) Operators are familiar with the maintenance and operation of traditional eMBB service, with 4G experience as a reference; (2) The technical standard of eMBB service scenarios was frozen by 3GPP first; so operators can, considering the process of standardization, launch commercial 5G in advance to better seize the market and establish brand influence; (3) 5G+ industrial application still needs cultivation. Introducing personal users by eMBB service first can help operators recover certain network investments to speed up the construction of a 5G network. Figure 1.15 shows the main service scenarios of 5G eMBB, including AR, VR, HD video, UAV video backhaul services, etc.

When applications of eMBB scenarios are being commercialized, those in URLLC scenarios are being cultivated. The performance of low latency and high reliability are required in URLLC scenarios mainly in remote or unmanned control services, such as Internet of Vehicles (IoV) and smart manufacturing. These services represent the direction for 5G service development. But there is no large demand for these services now. At the same time, the integration of MEC and vertical industrial applications is not mature at the technical level, so URLLC scenarios are not commercially available on a large scale. However, 5G pilot applications have been deployed worldwide in multiple industries, including healthcare, iron and steel, manufacturing, ports, etc.

Fig. 1.15 Main service scenarios of 5G eMBB

Rel-15 and Rel-16, frozen by 3GPP, didn't specify the related standard of mMTC. Therefore, the application of mMTC scenarios has not been officially commercialized. In the initial stage of 5G application, IoT technology will inherit the standards for air interface of NB-IoT and eMTC.

1.5.3 Process of Commercial 5G in Major Countries

China, the US, South Korea, the UK, and Switzerland are leading 5G commercialization.

Three telecom operators in South Korea officially launched commercial 5G services at 11 p.m. in local time on 3 April 2019, marking South Korea as the first country in the world to achieve 5G commercialization. One hour later, Verizon from the US followed suit. On 17 April 2019, Swisscom, the largest telecom operator in Switzerland, announced the launch of its commercial 5G network, which was the first 5G network for large-scale commercial use in Europe. On 30 May 2019, EE, a subsidiary of the operator "British Telecom" (BT), officially launched 5G services in London, Cardiff, Belfast, Edinburgh, Birmingham, Manchester, and other cities.

5G development in China has also achieved eye-catching results. On 6 June 2019, the Ministry of Industry and Information Technology issued 5G licenses to the four major operators: China Mobile, China Telecom, China Unicom, and China Broadnet. On 30 October 2019, China Mobile, China Telecom, and China Unicom all released commercial 5G contracts, which were officially launched on 1 November 2019, marking the commercialization of 5G networks in China. By April 2022, China has built 1.615 million 5G base stations, accounting for more than 60% of the total number worldwide. With more than 412 million 5G cell phone users and 5G networks carrying 30% of network traffic nationwide, China is already at the forefront of the world in aspects such as the scale of 5G network, the number of 5G core patents, the size of 5G users, 5G terminal shipments and the diversity of 5G applications.

1.5.4 Modes in Commercial 5G

In the 5G era, operators have three operating modes: super pipeline, ICT-enabled (Information and Communications Technology, ICT) basic lease, and ICT-enabled service market.

A super pipeline is made possible through the pipeline. The introduction of cloud infrastructure and network slicing enables 5G to have end-to-end service isolation and service quality guarantee. Meanwhile, the network design, deployment, operation, and maintenance are more intelligent. Therefore, it can be said that 5G is a super pipeline and a smart pipeline.

ICT-enabled basic lease means that operators build their own base platforms (such as IoT platforms and V2X platforms) and earn more by integrating services from other providers. For example, AT&T acquired a video platform through its acquisition of DIRECTV. After that, it can cooperate with some content providers to develop video services.

ICT-enabled service market means to open a channel for certain E2E services, so an operator can develop its content platform, for instance, AT&T's acquisition of Time Warner was in the hope of becoming a video service provider itself.

In these three operation modes, a super pipeline is the foundation. However, to support the basic lease and service market, operators must also deploy service and content platforms.

5G network slicing is the basis of a brand-new business mode for operators—network slicing is a service (Fig. 1.16). The operators provide an end-to-end network slicing to users, and multiple charging modes can be extended.

Operators can charge based on location, such as stadiums and smart factories. Meanwhile, in the same place, operators can also charge based on time. Take the stadium scenario as an example, and operators can provide VR live-streaming service based on time slice within 2 to 3 h after a concert or sports event begins. When the sport

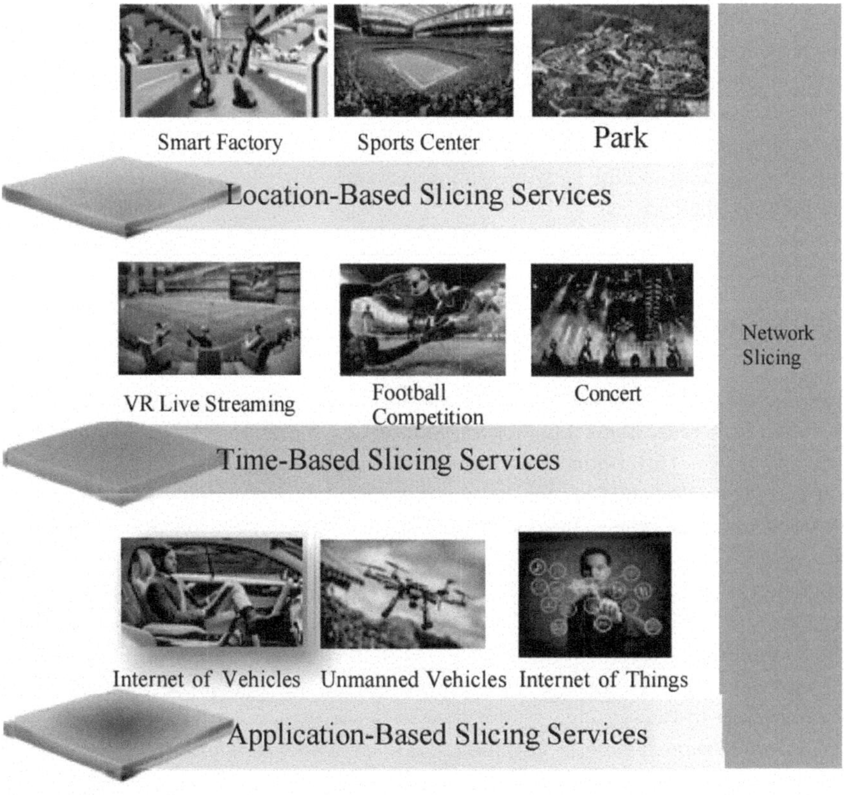

Fig. 1.16 Network slicing is service

Fig. 1.17 The evolution of 5G business modes

game is over, this network slicing disappears. This is a concrete example of future network slicing (service). Therefore, after the arrival of 5G, operators can change the traditional flat rate charging mode. By overlaying different services, operators can increase their charging rate reasonably, allowing 5G networks to bring greater value.

Currently, 5G is focused on eMBB services and services for personal users. Some industries have already started network slicing services, so the business mode of 5G should also be upgraded along with the network construction and the launch of commercial services. The evolution of 5G business model is shown in Fig. 1.17. In the early stage of 5G commercialization (Stage 1), there were no major changes at the architecture level and operators emphasized enhancing their network capacity more. In the second stage, the slicing architecture of 5G networks took shape and operators were committed to promoting the standardization of dispatch interfaces for cross-domain slicing capabilities. Operators also focus on integration with their management systems and piloting commercial slicing services. In the third stage, 5G networks enter a stage of maturity. Automated operation and management of network slicing is supported by the networks and their operation management systems. Operators provide end-to-end network slicing services to tenants and develop sound charging mechanisms.

1.6 Summary

This chapter first introduces the evolution of mobile communication network from its first to fifth generations and then explains each generation's main characteristics and capabilities. The continuous progress of science and technology as well as the ever-increasing demands in mobile communication, have become the main driving forces of 5G development. In addition, it also involves competition among countries and regions. Through the introduction, readers can get insight into the three major application scenarios (eMBB, URLLC, mMTC). The multi-dimensional enhancement has made it possible for mobile communication to support the informatization and the intelligentization of industries. 5G is characterized by ultra-large bandwidth, ultra-low latency and ultra-large connection which brings social development closer to the Internet of Everything.

The upgrading of mobile communication networks is a gradual process, and global standardization is inevitable for the development of the mobile communication industry. This chapter provides a detailed analysis of the process of 5G technical standards, presenting readers with the content and the significance supported by each

5G standard as well as its trend. In addition, this chapter illustrates the development of the 5G industry chain through the current process of 5G chips, terminals, and network equipment.

The year of 2019 initiated 5G commercialization globally, with South Korea and US being the first. China is also in the first echelon of commercial 5G. The huge mobile communication market and massive users provide a sound environment for 5G to prosper. Meanwhile, mobile communication manufacturers in China have achieved remarkable results through in-depth exploration over many years. As a result, China is leading in developing mobile communication in the 5G era. Unlike the more singular performance in traditional mobile communication networks, 5G attaches more importance to multi-dimensional information transmission. Moreover, it also achieves end-to-end resource isolation and SLA guarantees through technologies, such as network slicing and others, to bring more innovation to customers. Therefore, the traditional business model in mobile communication does not fit in the 5G era. This chapter introduces 5G commercialization worldwide and analyzes the stages of the three major scenarios of commercial 5G. Ideas on the business mode of 5G are also proposed.

By studying this chapter, readers should understand the overall history of mobile communication development, the background of 5G development, and the process of 5G industry chain and its commercialization.

Exercise 1

I. True or False

1.1 CDMA2000 is a 3G standard proposed by China. ()
1.2 The first official standard formulated by 3GPP for 5G is Rel-16. ()
1.3 The typical service in the URLLC scenario is Cloud VR. ()
1.4 5G SoC chip has more advantages than plug-in baseband chip. ()

II. Choice Questions

1.5 (Single choice) Which of the following is not a key performance target for 5G? ()

A. Peak download rate at 20 Gbps
B. Spectrum efficiency of over three times that of 4G
C. Minimum latency of air interface reaches 1 ms
D. Supports up to 350 km/h mobility

1.6 (Single choice) 5G can support up to 1 million devices per square kilometer. Which of the following belongs to this scenario? ()

A. eMBB
B. URLLC
C. mMTC
D. Massive MIMO

1.7 (Single choice) Which 3GPP protocol defines the SA networking for 5G? ()

A. Rel-14
B. Rel-15
C. Rel-16
D. Rel-17

1.8 (Single choice) Which of the following is incorrect about Kirin 9000? ()

A. Kirin 9000 is a Huawei product
B. Kirin 9000 is a SoC chip
C. Kirin 9000 uses a 14 nm process
D. Compared with a plug-in baseband chip, the Kirin 9000 is smaller and lower in power consumption

1.9 (Single choice) Which of the following is the main radio frequency module of the Huawei 5G outdoor macro station? ()

A. RRU
B. AAU
C. BBU
D. CPE

1.10 (Single choice) Globally, which year is the first year of 5G commercialization? ()

A. 2018
B. 2019
C. 2020
D. 2021

1.11 (Single choice) When did China officially launch commercial 5G? ()

A. 3 April 2019
B. 6 June 2019
C. 30 October 2019
D. 1 November 2019

1.12 (Multiple choice) In 2008, ITU publicly collected 4G standards. Which items are the alternatives to 4G standard? ()

A. LTE
B. UMB
C. WiMAX
D. NR

1.13 (Multiple choice) Which of the following can serve as 5G terminals? ().

A. VR glasses
B. Mobile phones
C. Cars
D. Industrial modules

1.14 (Multiple choice) What are the components of 5G network? ()

A. Local area network
B. Radio access network
C. Transport network
D. Core network

1.15 (Multiple choice) Which manufacturers have the top 5G chip design abilities in the industry? ()

A. Huawei
B. TSMC
C. Qualcomm
D. Samsung

1.16 (Multiple choice) Which of the following are defined in 3GPP Rel-16? ()

A. Enhanced URLLC
B. Industrial-class latency-sensitive web service
C. V2V direct communication
D. Precision positioning technology targeting vertical industries

III. Short Answer Questions

1.17 Please briefly describe the advantages of 5G compared with 4G in terms of capability

1.18 Please briefly describe the commercial stages of the three major scenarios of 5G

Chapter 2
5G End-to-End Network Architecture and Key Technologies

To adapt to various future service scenarios, the 5G end-to-end network will undergo profound changes, applying various new technologies with large bandwidth, high-density connectivity, and ultra-low latency. The architecture of 5G networks will also be more flexible and agile.

This chapter focuses on 5G end-to-end network architecture, 5G end-to-end network key technologies, and 5G security technologies.

Learning Objectives

- To know 5G end-to-end network architecture
- To master key technologies of 5G end-to-end network
- To master 5G security technologies

2.1 5G End-to-End Network Architecture

The FifthGeneration Mobile Communication System (5G) network architecture is divided into three parts: radio access, transport, and core, as shown in Fig. 2.1. Next, the three modules of the 5G radio access network, 5G core network, and 5G transport network are introduced.

Some new terms are annotated below.

CSG (Cell Sitc Gateway): a role in the mobile transport network located in the access layer, responsible for the access to the base station.

ASG (Aggregation Site Gateway): a role in the mobile transport network located in the aggregation layer, responsible for the massive CSG service flow aggregation's mobile transport network acccss layer.

RSG (Radio Service Gateway): a role in the mobile transport network in the aggregation layer, connecting the wireless controller.

L. Song et al., *5GtoB From Theory to Practice*,
https://doi.org/10.1007/978-981-97-8453-0_2

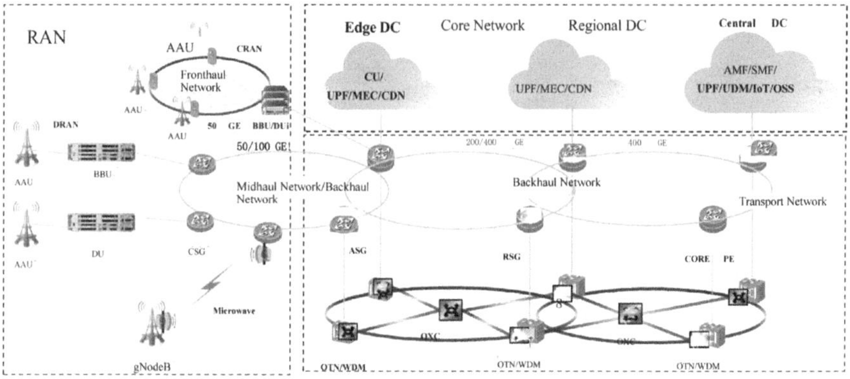

Fig. 2.1 The fifth generation mobile communication system (5G) network architecture

CORE PE (CORE Provider Edge Router): an edge device for service providers.
OTN (Optical Transport Network): a network that transmits information through optical signals.
WDM (Wavelength Division Multiplexing): a data transmission technology in which different optical signals are carried by different colors (wavelength frequencies) and then multiplexed for transmission on a single optical fiber.
OXC (Optical Cross-Connect): a technology for switching high-speed optical signals, usually used in optical mesh networks.

2.1.1 5G Radio Access Network Architecture

5G radio access network contains only one network element—the next generation NodeB, also known as gNB or gNodeB. It mainly interconnects with the transport network devices through wired media such as optical fibers. In special scenarios, wireless devices such as microwaves can interconnect with transport network devices.

Currently, there are two 5G radio networking methods: Distributed Radio Access Network (DRAN) and Centralized Radio Access Network (CRAN). The current strategy operators apply in China is based on DRAN and on-demand deployment of CRAN. In the CRAN scenario, the Base Band Unit (BBU) is centrally deployed and is then connected by optical fibers with Active Antenna Unit (AAU). Due to the far distance between BBU and AAU, optical fibers are highly required. In some scenarios, a transmitted wavefront is required. In the DRAN scenario, BBU and AAU are directly connected with optical fibers.

1. **DRAN Architecture**

In DRAN architecture, each station deploys an independent server room. BBU and RRU/AAU are co-located, while power distribution equipment and other supporting equipment are deployed independently, as shown in Fig. 2.2.

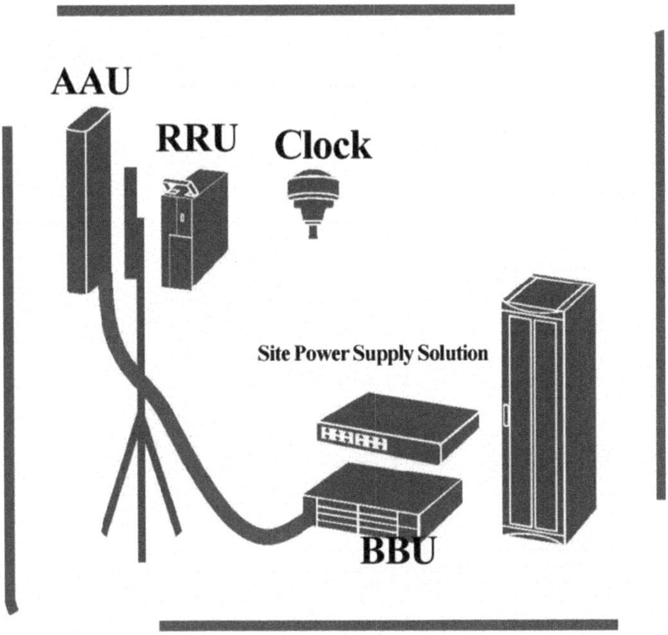

Fig. 2.2 DRAN site deployment

(1) In DRAN architecture, BBU and RRU/AAU are co-located, and the station backhaul can be flexibly grouped with microwave or fiber optic solutions according to the actual conditions of a server room.
(2) BBU and RRU/AAU are deployed in the same station. The fiber length of the Common Public Radio Interface (CPRI) is short and the overall fiber consumption is low.
(3) If power supply and transmission failures occur in a single station, it will not affect other stations.
(4) Benefiting from the long-term construction of 2G/3G/4G networks, the operators' existing network has many station server rooms or outdoor integrated cabinets. Although the higher frequency used in 5G will lead to more stations required for wireless coverage, operators will still use both the original station server rooms and the new ones for a long time in the future to deploy radio access networks with DRAN architecture.

2. **CRAN Architecture**

In CRAN architecture, BBU modules of multiple stations will be deployed in a central server room, as shown in Fig. 2.3, and each site radio frequency module is connected to BBU in the central server room through forward pull-out fiber.

(1) The ultra-dense site network of 5G will form more coverage overlap areas, making CRAN more suitable for the deployment of technologies, such as Carrier

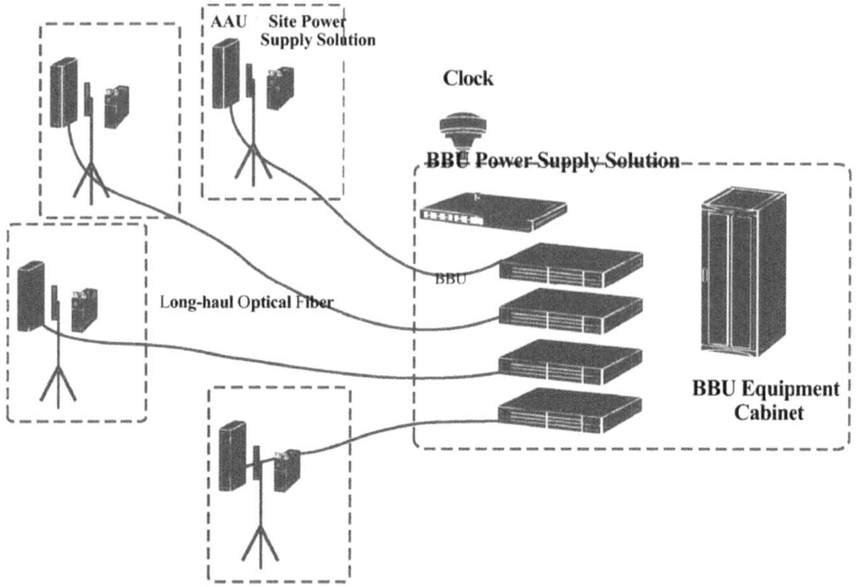

Fig. 2.3 CRAN site deployment

Aggregation (CA), the transmission and reception of Coordinated Multipoint (CoMP) and Single Frequency Network (SFN), etc., which can achieve efficient collaboration between stations and significantly improve wireless network performance.

(2) CRAN can simplify station acquisition. On the one hand, it can rapidly deploy radio access networks and shorten the construction cycle; on the other hand, it can achieve deep coverage in blind areas where it is not easy to deploy stations.

On the whole, because there is no need to build a server room at each station, through the measure of CRAN server room + remote holding pole, the radio access network station deployment can be completed quickly and form the overlay. Therefore, this solution is suitable for high-capacity, high-density call areas (densely populated urban areas, parks, shopping malls, residential areas, etc.), as well as other areas requiring the completion of base station deployment in a short period. Overall, the current proportion of CRAN stations is much lower than that of DRAN. However, in the future, in order to make it easier to deploy stations, various efficient and collaborative features can be started to improve the performance of wireless networks. Thus CRAN architecture in the deployment of 5G radio access networks will be more and more widely used.

3. **CloudRAN Architecture**

In the future, the wireless site will also evolve towards cloudization, and CloudRAN will introduce a structure that separates the Centralized Unit (CU) and Distributed

Unit (DU), as shown in Fig. 2.4. CU will be deployed in the edge data center after cloudization and is responsible for handling the high-level protocols of the traditional baseband unit. In contrast, DU can be centrally deployed in the edge data center or distributed near the AAU site and is responsible for handling the underlying protocols of the traditional baseband unit.

CU is typically deployed in an edge or regional data center. This data center can deploy User Plane Function (UPF) and MEC servers on demand in addition to CU network elements. For low latency services (such as autonomous driving), after DU sends the upstream data from the user plane to CU to complete the corresponding processing, CU needs to forward the data to UPF, which in turn forwards it to the corresponding autonomous driving MEC to generate control commands and then send them downstream to DU in the reverse direction. Therefore, CloudRAN architecture is more suitable for the flexible deployment of multiple services and scenarios in the future.

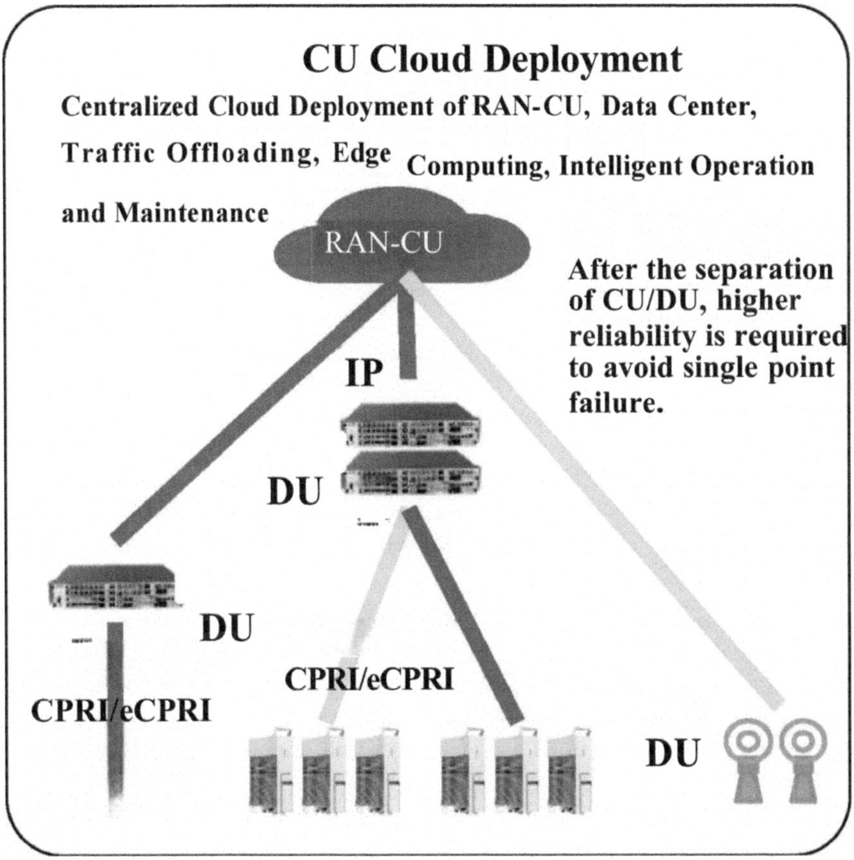

Fig. 2.4 CloudRAN architecture and names of each component

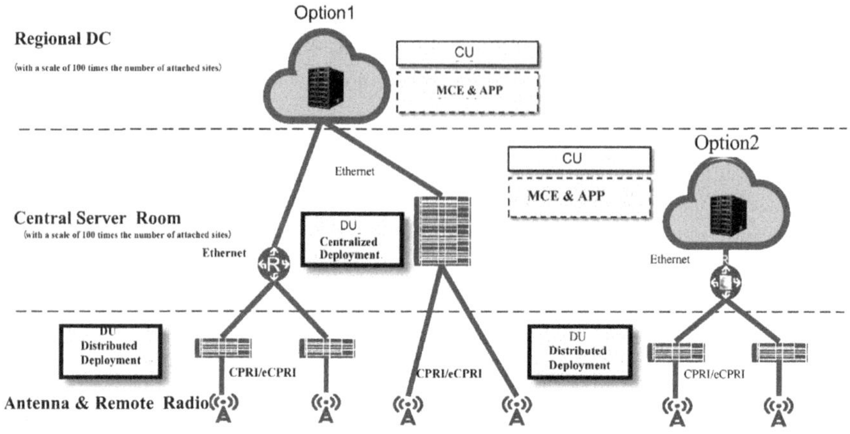

Fig. 2.5 CloudRAN overall deployment

DU remains in the baseband board and is deployed on the BBU site. DU can be deployed with either traditional DRAN architecture or CRAN architecture, which depends on specific installation scenarios.

The overall deployment of CloudRAN is shown in Fig. 2.5. In Option 1, CU is deployed in a regional data center (DC), which can realize highly centralized deployment of CU, and the corresponding DU can be deployed in the form of CRAN or DRAN/CRAN coexistence. CU is highly centralized and can achieve a greater range of control processing; however, the farther CU is from its users, the greater the service latency, and it is not suitable for latency-sensitive services. In Option 2, CU is deployed in an edge cloud data center, which corresponds to an ultra-low latency business scenario, with less BBU under the hang and less centralized CU; therefore, DU is generally suitable for DRAN deployment. It can be seen that CU in Option 2 is closer to users, has low latency, and can support latency-sensitive services well.

Implementing CloudRAN architecture will greatly increase the degree of collaboration and resource resilience of the radio access network and facilitate unified and simplified operations and maintenance. Overall, CloudRAN architecture serves the following purposes.

(1) To unify the architecture to achieve multi-dimensional integration of multi-system, multi-band, multi-layer, and super-dense networks.
(2) To utilize centralized control to reduce the complexity of the radio access network and facilitate efficient inter-system/inter-station service collaboration.
(3) To smoothly introduce 5G and achieve the ultimate user experience through dual connectivity.
(4) To decouple software and hardware and open the platform to promote agile service online.

(5) To facilitate the introduction of artificial intelligence and realize intelligent operation and maintenance management of radio access network slices to adapt to the diversity of future services.
(6) To realize resource pooling of cloud-based architecture so that networks can be deployed on demand and expanded/downsized elastically to improve resource utilization efficiency and protect investment.
(7) To adapt to various interface splitting options to ensure flexible networking under different transmission conditions.
(8) To achieve centralized deployment of network elements to save server rooms and reduce operating expenses.

2.1.2 5G Core Network Architecture

The core network is the control center of the mobile network and can implement various functions, including providing access and mobility management, session management, data forwarding, operator billing, and other related functions. In addition, it can target policy control functions (such as rate control, billing control, etc.) for different service scenarios. The core network is implemented by multiple logical entities, with different logical entities deployed on different dedicated hardware or common servers.

The 5G core network can be divided into the control plane and user plane. Both the control plane and user plane are deployed separately and at different locations according to different network requirements. And different logical entities are used to realize the control function and user plane forwarding function. The core network control plane elements are used to realize the control functions of the core network and generate corresponding control policies, such as billing control, access control, mobility management, policy control, etc.; the core network user plane elements are used to support the forwarding of user data messages. For example, the upstream/downstream data transmission of users.

The core network can be deployed in different locations according to different network demands, and it can be divided into three types of data centers (DC): central DC, regional DC, and edge DC. Among them, central DC is deployed in regional centers or provincial capitals, and regional DC is deployed in local city server rooms. Besides, edge DC is deployed in the transport network access server rooms or wireless server rooms. The core network control plane network elements are generally deployed in central DC rooms. In order to meet the needs of low latency service or service traffic diversion, core network user plane network elements will gradually move down to regional DC rooms or further down to edge DC rooms to shorten the distance from a base station to the core network and reduce the number of equipment involved in the forwarding process, so as to reduce service forwarding delay.

Before the birth of the 5G network, network elements defined by the 3GPP standard had multiple logical functions. Some logical function modules were similar or overlapped between different network elements, making it impossible to customize the control function combinations for a specific service type. Different services share the same set of network elements, and the coupling between many logical functions is very strong. Besides, the interface between network elements is particularly complex. These bring great difficulties to online service and network operation and maintenance, and it is not flexible enough to support multi-service scenarios in the 5G era.

To adapt to various service demands in the future, 5G network architecture has been placed with very high expectations. 3GPP standard combines IT cloud-native concept to change 5G network architecture in two ways: Firstly, control plane functions are abstracted into multiple independent network services, aiming at building the network in a software-based, modular-based and service-oriented way; secondly, Control and User Plane Separation (CUPS) enables user plane function to be free from "centralized" constraints, so that it can be flexibly deployed either in the core network or in the access network that is closer to users.

5G network architecture, which is a Service Based Architecture (SBA) as shown in Fig. 2.6, includes interfaces between network functions, as well as control plane network functions and user plane network functions. Among them, the core network control plane network has functions including Access and Mobility Management Function (AMF), Session Management Function (SMF), Network Repository Function (NRF), Unified Data Management (UDM), Policy Control Function (PCF), Authentication Server Function (AUSF), and Short Message Service Function (SMSF), etc. The user plane network of the core network has functions, including UPF. These network functions are used to implement different logical functions. Among them, AMF is used for user access and mobility management, including user authentication, user registration management, connection management, idle state mobility management, and connection state switching management. SMF is used for session management, including session creation, modification and release, IP address assignment, UPF control, etc. NRF is used for service registration and discovery. Core network functions can register their own supported service information into NRF for service discovery by other network functions. PCF is used for policy control, including providing mobility management policies for AMF, session management policies for SMF, and terminal policies for User Equipment (UE). UDM is a unified data management unit that holds a range of data, including user contract data.

5G core network uses service-oriented interfaces between control plane network functions to provide a series of services for other control plane network functions to invoke in the form of an open Application Programming Interface (API), which is defined based on HTTP2.0. Network functions register information about services they support into NRF, and other network functions can query NRF to obtain information about services that have completed registration with NRF.

The user plane network function of the core network provides forwarding of user messages, and the specific forwarding policy is controlled by the control plane network function. The user plane network function of the core network mainly refers to UPF with different roles (intermediate forwarding UPF, anchor UPF, etc.). SMF

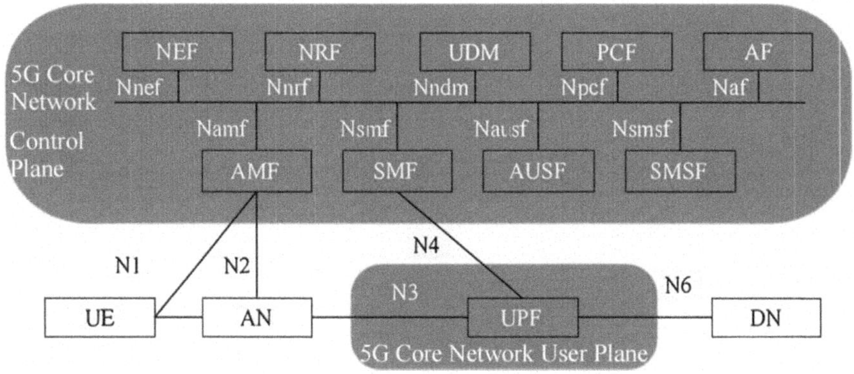

Fig. 2.6 Service based architecture (SBA)

passes message forwarding rules to UPF through N4 interfaces so that UPF can process user data messages according to the forwarding rules.

There are N2 and N3 interfaces between the core network and Access Network (AN). N2 interface is a control plane interface, connecting radio access networks AN to AMF, and is used to deliver control plane signaling, including mobility management, session management, and other control-related signaling. N3 interface is a user plane interface that connects radio access networks AN to UPF and is used to deliver user data messages.

In addition, there is a logical interface between UE and AMF for passing control plane signaling, such as registration and session management of UE. The interface between UE and AMF is called N1 interface or Non Access Stratum (NAS) interface.

In addition to the service-based interface architecture, 5G core network defines a reference point-based architecture, as shown in Fig. 2.7. The reference point-based architecture describes the interface relationships between network functions of 5G core network and interfaces between different network functions of core network. For example, in the reference point architecture in Fig. 2.7, reference point N2 exists between AMF and RAN. Besides, reference point N11 exists between AMF and SMF; Reference point N14 exists between AMF and AMF; Reference point N12 exists between AMF and AUSF; Reference point N8 exists between AMF and UDM; Reference point N15 exists between AMF and PCF, and reference point N1 exists between AMF and UE, etc. This indicates that in order to accomplish the control functions of AMF, AMF exchanges messages with these network functions with which it has reference points. The reference points (interfaces) between network functions on the control plane of the core network are still defined through services of the service-oriented architecture, which means that message passing and interaction between network functions on the control plane of the core network is performed through the invocation of services.

Each network function is operationally decoupled from other network functions and provides service-oriented interfaces to the outside world. The same network function can provide services to multiple other callers through the same interface, and

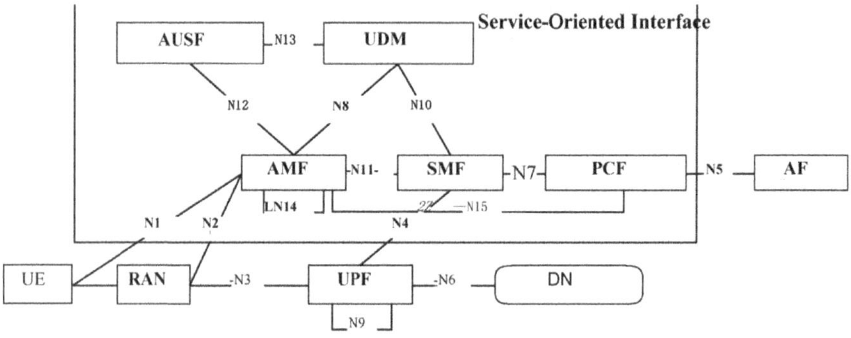

Fig. 2.7 Reference point-based architecture

multiple reference point interfaces can be considered as a unified service interface, thus reducing the number of logical interfaces to be configured. Take UDM as an example, in the reference point-based architecture, UDM has interfaces with AMF and SMF, and it needs to configure different docking data for these two interfaces in accordance to reference point architecture. In the service-based architecture, UDM can define a unified API, and both AMF and SMF can use API to obtain corresponding services from UDM.

Table 2.1 briefly describes network functions involved in 5G networks, including control plane network functions and user plane network functions.

5G core network design concept is similar to cloud-native and allows different network functions to be implemented through microservices. Network Function Virtualization (NFV) and Software Defined Network (SDN) technologies are used to enable different network functions to interact. These services are deployed on a shared, orchestrated cloud infrastructure and then designed accordingly to eventually match different service requirements.

2.1.3 5G Transport Network Architecture

5G transport network is composed of transport network equipment interconnected by fiber optic cable, such as Slicing Packet Network (SPN) equipment or IP Radio Access Network (IPRAN) equipment, and achieves the corresponding 5G service transmission function through an IP routing protocol, fault detection technology, protection reversal technology, etc. The main function of a 5G transport network is to connect a base station with another and a base station with a core network, provide a data forwarding function, and ensure data forwarding delay, rate, security, and other indicators to meet the requirements of relevant services. The 5G transport network structure can be divided from two dimensions: physical level and logical level.

In terms of the physical level, the transport network is divided into three parts: the front transmission network, which is used to connect RRU and DU, the middle

Table 2.1 5G core network function profiles

Network function	Function overview
ANIF	Access and Mobility Management Function executes registration, connection, reachability, and mobility management. It provides a message transmission channel for session management between UE (User Equipment) and SMF (Session Management Function), offering authentication, authorization function during user access, and serving as the control plane access point for terminal and wireless core network connectivity
SMF	Session Management Function is responsible for tunnel maintenance, IP address allocation and management, UP function selection, policy implementation, control within QoS, billing data collection, and roaming
AUSF	Authentication Server Function performs access authentication for both 3GPP and non-3GPP accesses
UPF	User Plane Function handles packet routing and forwarding, policy implementation, traffic reporting, and QoS handling
PCF	Policy Control Function offers a unified policy framework, providing policy rules for control plane functionalities
UDM	United Data Management Function includes 3GPP AKA authentication, user identification, access authorization, registration, mobility, subscription, and short message management, among other functions
NRF	Network Registration Function is a new feature that provides registration and discovery capabilities, enabling Network Function (NF) to discover each other and communicate via API interfaces
NSSF	Network Slice Selection Function determines LIE allowed network slice instances based on LIE slice selection assistance information contract information_ etc
NEF	Network Exposure Function exposes the capabilities of various NFs, transforming internal and external information
SMSF	Short Message Service Function is responsible for handling the forwarding and processing of short messages
AF	Application Function Entity processes business QoS authorization requests and more

transmission network, which is used to connect DU and CU, and the back transmission network which is used to connect CU and 5G core network (5G Core, 5GC), as shown in Fig. 2.8. Among them, the mid-transmission network is only available after the cloud evolution of BBU and the deployment separation of CU and DU. Suppose CU and DU are not separately deployed. In that case, the end-to-end networking architecture of the transport network will have only two parts: the front transmission network and the backhaul network, with the latter being generally known as the 5G transport network. The backhaul network will also achieve longer distance transmission with the help of wavelength division equipment, as shown in Fig. 2.1. The three upper rings are a 5G transport network. According to the location of equipment deployment and the requirements of service bearing, the 5G transport network is divided into three levels: access ring network, convergence ring network and core

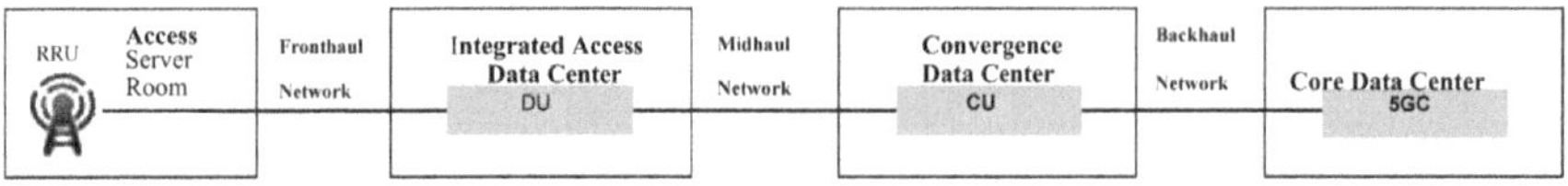

Fig. 2.8 Classification of carrier networks

ring network. The two lower rings are wavelength division network. Wavelength division rings are capable of long-distance transmission of large particles, while 5G transport networks have flexible forwarding capacities. The upper and lower rings are used together and can enable the flexible long-distance forwarding of large particles within the transport network. Generally, the front transmission network consists of 100 GE CPRI ports or 25 GE eCPRI ports. The access ring of the backhaul network, on the other hand, consists of 50 GE ports or 100 GE ports of Ethernet ports. The aggregation and core rings of the backhaul network consist of 200 GE ports or 400 GE ports of Ethernet ports.

Regarding the logical level, the 5G transport network is divided into three logical planes: management, control, and forwarding. Among them, the management plane realizes the basic management function of the transport network controller to the transport network equipment. The control plane calculates the transport network forwarding path (service tunnel). The forwarding plane realizes the function of forwarding user messages between different base stations or between base stations and core networks.

In the 5G stage, with the continuous improvement and wide application of NFV technology, the composition of the 5G core network will be decomposed into Control Plane (CP) and User Plane (UP). These two parts can be separated and deployed in a distributed manner, especially UP. It can be deployed downward according to service demand. Furthermore, it is closer to users, thus providing lower latency and better user experience.

With the downward shift of UP, the greatest change brought by 5G core network clouding to the transport network is the connection change. In the 4G era, the connection from a base station to a core network is convergence type. The network traffic is mainly S1 traffic, accounting for about 95%, and all S1 traffic is converged by thousands of base stations to several sets of core network deployed in the core layer. Whereas, after the downshifting of UP for the 5G core network, traffic from different core network UPs exists at individual base stations. For example, autonomous driving services are handled at MECs at the edge, and services like video classes are terminated at local data centers. Due to the need for content backup, VM migration, etc., traffic also exists between different layers of core network UPs, resulting in the mesh of traffic for the whole network. At the same time, the downshifting of core network UP is not a quick fix but should be planned and deployed according to the actual service development demand, considering multiple factors, such as network construction cost, user experience, etc. To cope with the mesh connection and its uncertainty, the 5G transport network needs to move down the three-layer network, at least to the location of the mobile edge computing MEC, to achieve flexible scheduling. The

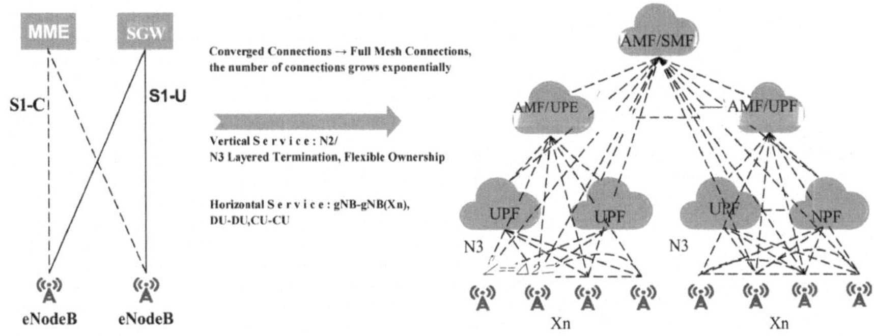

Fig. 2.9 The impact of 5G core network on the network architecture of 5G transport network

impact of the 5G core network on the networking architecture of the 5G transport network is shown in Fig. 2.9.

2.2 5G End-to-End Network Key Technologies

2.2.1 5G Radio Key Technologies

Compared to traditional 2G, 3G, and 4G networks, 5G networks can provide higher rates, lower latency, and larger connections. What are the performance goals of 5G air interfaces, and how are they achieved? With these two questions, the following will introduce the key technologies on the wireless side of 5G from the performance targets of 5G air interfaces, 5G new spectrum, 5G new coding technology, 5G new antenna technology, and super uplink.

1. **5G Air Interface Performance Targets**

The International Telecommunication Union defined three future application scenarios for 5G in June 2015, which are eMBB, URLLC and mMTC. Among them, eMBB refers to high-traffic mobile broadband services, such as augmented reality, virtual reality, and ultra-high-definition video. URLLC refers to services that require high reliability and low latency connections, such as unmanned driving, industrial control, etc. mMTC refers to large-scale IoT services, such as for smart cities, environmental monitoring and other application scenarios targeting sensing and data collection. In the eMBB scenario, the main focus is on peak rate and user experience rate, etc. In the URLLC scenario, the main concerns are latency and mobility. While in the mMTC scenario, the main focus is on the connection density.

To meet the future requirements of different service applications on network capacity, high requirements are set for 5G air interface rate, latency, connectivity and other performance, which are respectively 20 Gbps peak rate, 100 Mbps user

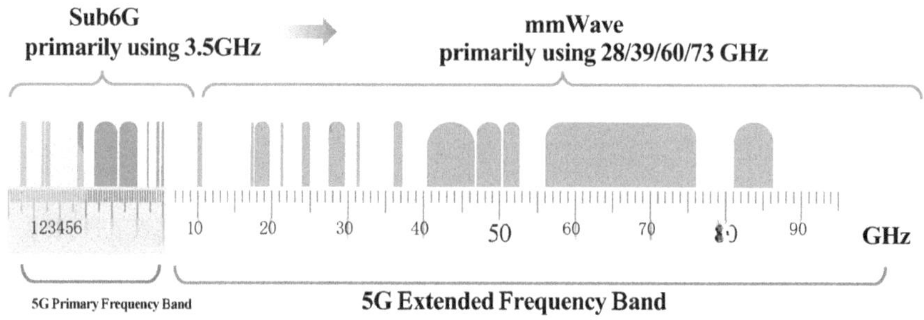

Fig. 2.10 5G spectrum resources

experience data rate, 1 ms air interface latency and the connection number density of 1 million devices per square kilometer.

2. **5G New Spectrum**

Spectrum resource is an important resource for wireless communication, and it is also a key factor affecting air interface rate and coverage. To enhance the air interface rate of 5G, its spectrum resource must also be very abundant.

In the 3GPP protocol, the overall spectrum resources for 5G can be divided into the following two frequency ranges (FR), as shown in Fig. 2.10.

FR1: It is the main band for 5G, also known as the low-frequency band, in which frequencies below 6 GHz are called Sub6G, while the band with frequencies from 3 to 6 GHz is usually called C-Band.

FR2: Millimeter wave, or what is called high-frequency band, is an extended band for 5G, and its spectrum resources are abundant.

FR1 is the main frequency band for 5G, with a maximum cell carrier bandwidth of 100 MHz. FR1 contains the traditional frequency bands below 3 GHz, as shown in Table 2.2. Many of these bands are spectrum resources previously used by2G, 3G, and 4G. 5G also supports these spectrum resources. As users continue to migrate to 5G networks, the spectrum resources used by 2G, 3G, and 4G can be gradually released, and some of them can be used for 5G deployment.

Sub3G spectrum resources are limited, and in order to meet the need for large bandwidth in 5G networks, the 3GPP protocol defines new frequency bands in FR1 for supporting 5G deployments, as shown in Table 2.3. In the world, 3 to 6 GHz is the main spectrum resource for current 5G deployment.

To enhance the unbalanced uplink and downlink coverage, 5G uses uplink and downlink decoupling techniques to enhance uplink coverage. Therefore, the 3GPP protocol defines Supplementary Uplink (SUL) bands as shown in Table 2.3 for uplink and downlink decoupling scenarios.

FR2 is the future expansion band for 5G, as shown in Table 2.4. With the development of 5G networks, the demand for spectrum resources will also increase as the demand for network capacity continues to increase. The spectrum resources

Table 2.2 Sub3G frequency bands

NR frequency band	Uplink frequency range	Downlink frequency range	Duplex mode
n1	1920–1980 MHz	2110–2170 MHz	FDD
n2	1850–1910 MHz	1930–1990 MHz	FDD
n3	1710–1785 MHz	1805–1880 1VIII2	FDD
n5	824–849 MHz	869–894 MHz	FDD
n7	2500–2570 MHz	2620–2690 MHz	FDD
n8	880–915 MI-Tz	925–960 MHz	FDD
n20	832–862 MHz	791–821 MHz	FDD
n28	703–748 MHz	758–803 MHz	FDD
n38	2570–2620 MHz	2570–2620 MITz	TDD
n41	2496–2690 MHz	2496–2690 MHz	TDD
n50	1432–1517 MHz	1432–1517 MHz	TDD
n51	1427–1432 MHz	1427–1432 MHz	TDD
n66	1710–1750 MHz	2110–2200 MHz	FDD
n70	1695–1710 MHz	1995–2020 MHz	FDD
n71	663–698 MHz	617–652 MHz	FDD
n74	1427–1470 MHz	1475–1518 MHz	FDD

Table 2.3 C-band and supplementary uplink bands

NR frequency band	Frequency range	Duplex mode
n75	1432–1517 MHz	SDL
n76	1427–1432 MHz	SDL
n77	3.3–4.2 GHz	TDD
n78	3.3–3.8 GHz	TDD
n79	4.4–5.0 GHz	TDD
n80	1710–1785 MHz	SUL
n81	880–915 MHz	SUL
n82	832–862 MHz	SUL
n83	703–748 MHz	SUL
n84	1920–1980 MHz	SUL

within FR2 are abundant and can provide spectrum resources for subsequent network expansion.

The lower the frequency, the better the coverage capability, but the more limited the available spectrum resources are. Millimeter wave resources are abundant, but their coverage capability is weak.

Combining available spectrum resources and coverage capability, the application scenarios of different spectrums of 5G are different, as shown in Fig. 2.11. Millimeter waves are mainly used in densely populated urban areas to take advantage

Table 2.4 FR2 bands

NR frequency band	Frequency range	Duplex mode
n257	26,500–29,500 MHz	TDD
n258	24,250–27,500 MHz	TDD
n260	37,000–40,000 MHz	TDD
n26I	27,500–28,350 MHz	TDD

of large bandwidth and absorb large service volumes. C-band takes advantage of both coverage and bandwidth and is used to achieve continuous coverage in urban areas. Sub3G takes advantage of coverage to achieve continuous coverage in suburban and rural areas.

3. **5G New Coding Technology**

In mobile communication, due to interference and fading, errors occur in the transmission of signals. Hence, error correction and error detection coding techniques need to be applied to digital signals to enhance the ability of data to resist various kinds of interference during channel transmission and improve the system's reliability. Error correction and detection coding of digital signals transmitted in the channel is channel coding technology.

The three major scenarios for 5G are eMBB, mMTC, and URLLC, and the performance metrics vary greatly across application scenarios. This places higher demands

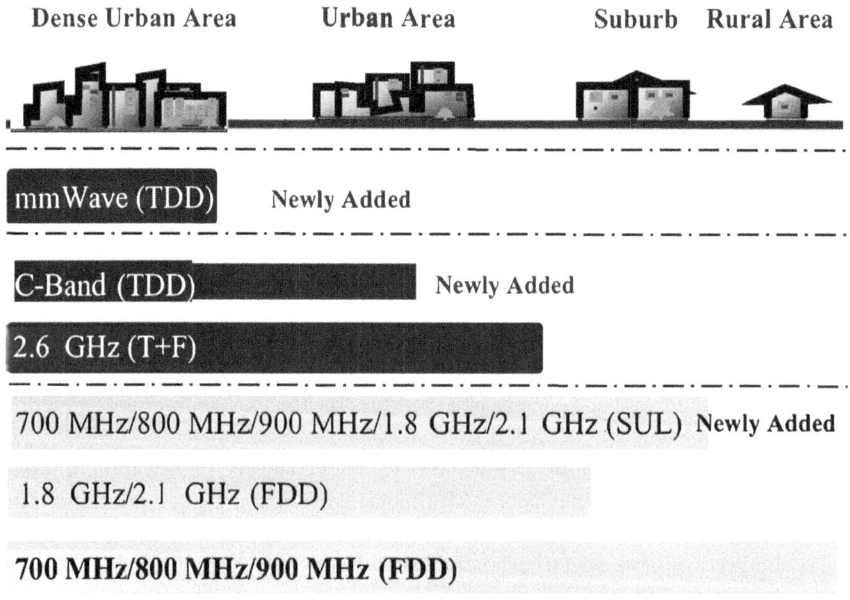

Fig. 2.11 5G spectrum applications

on 5G channel coding, which needs to support a wider range of code block lengths and higher coding rates. If the same coding efficiency is used for all application scenarios, it will waste data and thus spectrum resources.

The mainstream candidate coding schemes proposed by the industry are Polar codes, Turbo codes, and Low-Density Parity Check (LDPC) codes. The following describes the basic principles of channel coding selection.

(1) Coding performance: error correction capability and coding redundancy rate.
(2) Coding efficiency: complexity and energy efficiency.
(3) Flexibility: the size of encoded data blocks and the ability to support Incremental Redundancy-Hybrid Automatic Repeat Request (IR-HARQ).

It is found in comprehensive comparison that Turbo codes have been widely used and have good reliability. Still, the decoding latency is high, which makes it difficult to meet the requirements of 5G data transmission with low latency. LDPC codes have been verified for many years, with high theoretical and product maturity, and the latency of parallel decoding can support 5G low latency design. Polar codes are considered to be the coding technique closest to the Shannon Limit. Polar codes have higher coding efficiency and reliability than Turbo codes, which can meet service applications with high-reliability requirements for 5G, such as remote real-time control and unmanned driving.

In November 2016, in the 87th meeting of 3GPP RAN1, 3GPP selected Polar code as the coding scheme for the control channel of the 5G eMBB application scenario and LDPC code as the coding for the eMBB data channel. This is the final consensus reached by several companies worldwide after detailed evaluation of the performance, complexity, latency, and power consumption of multiple candidate coding schemes under unified guidelines.

4. **5G New Antenna Technology**

The rate of 5G has increased a lot compared to that of 4G, and because of this, 5G can realize large bandwidth services such as watching 4 K HD video or Cloud VR anytime and anywhere. The increase in 5G speed is mainly related to the increase of carrier bandwidth and spectrum efficiency, and the key new antenna technologies to improve spectrum utilization efficiency are introduced in detail below.

1) Massive MIMO

Massive MIMO is a large-scale antenna array, as shown in Fig. 2.12, which usually requires at least 16 transceiver antennas. At present, 64-channel transceiver antennas are mainly used in the network, which realizes 64T64R. With a greater number of antennas, a more flexible and accurate three-dimensional narrow beam assignment can be achieved, which enables more users to multiplex wireless time–frequency resources, thus achieving the purpose of improving coverage capability and system capacity and reducing system interference.

The implementation of Massive MIMO beam assignment is shown in Fig. 2.13. Massive MIMO uses the principle of wave interference and superposition so that

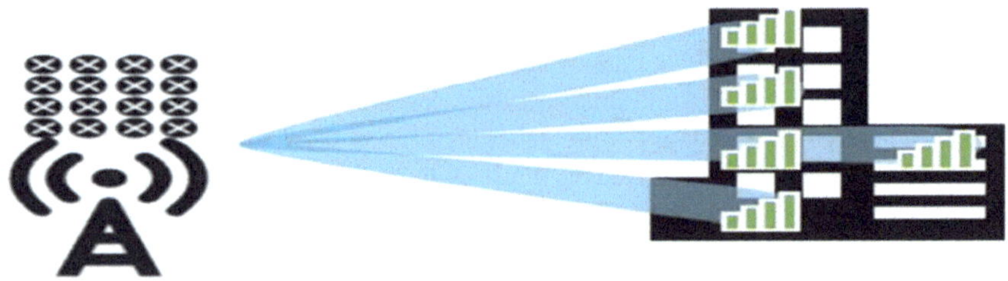

Fig. 2.12 Schematic diagram of massive MIMO transceiver antenna array

peaks are superimposed on peaks to enhance signals, and the wave crest is superimposed with the wave trough to weaken the signals. The base station estimates the downlink vector right from the uplink signal sent by the terminal or directly obtains it through the terminal reporting and finally uses this vector right to weigh the downlink signal to be sent, thus forming a directional beam.

Massive MIMO enables signal beam assignment not only in a horizontal direction but also in a vertical direction. In this way, the signal coverage of high-rise buildings is greatly improved, thus effectively enhancing the service experience of high-rise building users.

Due to narrower beam, Massive MIMO with 64T64R can send up to 16 data streams simultaneously in downlink through space division multiplexing, as shown in Fig. 2.14. This means that the base station can allocate the same time–frequency resources to 16 different users at the same time, thus significantly increasing the overall capacity of the cell, which is especially suitable for high-traffic scenarios, such as universities and urban central business districts. At the same time, the narrower beam can also reduce the interference between users in the cell.

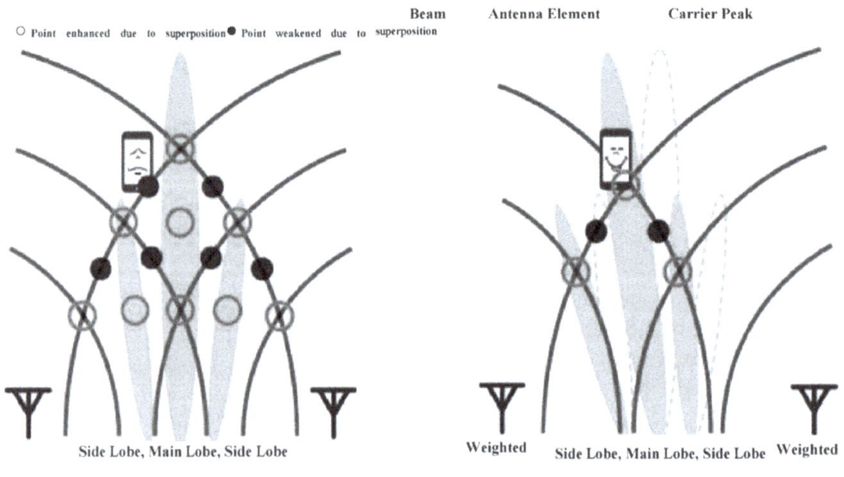

Fig. 2.13 Implementation of massive MIMO beam assignment

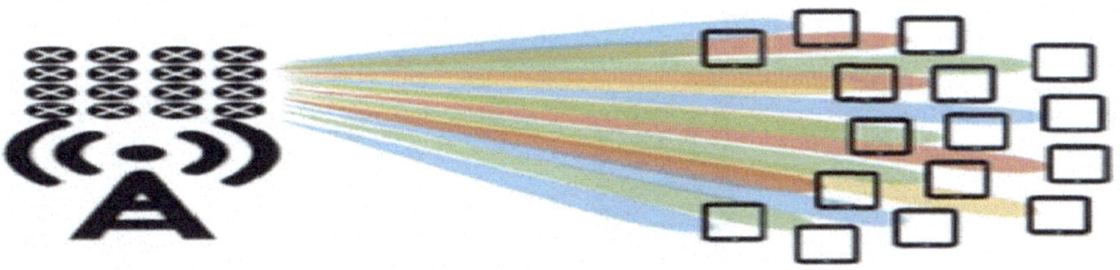

Fig. 2.14 64T64R antenna multi-stream transmission effect

The prerequisite for downlink multi-stream data transmission is that different terminals need to complete pairing in advance. For terminals that have completed pairing, the base station will schedule the same time–frequency resources for these end users, thus significantly improving spectrum resource utilization. At this stage, a 5G base station can theoretically pair up to 16 users in downlink and eight users in uplink. By realizing simultaneous transmission and reception of 16 streams in downlink and eight streams in uplink, the spectrum efficiency of air interface is improved, thus increasing the capacity of the cell uplink/downlink.

Massive MIMO effectively improves the capacity of a cell, so for a single user, is there the same gain in capacity improvement? If there are two antennas for the user terminal, Massive MIMO technology cannot improve the peak rate for a single user compared to the traditional 8T8R. Because the peak rate of the terminal is limited by the number of downlink receiving antennas, even if the base station side sends 16 data streams at once, the terminal can only receive up to two of the data streams simultaneously, so the peak rate will not increase. However, due to the use of Massive MIMO technology, the signal quality for a single user will be significantly improved compared to the traditional method, and then the user can adopt a more efficient coding scheme and higher-level modulation, so the average rate for a single user will also be increased.

When there are four or even more antennas for a user terminal, the peak rate for a single user will be increased exponentially because the user can receive four or even more data streams simultaneously in downlink at that point. At this stage, the mainstream 5G cell phone terminals support a 4-antenna configuration, and all four antennas support signal reception, while only 2 out of the four antennas support both signal reception and signal transmission, realizing 2T4R, as shown in Fig. 2.15. In this way, combined with the downlink multi-stream feature of Massive MIMO, the peak downlink rate of 5G terminals will be at least double compared to that of 4G terminals (1T2R).

2) Distributed-MIMO (D-MIMO)

Massive MIMO greatly enhances capacity and coverage in small areas by increasing the number of antennas at a given base station and enhancing beamforming techniques. If we integrate neighboring base stations in the area as super base stations and perform beam assignment and multiplexing with multiple antennas similar to

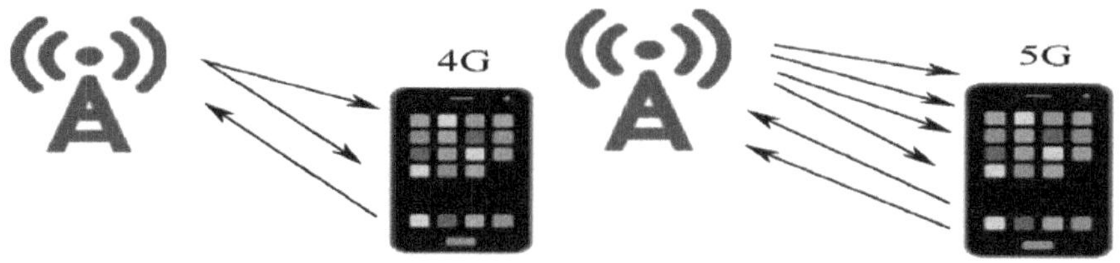

Fig. 2.15 Schematic diagram of 4G/5G terminal antenna transceiver mode

Massive MIMO, the capacity and the coverage of the whole area can be enhanced. This is Distributed-MIMO (D-MIMO) technology.

The traditional way of network construction is network-centered, and user experience is limited by network planning and design. Besides, strong interference and weak coverage inevitably affect user experience. D-MIMO technology, on the other hand, unites base stations in a certain area to provide services to users, turning co-channel interference in traditional networks into useful signals, improving edge coverage, and achieving joint beam assignment and multiplexing in the spatial dimension to improve system capacity. Therefore, D-MIMO can be regarded as a new user-centered network construction method, which greatly reduces the reliance on network planning and optimization. Looking ahead, Massive MIMO will achieve the ultimate beamforming capability within the station, while D-MIMO will enhance spatial coordination between stations, both of which are naturally complementary and will become the core technology of future 5G.

5. **Super Uplink**

5G new services have an increasing demand for network bandwidth, where toC services require the high speed of network downlink and also need further improvement of network uplink rates, such as ultra-high definition video and VR services shown in Fig. 2.16. The applications in the toB industry are more about transferring information from the device side to the network side, so they put higher demands on the network uplink rate. As shown in Fig. 2.17, backhauls of HD medical images in remote medical care, HD inspection video in smart grid, machine vision in smart manufacturing, and images taken by HD cameras in autonomous driving vehicles, all require extremely high uplink rates.

Currently, 5G mainly adopts Time Division Duplex (TDD) method of networking, that is, uplink and downlink time divisions multiplexing a section of spectrum resources, and the actual time and frequency resources for uplink are limited, resulting in poor user uplink experience.

Super Uplink greatly increases the available uplink time–frequency resources for users by sending uplink data in the SUL spectrum and NR TDD spectrum in a time-sharing manner, as shown in Fig. 2.18. NR TDD spectrum is used in the uplink time slot for uplink data transmission. In the downlink time slot of the NR TDD spectrum, the free SUL spectrum is used to supplement uplink data transmission, ensuring that uplink data can be sent in the full-time slot, as shown in Fig. 2.19.

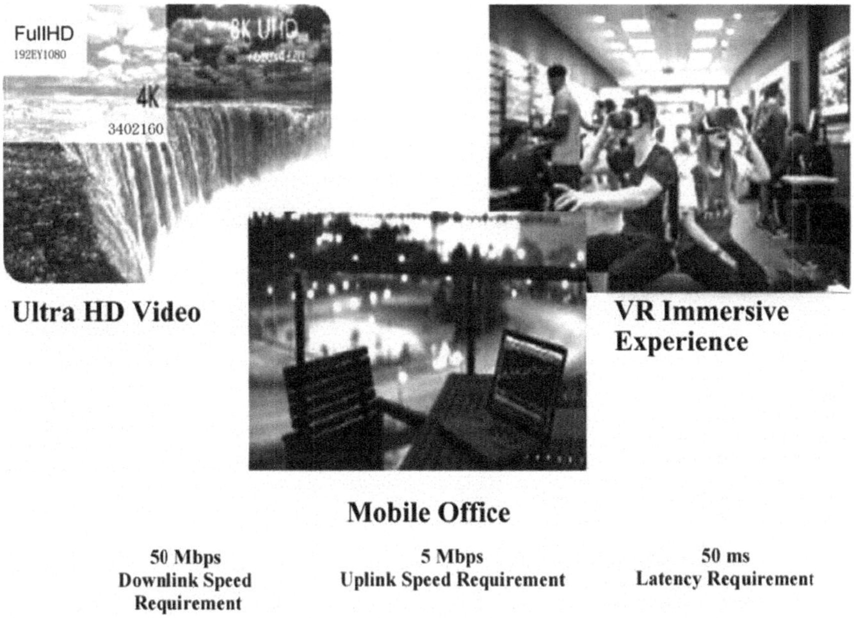

Fig. 2.16 Network bandwidth requirements for typical toC services

How much improvement does Super Uplink have on the uplink rate? The following is an example of C-Band using 100 MHz to support uplink 2-antenna transmission and Sub3G using 20 MHz to support 1-antenna transmission, as shown in Fig. 2.20. Referring to Table 2.5, the theoretical peak uplink rate equals the rate on C-Band plus 80% of the rate on Sub3G (because the terminal only occupies 80% of the resources on Sub3G), which is about 349 Mbps.

2.2.2 *Key Technologies of 5G Core Network*

With the rapid growth of mobile networks, the service scenarios of 5G networks have changed dramatically, from mainly toC service scenarios to mainly toB service scenarios. 5G end-to-end network is undergoing profound changes, and different applications have great variability in network requirements, putting forward demands for large bandwidth, high-density connectivity, and ultra-low latency on network side. In order to adapt to various service scenarios, core network architecture needs to become more flexible and agile to adapt to the demands of various services.

Currently, as 5GtoB is developing rapidly, industry customers are putting forward more stringent requirements for network maintenance and network quality, as well as specific index requirements for network isolation, security, and convenience.

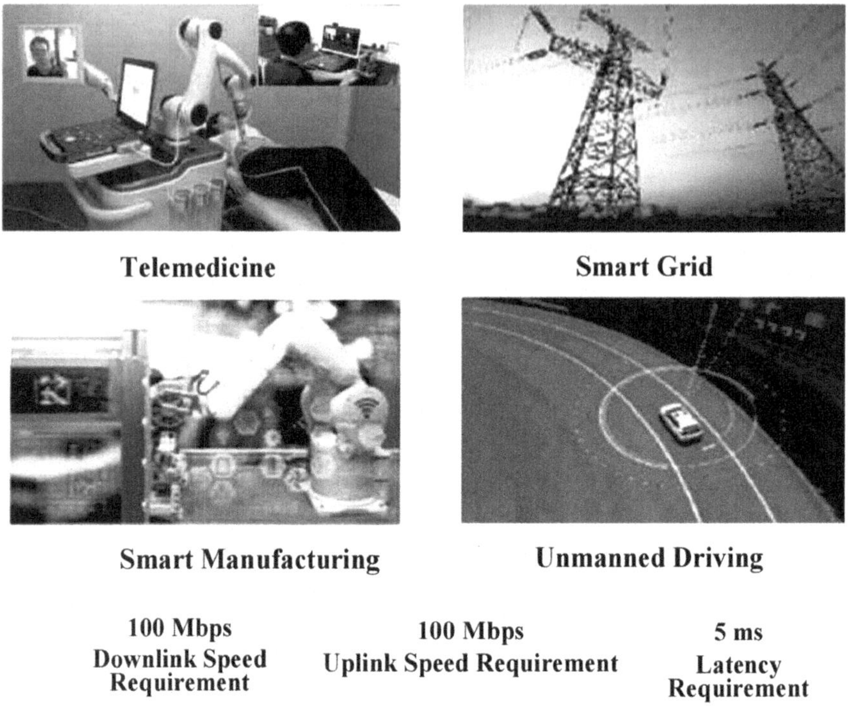

Fig. 2.17 Network bandwidth requirements for typical toB services

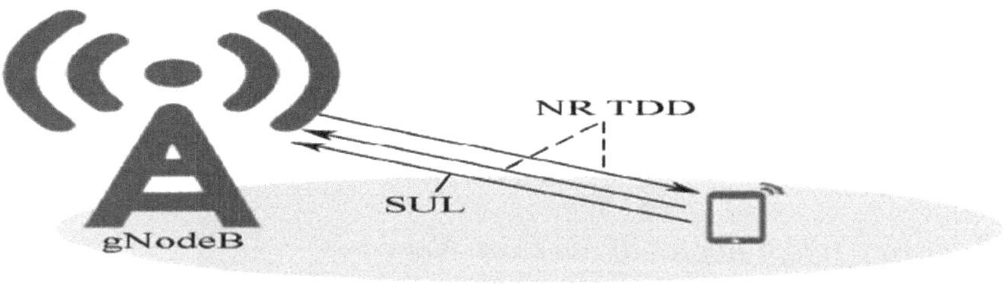

Fig. 2.18 Super uplink transceiver diagram

To address various concerns and challenges from toB customers, different solutions are proposed for the core network. The following will introduce key technologies on 5G core network side in terms of data not leaving the park (UPF sinking), service isolation (5G slicing solution), 5G Local Area Network (5G LAN), and service non-disruption (edge high availability gateway solution).

1. **Data not Leaving the Park (UPF Sinking)**

Before the birth of 5G networks, the control and user planes of the core network forwarding gateway were intertwined and difficult to divest. In other words, data

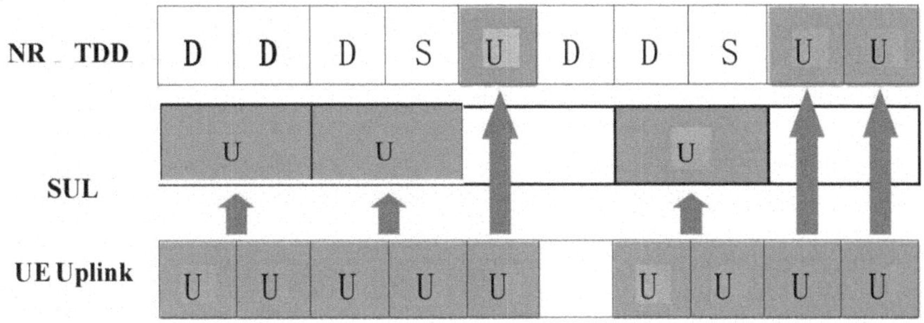

Fig. 2.19 Super uplink time slot usage diagram

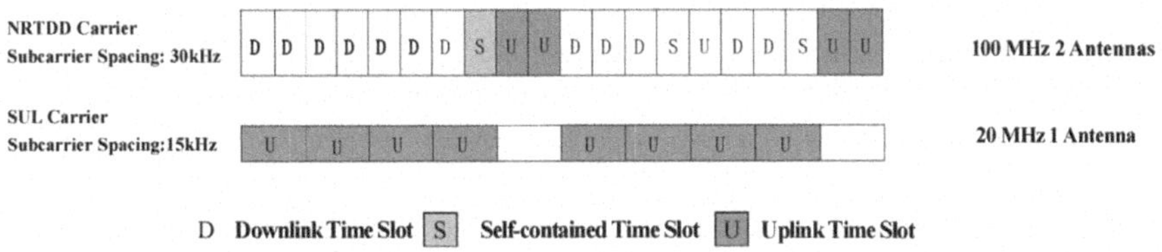

Fig. 2.20 Example of super uplink scenarios

Table 2.5 Theoretical peak of uplink rates

Frequency band	Antenna number	Configuration	Bandwidth	Time slot ratio	Theoretical uplink rate
C-Band	Uplink 2 T	256QAM	100 MHz	8:2	254 Mbps
Sub3G	Uplink 1 T	256QAM	20 MHz	FDD	119 Mbps
Total uplink rate of supplementary uplink terminal = 254 + 119 × 0.8 ≈ 349 Mbps					

forwarding devices were responsible for both control plane forwarding and user plane forwarding. In 4G networks, the 3GPP Rel-14 standard defines Control and User Plane Separation (CUPS) architecture, which separates network functions of the Serving GateWay (SGW) and Public Data Network Gateway (PDN GateWay, PGW) into the control plane and user plane. A new logical interface between the control plane and user plane is added to support the service control of the control plane on the user plane. 5G core network completely separates the control and user planes through technologies such as SBA. The control plane function is carried by multiple Network Function (NF), and the user plane function is carried by the user plane network element UPF, which can be flexibly deployed in the centralized control room of the core network or in the access network server room of the transport network closer to users.

Once the user plane gateway function is independent, it can be deployed in a distributed manner according to the service scenario, either in a central Data Center

(DC), in a local DC, or even in an edge DC closer to users. This depends on the vertical industry's network requirements, such as latency, bandwidth, reliability, etc.

As mentioned above, the rapid growth of 5GtoB has put forward higher requirements on the network, such as lower latency, traffic not leaving the park and other service demands. After the complete separation of the control plane from the user plane of 5G core network, the control plane can be collectively deployed and the user plane can be sunk to different DCs as needed.

A more flexible network architecture can be constructed with CUPS, and the unique value can be realized as follows.

Reduce user plane latency: The user plane is deployed in the user access area, which can access the enterprise network through a nearby user plane, reducing the transmission distance and transmission latency.

Save backbone transmission resources: The user plane is separated and sunk to local city deployment or even enterprise park, and direct access to the enterprise intranet or local Content Delivery Network (CDN) reduces the consumption of backbone transmission resources and transmission cost.

Simplified operation and networking: The control plane is centrally deployed and the signaling interface is unified, which has no impact on the docking of peripheral network elements before and after CUPS transformation, thus simplifying control plane operation and networking with less change in personnel and architecture of the operation and maintenance.

Flexible expansion: Under CUPS networking architecture, the user plane can be expanded independently to address the increasing throughput of operators. The control plane can be expanded separately to increase the number of user accesses.

In the actual deployment of 5GtoB incumbent networks, UPF is typically co-located with MEC, with UPF acting as a data-plane input to MEC. The European Telecommunications Standards Institute (ETSI) defines MEC as the provision of IT service environments and cloud computing capabilities at the edge of the mobile network. Thus, MEC can be understood as a cloud server running at the edge of the mobile network that can handle tasks that cannot be dealt with by the traditional network infrastructure.

There are generally two types of MEC deployed by operators: shared MEC and On-site MEC. The difference between these two deployment methods is that a shared MEC is one in which multiple customers share one MEC resource, and the deployment location is generally in a local city or region, supporting multiple services for multiple tenants. On-site MEC is generally deployed inside an enterprise park, mainly for multiple services of one enterprise, with exclusive use of resources, as shown in Fig. 2.21. Different services of different enterprises can access different enterprise intranets through the same MEC.

In the process of MEC deployment, UPF mainly functions as triage. Currently, there are two main core network triage scenarios. One is a service level triage approach, where local services are identified and triaged through Uplink Classifier (UL CL) service. The other is Access Point Name (APN) approach, where triage is performed by signing different Access Point Names/Data Network Name (DNN). Service level triage can be done in the form of UL CL for service identification and

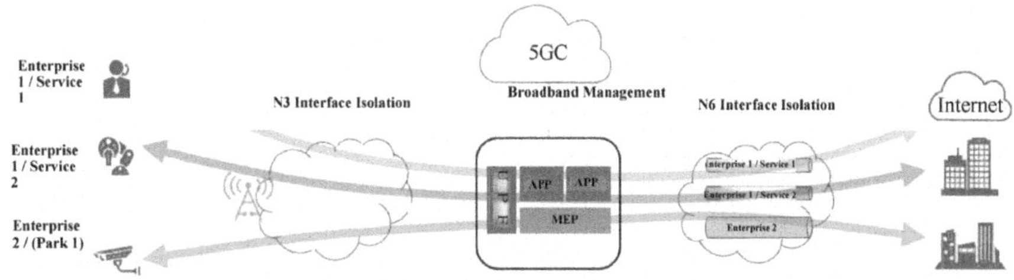

Fig. 2.21 MEC shared by different enterprises

triage for local traffic offloading. This type of triage network performs real-time triage based on actual access traffic and service identification by UL CL UPF to offload or redirect traffic. APN-level triage points to different data networks by signing terminal users to different APNs. The main application scenario of APN-level triage is that the user identity can be known in advance and the contract can be signed in advance when all traffic is sent to the unified data network through the UPF selected by the APN. The main application scenario of service-level triage is that service-level triage is mainly applied to scenarios that cannot be distinguished through APN-level triage or scenarios where access users cannot be obtained in advance. When accessing certain services, triage is implemented on the basis of determining APN, and the services are processed locally. When other services are carried out, no triage is performed and the services are processed in the remote central DC. Location identification is generally achieved through APN and tracking areas. Service matching is realized through UL CL UPF, and successful matching is directly forwarded locally. Figure 2.22 shows the difference between APN-level triage and UL CL service triage. In APN-level triage, all traffic is delivered to a network without distinction, and dedicated DNN triage is achieved by sinking the inbound UPF to the park and using the main anchor point sink method (requiring access data to be configured at UPF sunk to the park). The UL CL service triage mode will shunt to different target networks depending on the services. Users will first select UPF of larger networks, then trigger the insertion of the sunken UPF, and finally, the sunken UL CL UPF will realize service-level triage. APN-level triage is generally achieved in the form of a custom dedicated APN/DNN, while service-level triage can be achieved by a dedicated or large network APN/DNN.

The scenario of triage through UL CL UPF, as shown in Fig. 2.23, requires a decision on whether to use UL CL shunting through the core network control plane network element SMF and other network elements and then service triage is realized through UL CL UPF according to the configuration. The UL CL UPF acts as a triage to decide whether to shunt packets to the secondary anchor UPF PSA2 and the local network or to the primary anchor UPF PSA1 according to the configured shunt rules. Users accessing local services inside the park can directly access the local DN through PSA2, and if the access destination is the Internet, they can access via forwarding through PSA1.

Fig. 2.22 Triage mode

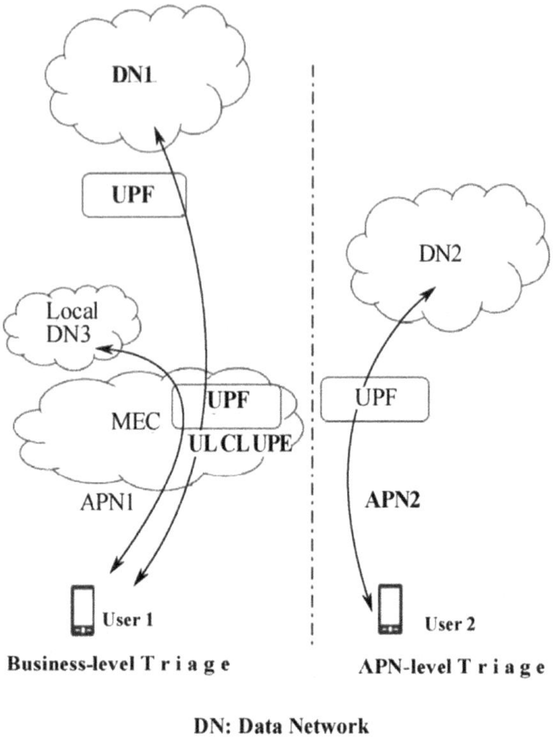

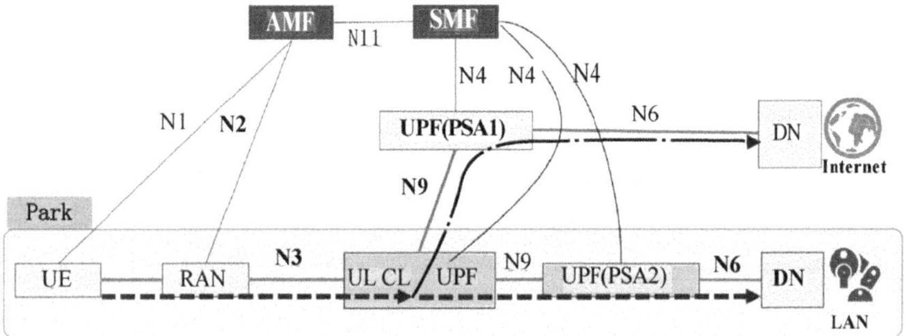

Fig. 2.23 Scenario of triage through UL CL UPF

Different MECs can provide different levels of service experience for industry customers. Inbound MEC has exclusive access to resources, while shared MEC can achieve service slicing isolation through UPF interface binding slices.

The UPF sinking meets the enterprise traffic need to detour from the Internet, reduces latency, improves security, and enhances user experience.

So, where exactly does UPF sink to? Should a shared MEC or an inbound MEC be chosen? The sinking location depends on the service scale and latency requirements. The larger the service scale, the more UPFs are needed, and UPFs can be sunk to a server room closer to users. The lower the latency demand, the closer UPF should be to users. Shared MEC is cheaper than inbound MEC, and service security and isolation are not as good as inbound MEC, so it is recommended to use shared MEC if cost is considered. If the security requirement is higher and the data does not leave the park, then an inbound MEC will be chosen. Because of the exclusive use of resources, it is more convenient to update subsequent services, but the cost is relatively high.

2. **Service Isolation (5G Slicing Solution)**

In implementing 5GtoB service, industry customers may have multiple service requirements in the same region, with different service priority configurations. Their consumption of resources also varies greatly. For example, there are two services in a steel plant remote control scenario: video monitoring backhaul and Programmable Logic Controller (PLC) remote control. Video consumes more bandwidth resources, while PLC consumes less bandwidth but has high requirements for latency. In 3G/4G networks, different services may interfere with each other by occupying network resources, or network resources may be wasted due to unreasonable utilization. Can 5G networks achieve strict isolation for such scenarios and meet the challenge of huge differential demand? The answer is yes. 5G slicing solutions can achieve end-to-end resource isolation of services and meet the isolation needs of different services.

Different business scenarios have different requirements for 5G networks, such as ultra-low latency, inter-network isolation, service customization, etc. These requirements pose multiple challenges for 5G networks.

(1) Guaranteed Service Level Agreement (SLA): It can provide reliable SLA assurance. For example, the network can meet the requirements of large bandwidth, low latency, low packet loss rate and delay shaking.
(2) Isolated network: Due to the exclusive use of network resources, network congestion or resource insufficiency will never arise to affect service operations.
(3) Self-operated and self-maintained Network: Users of a slice can track and manage the operation status of the slice in real-time to ensure that the slice and the services provided by the slice are in normal working condition. The network slicing can effectively deal with the above challenges.

Network slicing is essentially the division of the same physical network of an operator into multiple virtual networks. Each virtual network is supposed to meet different service requirements, such as division of latency, bandwidth, security, and reliability requirements, so as to respond to different network application scenarios flexibly.

Network slicing is an end-to-end system engineering that requires end-to-end network cooperation, and slicing runs through the three sub-domains of radio access network, transport network, and core network. Slicing can also meet a variety of customer service experience requirements, as shown in Fig. 2.24. Different vertical

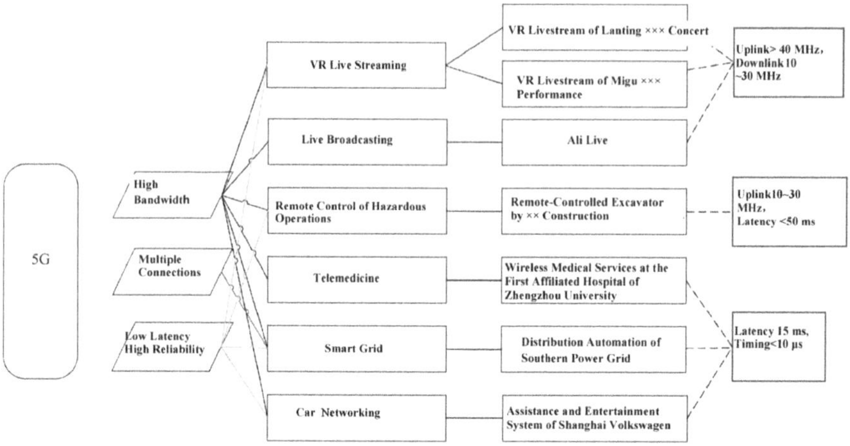

Fig. 2.24 5G network slicing

industries have different needs for the network, and to meet the requirements, 5G network can be sliced to achieve the effect of meeting different service needs of users through one network.

To achieve different service isolation requirements, network slicing technology needs to have functions, such as end-to-end, resource isolation, customization, virtual network, private network, capability opening, etc., to meet various service challenges.

(1) End-to-end: Each slice needs simultaneously the support of a radio access network, transport network and core network functions, so as to achieve end-to-end service isolation.
(2) Resource isolation: Slices will not interfere with each other, and each slice will have exclusive access to network resources. They are of high-security level. Figure 2.25 shows the core network's fully exclusive, partially exclusive, and fully shared deployments.
(3) Customization: Each slice can choose its corresponding service metrics, such as bandwidth size, delay size, etc., according to user needs.
(4) Virtual network: Slices are all virtual networks logically divided based on a physical network, rather than a splitting of a real physical network.

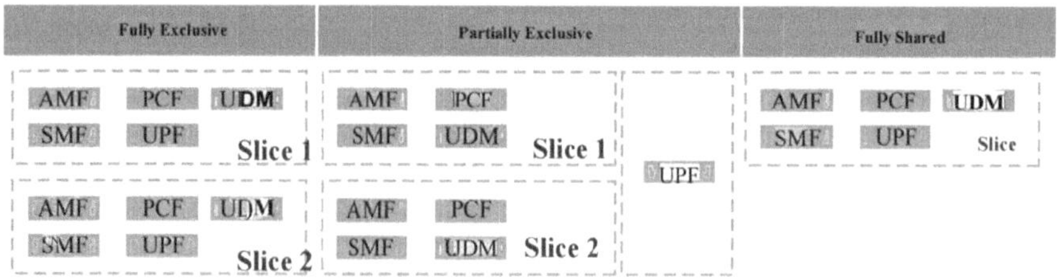

Fig. 2.25 Resource isolation model for core network slicing

(5) Private network: A slice provides a private network for a particular service and does not allow other users to access it without permission.
(6) Open capability: Slices can be developed based on the API interface of 5G to meet the customized needs of different users.

5G network introduces network slicing technology, enabling a network to support many different service scenarios simultaneously. Slicing provides more security measures based on the inherited security features of 5G networks: service and resources are isolated between different slices; slice access security can be achieved through user-contracted slice authentication and authorization; slice API interface authentication management is conducted to prevent slices from overstepping operation and maintenance, and so on.

Whether or not the final network chooses to deploy slices depends on industry customers' demand for resource exclusivity and service isolation, and the type and the number of slices deployed can be selected according to actual service needs.

3. **5G Local Area Network (5G LAN)**

5G industry modules are used less, and most industrial terminals do not have 5G communication capability. Many scenarios need to be relayed through Customer Premise Equipment (CPE), which greatly limits the development of 5GtoB. Enterprises face many troubles and challenges in deploying 5G networks. For example, many production networks need to be relayed through AR routers and other equipment to adapt to 5G networks, making networking complex and increasing costs. Interconnections between different regions or different branches also require the rent of carrier Internet private lines. When roaming, if temporary access to a corporate intranet is needed, APP installation is required. 5G LAN solutions are trying to solve these troubles for enterprises.

So, what is 5G LAN?

In brief, 5G LAN is a kind of network or slice service, a Layer 2 switching service mode provided on a network or a slice, as shown in Fig. 2.26, which connects cell phones, or cell phones and servers, through Layer 2. In practical application scenarios, the 5G LAN service mainly supports Layer 2 switching, and the original 5G DNN service supports Layer 3 routing. 5G LAN service is one of the service forms of network slicing. The application of slicing focuses on resource management and isolation of different domains, including reservation and isolation, etc. Corresponding to slicing is a logical network. Slicing can further achieve differentiation and deterministic SLA guarantee among different slices by enhancing features, such as supporting 5G LAN and 5G latency-sensitive networks. 5G LAN corresponds to a Virtual Local Area Network (VLAN) inside a logical network.

5G LAN terminals support access to an intranet from any location, and branches can quickly access the intranet without deploying dedicated Internet lines, allowing secure and isolated Layer 2 interoperability. Industrial applications have various scenarios that require pure Layer 2 communication or mixed Layers 2/3 communication. Many industrial Ethernet technologies do not define network layer and transport layer and can only be switched through Layer 2. Before the birth of 5G

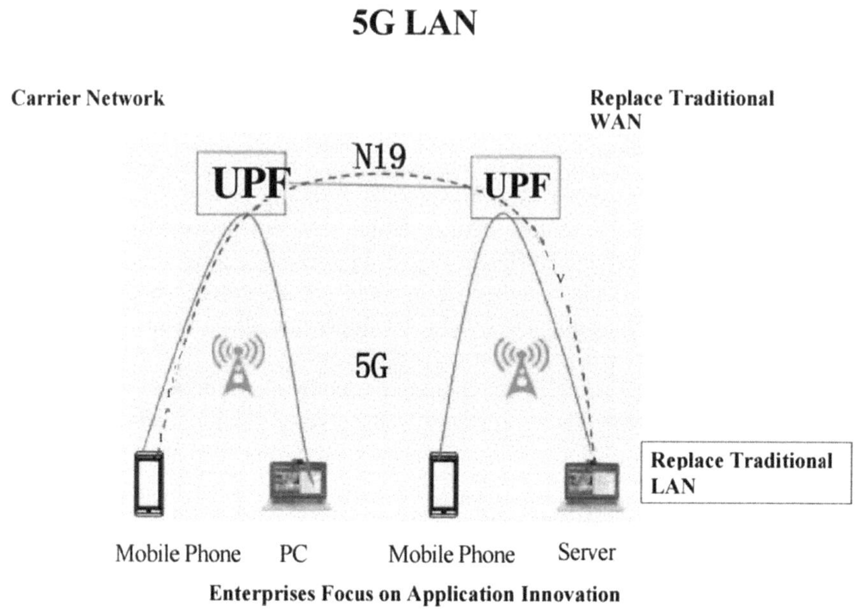

Fig. 2.26 5G LAN networking

LAN, this problem could only be solved by configuring Layer 2 tunnels through CPE or routers. With 5G LAN, simpler networking and lower deployment costs can be achieved.

3GPP defines the 5G LAN access method, as shown in Fig. 2.27. 5G LAN terminals are accessed via 5G, and the network side provides LAN and services with similar functions to Virtual Private Network (VPN). The 5G LAN terminal needs to support the corresponding features. For example, UE must support 5G network access, 5G-related link layer protocols, and 5G LAN type services; UE needs to support the creation of PDU sessions with Ethernet (Ethernet) type. The PDU part of the 5G LAN service is an Ethernet Frame, with UE needing to support Ethernet-related switching protocols. The Application part, depending on the specific application requirements, does not require 5G LAN for adaptation. UE and 5GC supports are required in the end-to-end process, and the wireless side is not aware of these upper layer service applications. There are no additional requirements by 5G LAN services for 5G QoS mechanisms, and end-to-end QoS mechanisms based on 5G systems need to ensure smooth service.

5G LAN networks are generally controlled and adapted to subscribers through different 5G Virtual Network groups (5G VN groups), where a 5G VN group is a collection of UE using 5G LAN-type dedicated communication. A 5G VN group has a unique 5G virtual network group identity. Each 5G VN group member has a unique Generic Public Subscription Identifier (GPSI). The 5G network needs to support the management of the 5G VN group, including both the management of the group

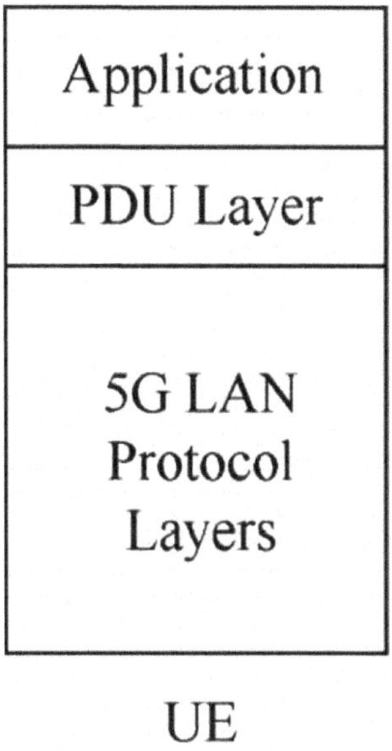

Fig. 2.27 UE protocol stack

and the management of the group members. 5G VN groups can be managed by the network administrator, or management permission can be opened to the enterprise administrator to set or delete them dynamically.

4. **No Service Interruption (Edge Gateway High Availability Solution)**

In the 5GtoB scenario, because the network involves production areas, industry users have higher requirements for network stability. For example, it is necessary to recover from failure within 1 h and promise continuous uninterrupted service, etc. There are a series of solutions on the network side to achieve reliable network operation. Core network (5GC) control plane mainly adopts the solution of edge gateway high availability to achieve high availability in local scenarios, and user plane utilizes MEC load sharing to achieve service continuity.

The solution of edge gateway high availability is proposed for the network to change from "best effort" mode to "stable industry connectivity" mode, which can meet the demand for emergency access of local data services when the 5GC control plane is disconnected in the 5GtoB scenario.

The commonly mentioned edge gateway high availability solution generally refers to adding single boards through park UPF to realize sinking AMF/SMF/UDM services and act as a temporary emergency module. An example of an edge gateway high availability solution is shown in Fig. 2.28. In an emergency scenario, the core network control plane fails, and if new users are connected to the park, users will choose local emergency control plane equipment. Users who are already online (stock services) are not affected by inertia operation. In the scenario of large network control plane network element loss (full break of N2, N4) an emergency control plane can be used to handle user access.

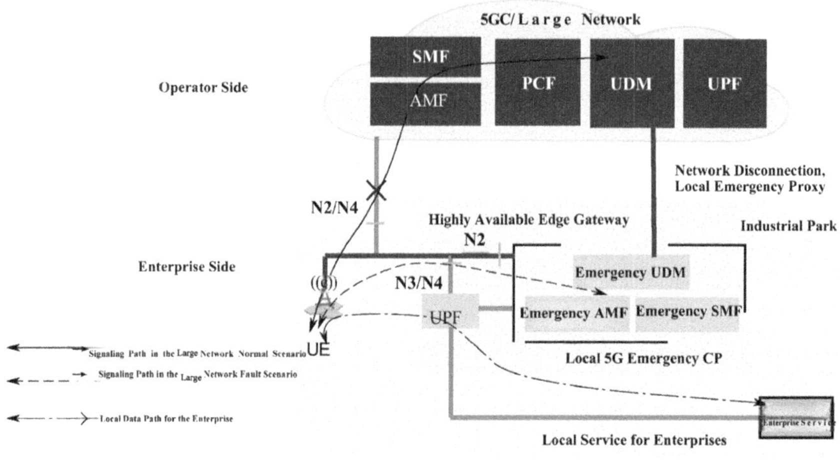

Fig. 2.28 Example of edge gateway high availability solution

In the scenario where no edge gateway high availability solution is deployed, if N4 is disrupted, UPF releases the session and users re-select the user plane. When AMF N2 breaks, steady-state service cannot be maintained, and a new AMF needs to be selected and re-accessed. When N2/N4 is fully disconnected, users cannot re-access. After deploying edge gateway high availability solution, network stability is greatly improved. In the scenario of large network loss, it supports steady-state users (online users) without dropping and inertia operation. It supports user mobile switching, and in control plane, the emergency control plane needs to be selected. It supports user re-access after disconnection. In the control plane, sinking emergency control plane is selected; and in the user plane, park UPF is selected.

The following is a review of the access situation of a large network loss scenario after the deployment of the edge gateway high availability solution, as shown in Fig. 2.29.

Type 1. New start-up user: gNB selects park emergency AMF access, and service access is not affected.

Type 2. Online service users (service is not interrupted): park gNB keeps the connection state; large network control plane network element and park UPF keep the service registration state of users, and service is not affected (inertia operation).

Type 3. A user of new signaling: park gNB selects emergency AMF processing, and the existing service will be interrupted, at which time emergency AMF guides users to re-register, and service is quickly restored.

After the recovery of large network loss, the network is handled as shown in Fig. 2.30.

Type 1. Newly powered-on subscriber: select large network toB AMF access, and service is unaffected.

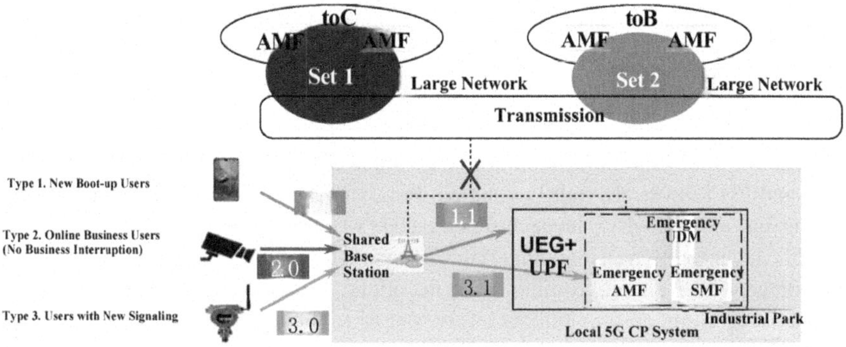

Fig. 2.29 Large network loss

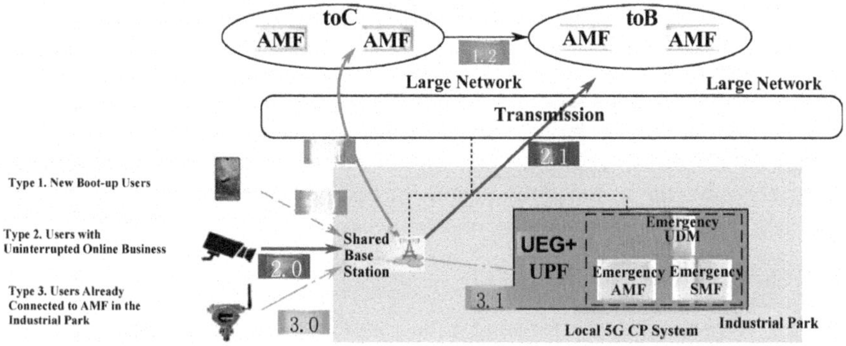

Fig. 2.30 Large network loss recovery

Type 2. Users whose online service is not interrupted: users initiate a mobility signaling process to resume communication with big network AMF and carry Globally Unique Temporary Identity (GUTI) assigned by big network AMF, and service is not affected.

Type 3. A user who has accessed park AMF: user's GUTI contains globally unique AMF ID of emergency AMF and if emergency AMF is selected, service is not affected (residual registered users of big network AMF will be cleared through SMF verification mechanism).

2.2.3 5G Key Technology of Transport Network

5G transport network is a new mobile transport network for 5G services, realizing user access from the next generation base (gNB) station to the 5G core network. Compared with the traditional 4G transport network, 5G transport network mainly introduces network slicing, SRv6 and other technologies, to meet the 5G service needs for large bandwidth, low latency, and wide connection.

1. **5G Slicing Technology of Transport Network**

In the traditional Ethernet, the interface includes MAC layer and PHY layer. MAC layer and PHY layer are tightly coupled. For example, when the interface is transporting service, the MAC layer makes a packet of service flow, wears a "coat", and the PHY layer then gives it another outer wrapper. Traditional Ethernet interfaces are confronted with technical limitations in the application. Firstly, the cost of interface production is high, and the interface rate is slow to increase. The current maximum rate is 200 Gbps, while the 400 Gbps interface rate is still to be commercialized. Secondly, bandwidth grabbing takes place between multiple services. An interface carries multiple services that share a MAC layer, so each service will grab bandwidth, which is determined by the "best effort" of traditional Ethernet and cannot be avoided. For example, if a long packet comes from one service, all messages from other services will be blocked at the interface during the delivery of the long packet. Flexible Ethernet (FlexE) is the main network slicing technology in the 5G transport network. It is an Ethernet-based transport technology with multi-rate sub-interfaces over multi-PHY links, supporting bundling, channelization, and sub-rate. It can also effectively solve the technical limitations of traditional Ethernet interfaces. Next, how FlexE in 5G transport network realizes network slicing is introduced.

1) Technical Architecture

Figure 2.31 shows that FlexE includes three parts: FlexE Client, FlexE Shim, and FlexE Group.

FlexE Client: denotes the user interface for 64 bit/66 bit Ethernet streams, currently supporting n × 5 Gbps rate.

FlexE Shim: indicates the new sub-layer between MAC/RS and PCS/PHY layers, completing the multiplexing and demultiplexing between the contents carried from FlexE Client to FlexE Group.

FlexE Group: indicates a set of bound Ethernet PHYs.

FlexE interface has been changed from the two-layer architecture of traditional Ethernet to a three-layer architecture: Client, Shim, and Group. Client corresponds to

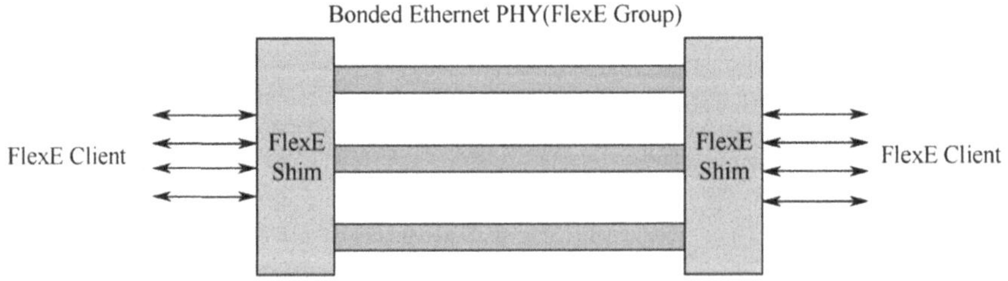

Fig. 2.31 FlexE technical architecture diagram

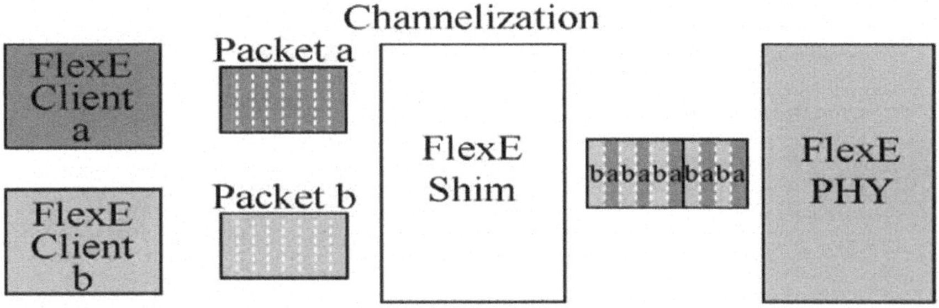

Fig. 2.32 FlexE port channelization diagram

customers' service flow and is accessed through the FlexE interface. Multiple Clients can be encapsulated into one FlexE interface, which can support non-standard rates. The shim layer is a time slot slice of PHY bandwidth. For example, when the 50 GE interface is divided into ten time slots, each time slot is 5 GE. The administrator can configure multiple time slots. For example, if 15 GE is required, three time slots can be bound so that there is a 15 GE interface. Group is the binding of several ports through FlexE interface. For example, there are eighty 10 GE time slots for two 400 GE interfaces. They transmit data in a unified manner, called a Group.

2) Main Features of FlexE

FlexE's new Ethernet port has the following three main functions.

(1) Channelization function. The FlexE port channelization diagram is shown in Fig. 2.32, which places data from different Clients in different time slots of the same PHY for transmission.

    Hard channelization of Ethernet mainly uses the Synchronous Digital Hierarchy (SDH) method to flexibly divide traditional Ethernet interface rates to carry different sliced services. The physical bandwidth of each interface is divided based on time slots and will not be seized and shared, solving the problem of sharing all the bandwidth of traditional Ethernet. For example, a 100 GE port can be divided into twenty 5 GE, and then time slots are configured so that one flexible interface bandwidth is 30 GE, one is 35 GE, and the other is 25 GE. The bandwidth is not shared among these three interfaces to achieve Ethernet hard pipe.

(2) Port binding function. FlexE port binding diagram is shown in Fig. 2.33, which bundles multiple PHYs through FlexE Shim to achieve larger capacity ports.

    Bandwidth expansion allows the bundling of multiple physical interfaces to expand bandwidth. Compared to Eth-Trunk, bandwidth expansion can solve the problem of uneven load. For example, when there are two 100 GE ports on a device, but the actual traffic requires 200 GE, the ports can be bundled via FlexE to expand the bandwidth.

(3) Subrate function. The FlexE port subrate diagram is shown in Fig. 2.34, which allocates part of PHY time slots to the Client.

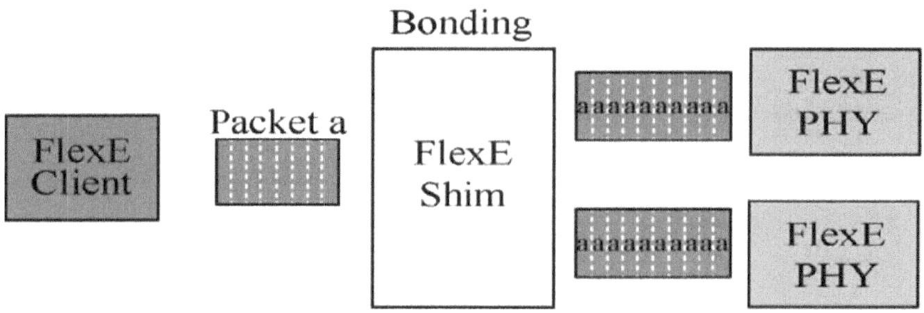

Fig. 2.33 FlexE port binding diagram

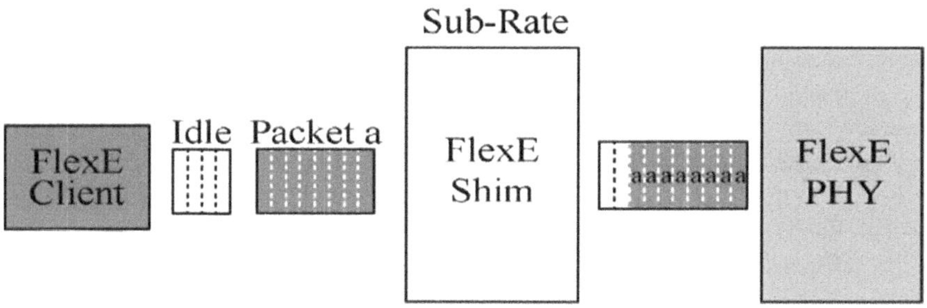

Fig. 2.34 FlexE port subrate diagram

FlexE subrate function is mainly used when interfacing with WDM devices. When router device port utilization is too low, time slots can be allocated. Part of the time slots are used to transmit valid bandwidth, and invalid bandwidth will not have to occupy OTN side bandwidth.

3) Crossover Principle of FlexE

FlexE crossover technology (also known as FlexE +) refers to the extension of networking technology based on standard FlexE interface technology. Through the crossover of FlexE bit blocks, an end-to-end FlexE networking is constructed, which can reduce FlexE's switching latency. FlexE crossover module has mainly three functions.

(1) Cross-connections between time slots. It supports the cross-connection of FlexE channel layers, that is, 66-bit code block exchange. Due to the smooth evolution of the existing network, some single boards support fixed packet length exchange mode.
(2) Operation, Administration, and Maintenance (OAM). The FlexE crossover module cooperates with OAM module to conduct OAM extraction, alarm processing and performance monitoring of FlexE channel layer.

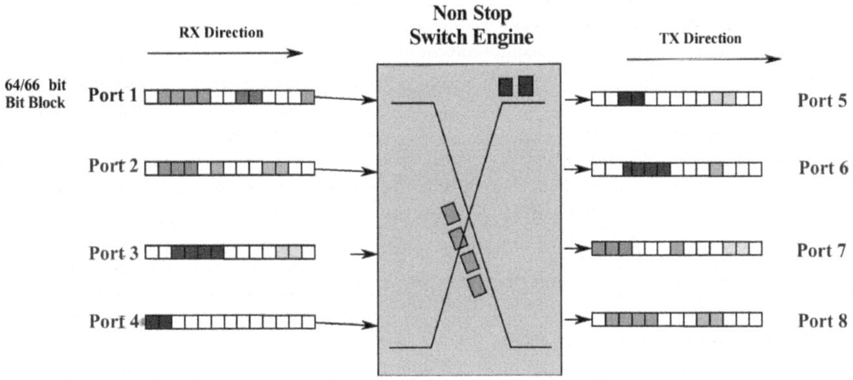

Fig. 2.35 Fundamentals of FlexE client crossover

(3) Protection function. It cooperates with the protection module to modify the cross-connection relationship and realize the Automatic Protection Switching (APS) of the FlexE channel.

The basic principle of FlexE Client crossover is shown in Fig. 2.35.

The following describes the process of FlexE Client cross-forwarding.

(1) Inbound port processing: The Client number where the bit block is located determines the Client's outbound port.
(2) Non-Stop Switch Engine processing: The Non-Stop Switch Engine module swaps the bit block of this Client to the outgoing port based on the previously configured outgoing port information.
(3) Outgoing port processing: Swap to the corresponding Client of the outgoing port. For example, in Fig. 2.35, Client1 of Port 1 crosses to Client1 of Port 8.

2. SRv6 Technology

Segment Routing IPv6 (SRv6) based on the IPv6 forwarding plane is a protocol designed to forward IPv6 packets over the network based on the source routing concept. SRv6 accomplishes hop-by-hop forwarding by inserting a routing extension header into an IPv6 message, pressing an explicit IPv6 address stack into SRH, and having intermediate nodes continuously perform operations to update destination addresses and offset address stacks.

The advantages of SRv6 technology are as follows.

Firstly, it simplifies network configuration and makes implementing a virtual private network (VPN) easier. SRv6 is based on IPv6 forwarding and is fully compatible with existing IPv6 networks without using Multiprotocol Label Switching (MPLS) technology. Intermediate forwarding nodes do not necessarily support SRv6, but only need to be able to forward IPv6 messages normally.

Secondly, it provides higher protection capability. Based on SRv6 technology, it combines Remote Loop-free Alternate (RLFA) Fast Reroute (FRR) algorithm to form an efficient Topology-Independent Loop-free Alternate (TI-LFA) FRR algorithm. Alternate (TI-LFA) FRR algorithm, which supports arbitrary topology protection in principle, can compensate for the shortages of traditional RFLA FRR protection technology.

Finally, it facilitates the traffic tuning of IPv6 forwarding paths. By using the Segment ID (SID) of various service types, the explicit path can be flexibly planned at the head node of the forwarding path to adjust the corresponding service traffic and achieve traffic tuning.

Next, some key technologies of SRv6 are introduced.

1) Definition of SRH

The IPv6 message comprises IPv6 standard header + extension header + load data. In order to implement Segment Routing (SR) based on the IPv6 forwarding plane, a new type of IPv6 routing extension header is added, namely SRH. SRH specifies an explicit path of IPv6 and stores Segment List information of IPv6, which serves the same purpose as Segment List in SR-MPLS. The head node device of the forwarding path adds an SRH to the IPv6 message, and the intermediate nodes only need to forward the path information contained in SRH, and the SRH format is shown in Fig. 2.36.

2) Processing of SRH

In SRv6, the Segments Left (SL) field decreases by 1 for each passing SRv6 node, and the IPv6 destination address information is transformed once. Segment Left and

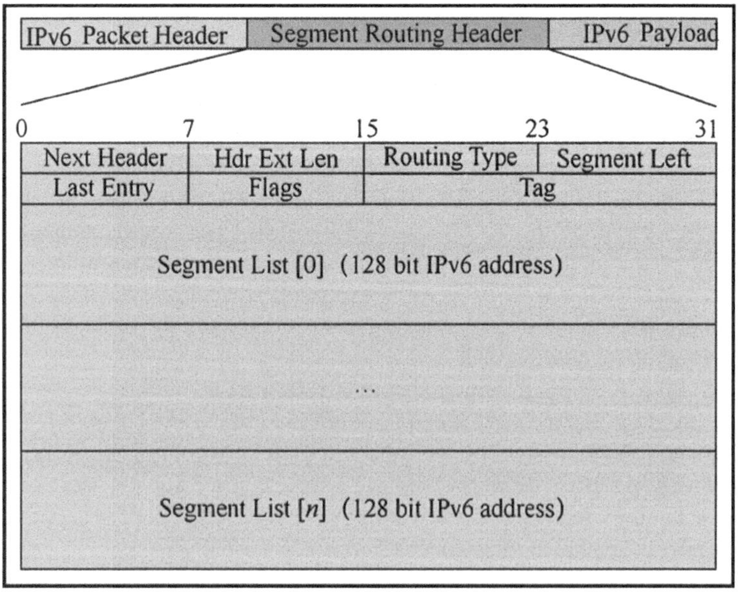

Fig. 2.36 SRH format diagram

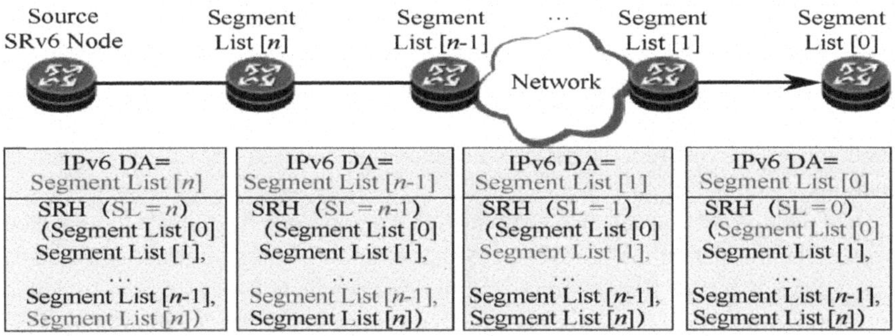

Fig. 2.37 Processing diagram of SRH

Segment List fields determine IPv6 destination address information together. The detailed process is shown in Fig. 2.37.

The specific processing is described as follows.

If the Segment Left value is n, the IPv6 DA (Destination Address) takes the value of Segment List [n].

If the Segment Left value is n − 1, the IPv6 DA takes the value of Segment List [n − 1].

......

If the Segment Left value is 1, the IPv6 DA takes the value of the Segment List [1].

If the Segment Left value is 0, the IPv6 DA takes the value of Segment List [0].

3) Definition of SRv6 Segment ID

SRv6 Segment ID is a form of IPv6 address totaling 128 bits, also commonly referred to as SRv6 SID, as shown in Fig. 2.38. SRv6 SID consists of two parts, Locator and Function, in the format of "Locator:Function". The locater occupies the high bit portion of the IPv6 address, and the function occupies the remaining low bit portion of the IPv6 address.

The locator has a network location function and is generally unique within the SR domain. However, in special scenarios, such as Anycast protection scenarios, multiple devices may be configured with the same Locator. After a node configures a Locator, the system generates a Locator segment route and spreads it within the SR domain through Internal Gateway Protocol (IGP), so that other nodes in the network

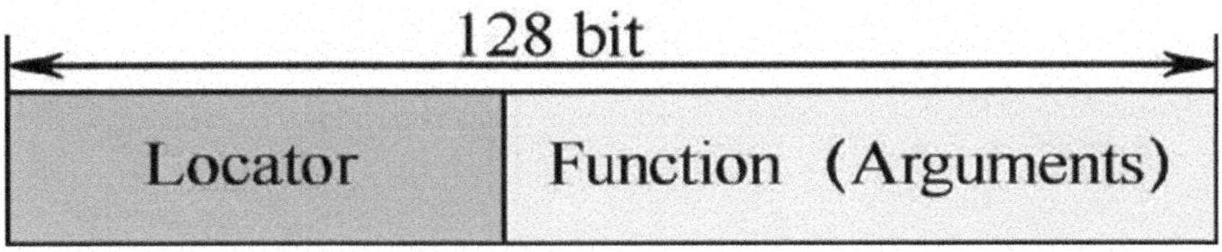

Fig. 2.38 SRv6 SID diagram

can locate this node through the Locator segment route. At the same time, all SRv6 SIDs published by this node can also be reached by this Locator segment route.

The function represents the device command, also known as Opcode, which can be dynamically assigned via IGP or statically configured via Opcode command. A function is used to instruct the SRv6 SID generation node to perform corresponding functions.

The Function section can also be divided into an optional Arguments field, which changes the format of SRv6 SID to "Locator:Function:Arguments". Arguments occupy the low bits of IPv6 address, and the Arguments field can be used to define information such as the flow and service of a message. Currently, an important application is the Customer Edge (CE) multi-homing scenario for Ethernet Virtual Private Network (EVPN). Arguments can be used to achieve horizontal segmentation when forwarding Broadcast Unknown-unicast and Multicast (BUM) traffic.

Both Function and Arguments are customizable, which reflects the advantage of SRv6 in facilitating network programming.

An SRv6-enabled node will maintain a local SID table containing all SRv6 SIDs generated at the node, and an SRv6 forwarding table can be generated based on the local SID table. Therefore, the local SID table has the following purposes.

(1) Define locally generated SIDs, such as End.X SID.
(2) Specify commands bound to these SIDs.
(3) Store forwarding information associated with these commands, such as outgoing interfaces and next hops.

4) Introduction to L3VPNv4 over SRv6 BE

L3VPNv4 over SRv6 BE mainly refers to the SRv6 BE tunnel based on which Layer 3 Virtual Private Network (L3VPN) services are carried and used to deliver L3VPNv4 data. A typical networking example of L3VPNv4 over SRv6 BE is shown in Fig. 2.39.

The following describes the device roles involved in Fig. 2.39.

CE: Customer Edge.

P: Provider, backbone equipment in the service provider network.

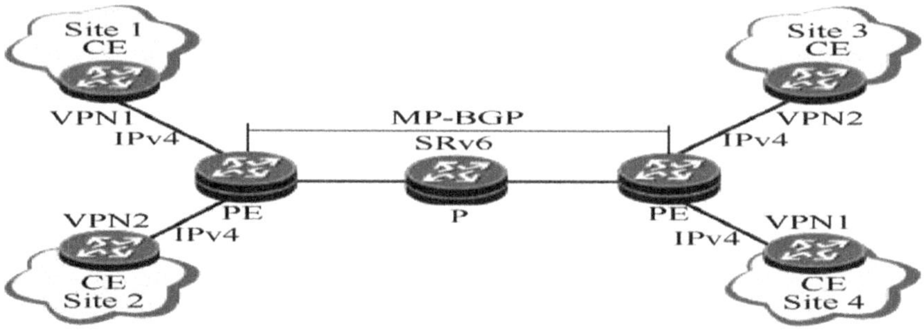

Fig. 2.39 Typical networking example of L3VPNv4 over SRv6 BE

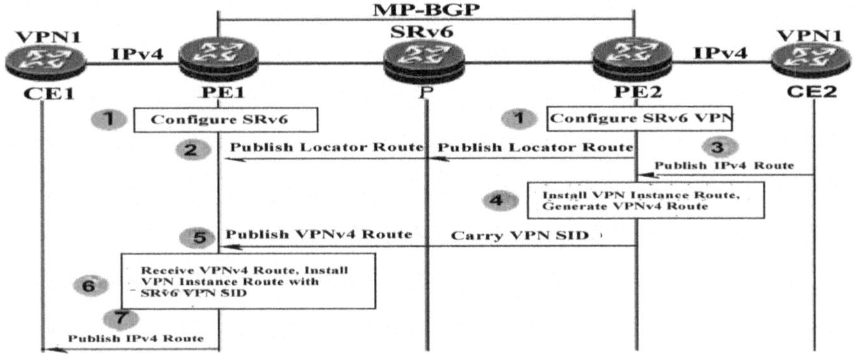

Fig. 2.40 Detailed process of establishing L3VPNv4 over SRv6 BE service

PE: Provider Edge.

L3VPNv4 over SRv6 BE has the following features:

(1) Transmit routing messages via extended BGP.
(2) Use SRv6 BE to encapsulate private network data messages for transmission.

The key implementation steps of L3VPNv4 over SRv6 BE include SRv6 BE path establishment, VPN route intercommunication, and data forwarding, etc. The detailed process of service establishment is shown in Fig. 2.40.

The detailed process is described as follows.

(1) PE configures SRv6 and SRv6 VPN, and the intermediate node P device needs to support IPv6.
(2) PE2 publishes SRv6 Locator routes to PE1.
(3) Routing information exchanges from CE to PE, i.e., CE2 publishes IPv4 routes from this site to PE2. Static routes, Routing Information Protocol (RIP), Open Shortest Path First (OSPF), Intermediate System-to- Intermediate System (IS-IS), or Border Gateway Protocol (BGP) can be used between CE and PE.
(4) PE2 learns VPN routing information from CE2 and stores it in VPN instance routing table while converting it to VPNv4 routes.
(5) Route publication between PEs, that is, PE2 publishes VPNv4 routes to egress PE1 through MP-BGP, and the Update message also carries the RT attribute of VPN and SRv6 VPN SID attribute.
(6) PE1 receives VPNv4 route, i.e., PE1 receives VPNv4 route and, based on the ingress policy of BGP, decides whether to add the route to VPN instance routing table by performing private network routing crossover, route iteration SRv6 BE path, route preference, etc., if the next hop is reachable. The VPN route issues and associates with SRv6 VPN SID simultaneously.
(7) Routing information exchange from PE to CE, that is, CE1 has multiple ways to learn VPN routes from PE1, including static routes, RIP, OSPF, IS-IS, and BGP, the same as the routing information exchange from CE2 to PE2.

5) Introduction to SRv6 TE Policy

SRv6 TE Policy is a new tunnel diversion technology based on SRv6 technology. SRv6 TE Policy paths can be represented as a segment list of a specified path, called a Segment ID List (SID list). Each SID list is an end-to-end path from the source to the destination and instructs devices in the network to follow the specified path instead of following the shortest path calculated by IGP. If a packet is imported into SRv6 TE Policy, an SID list is added to the packet by the head end, and the rest of the network implements the instructions embedded in the SID list to forward the data.

SRv6 TE Policy consists of three parts.

HeadEnd: The node where SRv6 TE Policy is generated.

Color: The extended group attribute carried by SRv6 TE Policy. BGP routes carrying the same Color attribute can use this SRv6 TE Policy.

Endpoint: The destination addresses of SRv6 TE Policy.

Color and Endpoint information is added to the SRv6 TE Policy through configuration, and the service network head-end node matches the corresponding SRv6 TE Policy with the Color attribute and next-hop information carried by the route to achieve service data forwarding. Color attribute defines the application-level network SLA policy, which can be used to plan the network path based on the SLA of specific services, realize the segmentation of service value, and build new business models.

2.3 5G Security Technologies

2.3.1 5G Security Standards

In the 5G era, mobile communication will not only provide high-speed Internet connection for toC users and build a new Internet form but will also become a new basic network facility for the connectivity of everything. New toB services, such as Industrial Internet, Telematics, Smart Grid, Smart Medical, Smart City, and so on, will be built on a 5G network. Therefore, the security of the 5G network will face higher requirements.

1, **Threats to 5G Security**

5G network security threats are divided into extra-domain security threats to the operator network and inter-domain security threats to the operator network. Extra-domain security threats are shown in Fig. 2.41, including air interface, Internet, network roaming, and external access EMS security threats.

(1) Security threats at the air interface (such as ❶ shown in Fig. 2.41): user data eavesdropping or tampering, Distributed Denial of Service (DDoS), attacks to deny user access, unauthorized, illegal terminal access to the network, fake base stations, malicious interference at the air interface, etc.

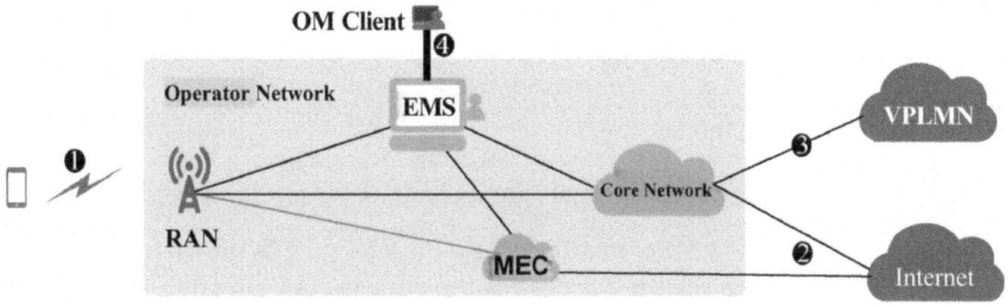

Fig. 2.41 5G extra-domain security threats to operator network

(2) Internet security threats (such as ❷ shown in Fig. 2.41): user data transmission leakage or tampering, counterfeit network applications to deny specific services, Internet-side DDoS attacks to deny data services, unauthorized access to capability open APIs, etc.
(3) Network roaming security threats (such as ❸ as shown in Fig. 2.41): user-sensitive information transmission leakage or tampering, impersonation of forwarding operators to deny service, etc.
(4) External EMS access security threats (such as ❹ as shown in Fig. 2.41): user sensitive information transmission leakage or tampering, access by unauthorized users, malicious operations by legitimate users, etc.

Security threats within the operator domain include service-based architecture (SBA) threats, MEC inter-module threats, inter-element interfaces and inter-module interface threats within network elements, as shown in Fig. 2.42.

1) SBA service-based architecture threats

The main SBA service-based architecture threats are:

(1) DoS attacks on NRF, resulting in unregistered/undiscovered services, such as ❶ shown in Fig. 2.42.
(2) An attacker's illegal access to the core network, such as ❷ shown in Fig. 2.42.

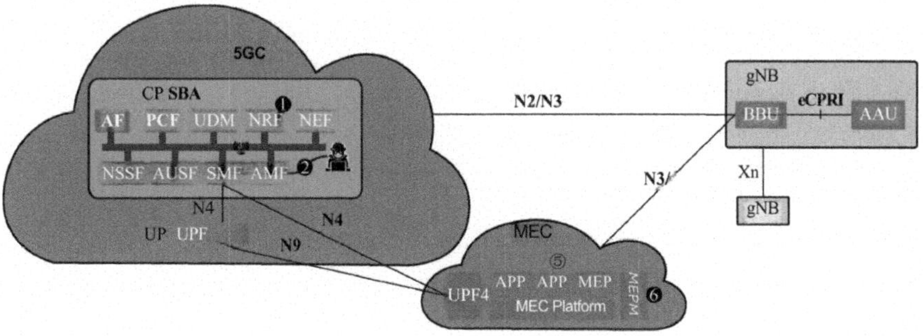

Fig. 2.42 Security threats within 5G operator network domain

(3) Communication data transmitted between NFs is eavesdropped or tampered with, as ❸ shown in Fig. 2.42.

2) MEC inter-module threats

The main MEC inter-module threats are:

(1) Malicious attacks on MEC platforms or UPF VNFs, as ❹ shown in Fig. 2.42.
(2) Inter-APP resource seizure (compute/storage/network) affecting other APPs, such as ❺ shown in Fig. 2.42.
(3) Overstepping the authority to manage the operation and maintenance of third-party applications, such as ❻ shown in Fig. 2.42.

3) Inter-element interface and inter-module interface threats

The main threats to inter-element and inter-module interfaces within network elements are eavesdropping or tampering of transmission data, illegal access to network element modules, etc.

2. **5G Security Standard Objectives**

3GPP has set 5G security standard objectives to ensure legitimate access to the network, safeguard the confidentiality and integrity of the air interface, and ensure the security of the connection between 3GPP network elements.

To ensure legal access to the network, two-way authentication between UE and the network is needed to prevent fake base stations. Regarding key architecture, 5G inherits NAS signaling encryption and integrity protection of 4G. Furthermore, 5G is enhanced in access authentication by developing a unified access authentication framework for 3GPP and non-3GPP, supporting the Extensible Authentication Protocol Method for Third Generation Authentication and Key Agreement (EAP-AKA) and 5G-AKA authentication mechanism to enhance the flexibility of authentication.

To safeguard confidentiality and integrity of air interface: enhance air interface encryption algorithm, add International Mobile Subscriber Identity (IMSI) encryption to protect subscribers' privacy, and add subscriber plane integrity protection.

To secure the connection between 3GPP network elements, IPSec and/or Transport Layer Security (TLS) security mechanisms are used between 3GPP network elements to protect the security of transmission information. Security between 5GC home and roaming domains is ensured by Security Edge Protection Proxy (SEPP).

5G SA security architecture was defined in Rel-15, with continued enhancements in subsequent versions, as shown in Fig. 2.43.

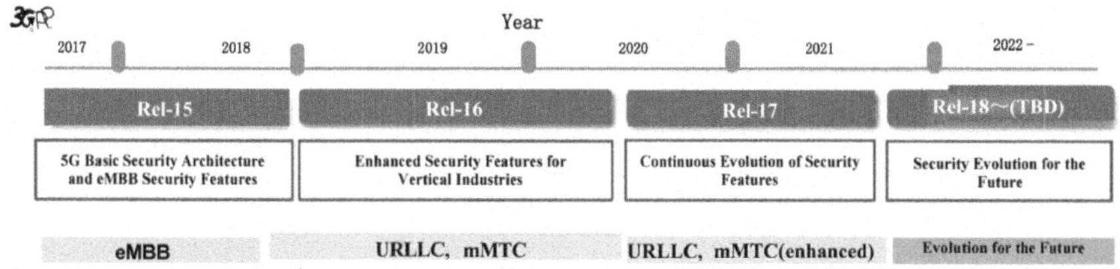

Fig. 2.43 Evolution of 5G security standards

2.3.2 *Key Features of 5G Security*

Key features of 5G security include six aspects, namely, SUPI encryption, 5G key enhancement, user plane integrity protection, slice security, URLLC security, and mMTC security.

1. **SUPI Encryption**

Before 4G terminal registration authentication, IMSI plaintext transmission of registration information is shown in Fig. 2.44. There is a possibility of registration information leakage.

The Subscription Permanent Identifier (SUPI) is equivalent to IMSI in 4G. With newly added UE identity encryption protection in 5G, SUPI is not transmitted directly in the network but in the form of encrypted SUPI, Subscription Concealed Identifier (SUCI). SUCI can be considered as an encrypted form of SUPI. Figure 2.45 shows

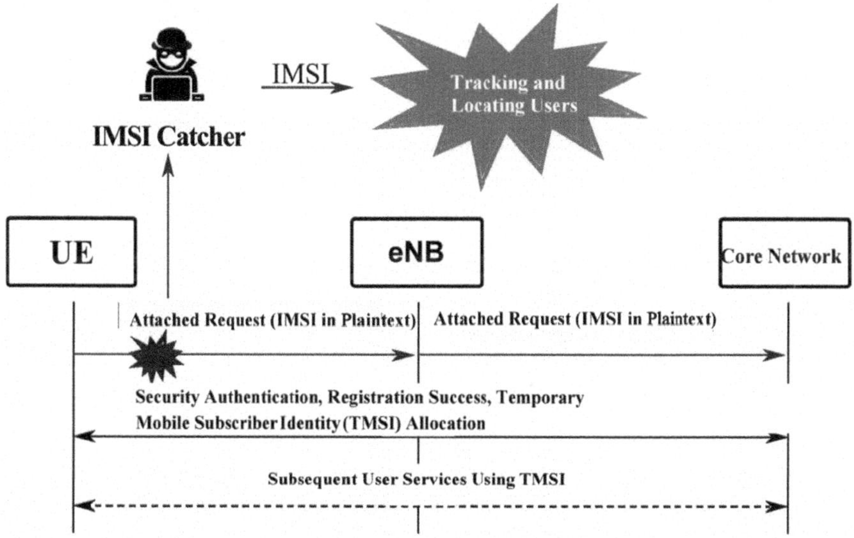

Fig. 2.44 IMSI plaintext transmission of registration information

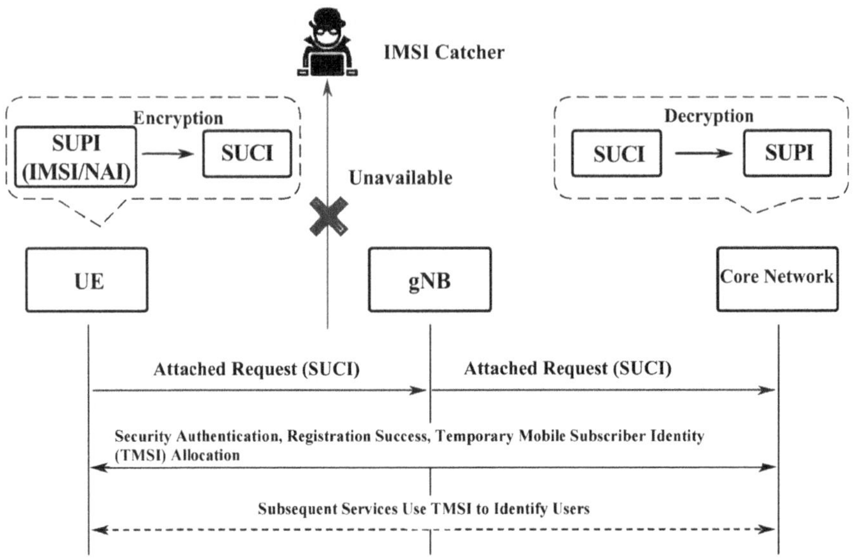

Fig. 2.45 SUCI registration after encryption

that a USIM card contains a public key, and USIM or UE encrypts SUPI to obtain SUCI. UE actually uses SUCI when it interacts with the network for signaling. After the security process, the core network will assign TMSI in 5G-GUTI to users. The whole process avoids direct interaction between air interface and SUPI, so as to protect user information security.

2. **5G Key Enhancement**

In the current network, the length of 2G, 3G, and 4G encryption keys is 64 bits or 128 bits, as shown in Fig. 2.46. Even with a supercomputer, the 128-bit key cannot be cracked, and only a few national laboratories in a few countries have supercomputers. So far, 128-bit keys are relatively safe.

In the future, quantum computers can halve the complexity of cracking symmetric cryptography algorithms and thus may be able to crack current symmetric cryptography algorithms based on 128-bit keys in a relatively short time. Therefore, 5G has upgraded the key to 256 bits to enhance protection, as shown in Fig. 2.47.

3. **User Plane Integrity Protection**

Air interface encryption means that a sender encrypts a message with a negotiated encryption algorithm and then sends the encrypted message to a receiver, who decrypts the encrypted message with the negotiated encryption algorithm.

Integrity protection means that a sender uses a negotiated integrity protection algorithm to calculate the Message Authentication Code for Integrity (MAC-I). The message is then sent to a receiver along with its MAC-I, and the receiver uses the

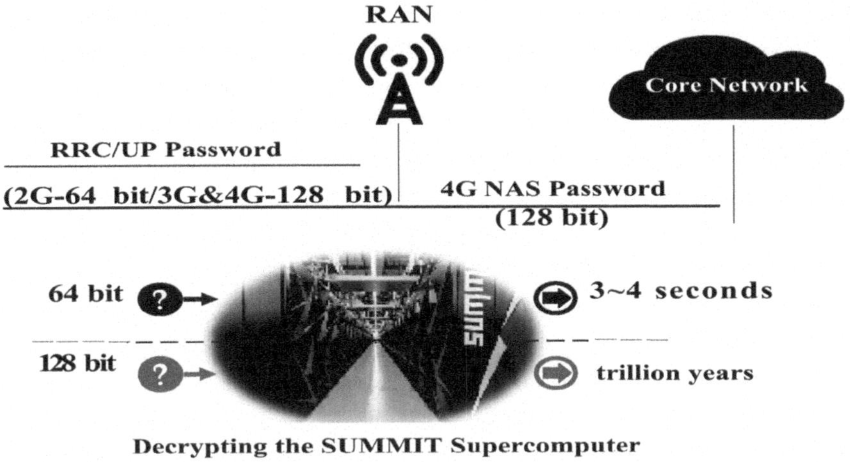

Fig. 2.46 Current key length

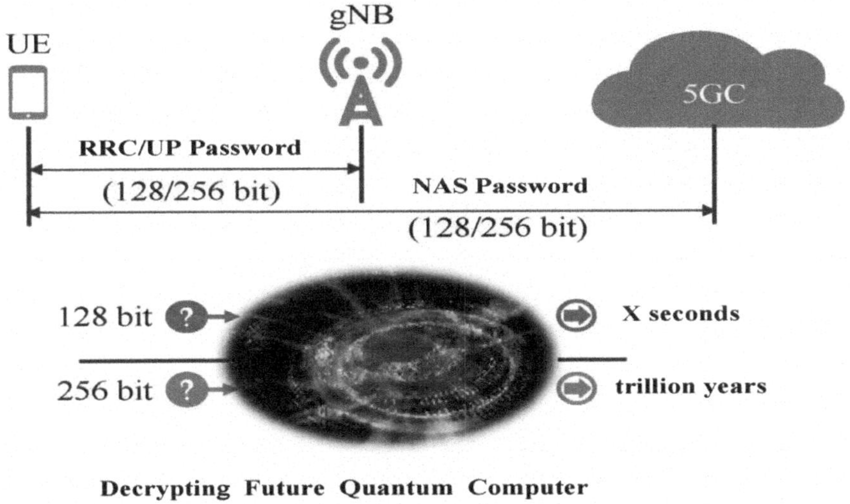

Fig. 2.47 Key enhancement

negotiated integrity protection algorithm to calculate the X-MAC of the message and compares MAC-I with X-MAC.

The 4G network air interface user plane employs encryption, but integrity protection is not enabled. Based on such facts, the research team demonstrated in a lab environment that an active attacker redirects DNS (Domain Name System) requests and then performs a DNS spoofing attack that causes the victim mobile device to use a malicious DNS server, ultimately redirecting the victim to a spoofed malicious website, as shown in Fig. 2.48. This dangerous attack is possible in specific scenarios in lab. It is relatively difficult to execute in a commercial 4G network because it also requires an expensive device like an IMSI catcher to operate within a 1-mile (about 1609 m) range.

To circumvent this vulnerability, 5G adds user plane integrity protection, as shown in Fig. 2.49, to prevent information from being tampered with.

4. **Slice Security**

5G enables vertical industries, and different industry applications have different demands on the network. To meet the diverse network requirements of vertical industries, 5G proposes the concept of network slicing, in which a network slice meets a certain type or a certain case of demand for connected communication services, and the entire 5G network consists of multiple network slices.

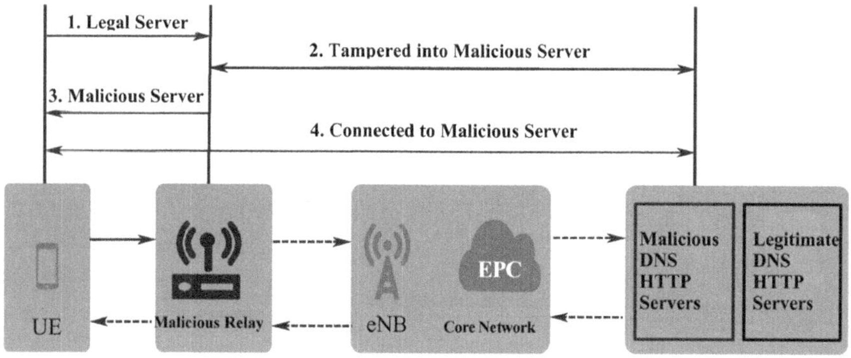

Fig. 2.48 DNS spoofing attack

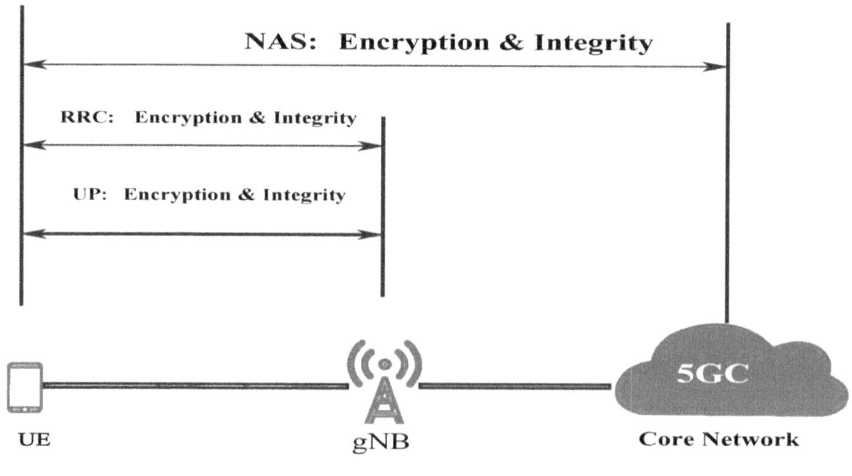

Fig. 2.49 User plane integrity protection

Users can access multiple slices; slice management authority is open to third parties. The security risks that confront network slices due to illegal access and overstepping management are shown in Fig. 2.50.

❶ Authentication and authorization of user (UE) access to slices is at risk, with UE accessing unauthorized slices.

❷ Slice management privileges are open to third parties, and attackers obtain slice management functions; legitimate third parties are not authorized to provision and obtain other slice resources and data opportunity points.

❸ Abnormal behaviors of users with legitimate access privileges.

To deal with illegal access and unauthorized management, the network adopts multiple security measures to ensure the security of slices, as shown in Fig. 2.51.

① Slice ID authentication is conducted to make slice access security visible.

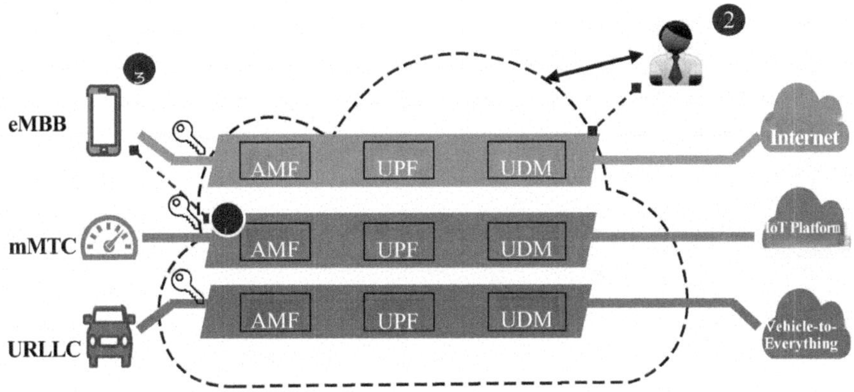

Fig. 2.50 Security risks to network slicing

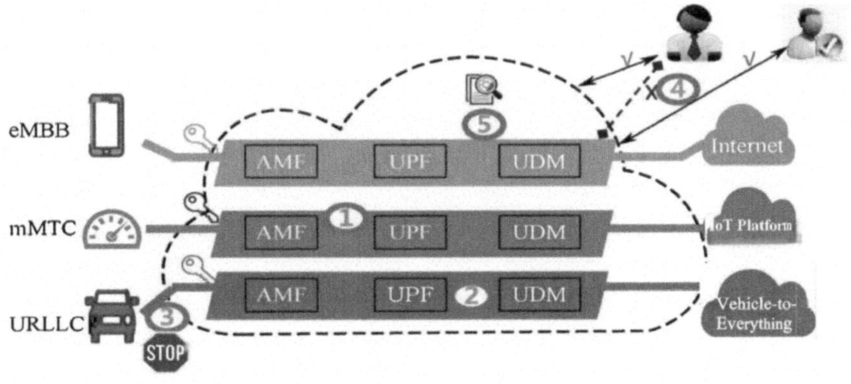

Fig. 2.51 Slicing security

② Statistics monitoring is integrated into the slice to avoid unauthorized slice access.

③ Restricted access and forced offline are supposed to suppress abnormal behavior of legitimate users.

④ System is separated from slice operation and maintenance, and slice operation and maintenance authority is controlled.

⑤ Operation and maintenance are audited, and the operation is traceable.

Compared with the traditional network, the physical resources of network slices can be shared, and the illegal occupation of a certain resource will affect other slices. If a large number of terminals or network sides launch DDoS attacks on slices, it will lead to excessive resource consumption and affect the experience of other users in the same slice, as well as the resource availability, latency, and throughput of other slices.

Through resource reservation on demand and Key Performance Indicator (KPI) monitoring, effective slice resource isolation is ensured to reduce the mutual impact between slices, as shown in Fig. 2.52.

① Generic cloud-based security solution with slice resource isolation ensures the security of NFVI and NFV.

② Slice-based flow control restricts the flow of users within the slice and realizes reasonable utilization of resources within the slice.

③ Slice resources should be reserved, and this slice does not infringe on neighboring slice resources.

④ Slice KPI (throughput, delay, etc.) should be detected and monitored, and the operation status should be controllable. Alarms should be generated when abnormalities occur.

5. **URLLC Security**

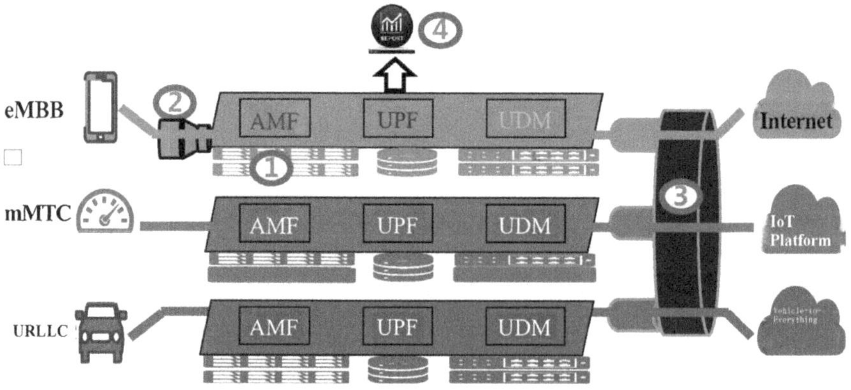

Fig. 2.52 Slicing resource isolation

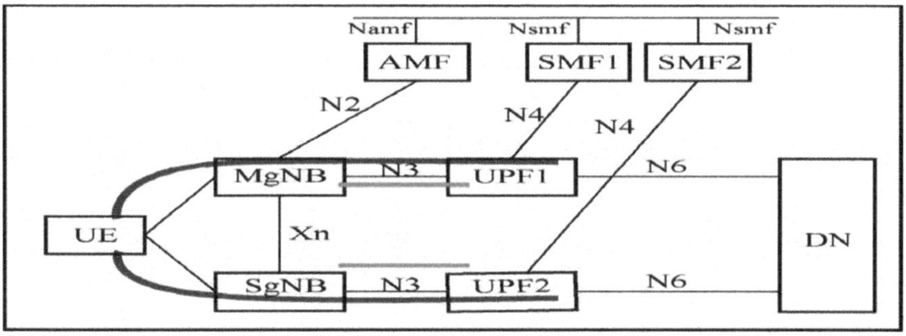

Fig. 2.53 Dual channel connection

Two paths can be used to transmit the same data to guarantee low latency and reliability of URLLC service data, as shown in Fig. 2.53. To ensure the reliability of the whole session, two sessions are established for the same service to be transmitted through the dual connection of Master gNodeB (MgNB) and Secondary gNodeB (SgNB).

There is a risk that the two redundant channels may have inconsistent security protection, and for this reason, 5G introduces on-demand user plane security policies. Suppose the user plane data protection methods between UE and MgNB are different from those between UE and SgNB. In that case, an attacker can access the message content by eavesdropping on the weaker protected connection (e.g., with encryption between UE and MgNB while without encryption between UE and SgNB). To avoid such risks, the same user plane security protection method can be used for both the primary base station and secondary base station user planes and MgNB will inform SgNB of the activated security protection method to make SgNB and MgNB use the same user plane security protection method.

6. **mMTC Security**

With the rapid growth of wireless services, base stations still need to support massive IoT connections. Yet, the security capabilities of many IoT terminals cannot meet the needs.

Attackers may eavesdrop and tamper with IoT data. To avoid such risks, NAS signaling can be used to transmit non-frequent small data (such as electricity meters, etc.), and based on the NAS protection method between UE and AMF, the transmission of non-frequent small data is protected.

Some IoT terminals are deployed in unattended areas, which are easily accessed and hijacked by attackers who then launch distributed denial-of-service (DDoS) attacks on the base station, as shown in Fig. 2.54. An attacker manages to attack the system resources of a base station through vulnerability, causing the resources of the attacked base station to be exhausted, which results in the base station being unable to provide normal services or paralyzed.

To avoid such risks, a flow control-based mechanism can be used to reduce the risk of air interface DDoS attacks, as shown in Fig. 2.55. The overload mechanism

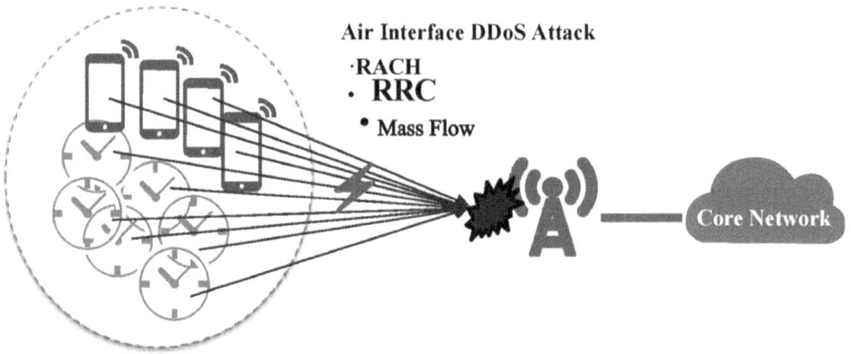

Fig. 2.54 Air interface DDoS attack

is triggered when the core network is overloaded, and the core network sends an Overload Start message back to the base station, requesting the base station to deny the accessed services or flow control the accessed services proportionally according to specified Radio Resource Control (RRC) established reason value. After the overload is eliminated, the base station is notified to stop the flow control through the Overload Stop message.

Resource monitoring is also performed within the base station equipment. When the load of a base station is too high or congested, the base station directly rejects or discards some of the initial access messages based on CPU occupancy of the main control board or the base band board. When the load of the base station is persistently too high, it is necessary to reduce the number of messages from terminal devices accessing the base station to reduce the system load and to reduce the frequency of terminal access through cell Access Barring to reduce the load of the base station.

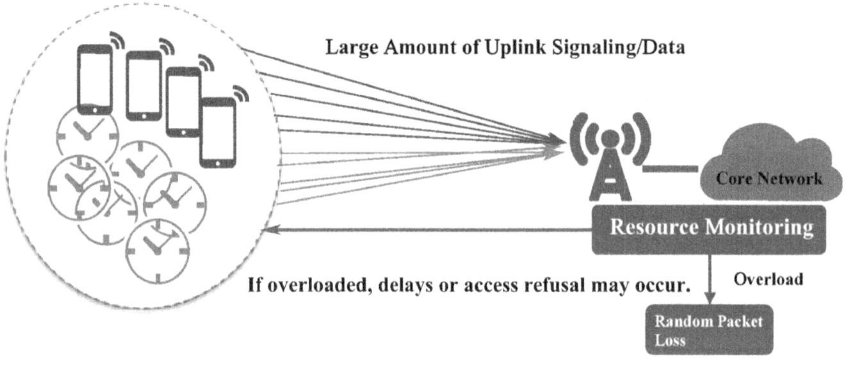

Fig. 2.55 Flow control mechanism

2.4 Summary

To meet the future demand for services on a network, the 5G end-to-end network has undergone profound changes. This chapter first explains the 5G end-to-end network architecture, in which 5G radio access network architecture undergoes continuous evolution from DRAN architecture to CRAN architecture and then to the future CloudRAN architecture. 5G transport network architecture evolves mainly to SDN architecture. The 5G core network architecture section introduces overall core network functions, functions of each NF, service-oriented architecture, etc.

To meet 5G network performance targets, 5G end-to-end network employs many new key technologies. Key technologies for wireless networks include new channel coding technologies of LDPC code and Polar code; Massive MIMO and D-MIMO, technologies that improve spectrum utilization rate and coverage at the air interface; and Super Uplink, a technology that increases uplink rates. Key technologies for carrier networks include FlexE network slicing, SRv6 segment routing, etc. Key technologies for the core network include data not leaving the park (UPF sinking), service isolation (5G slicing solution), 5G local area network (5G LAN), and service non-disruption (edge gateway high availability solution).

Finally, 5G security technologies are explained, including security standards and security key features. The section on key features of 5G security focuses on SUPI encryption, 5G key enhancement, user plane integrity protection, slice security, URLLC security, and mMTC security.

This chapter requires readers to understand 5G end-to-end network architecture, master key technologies of 5G end-to-end network, and understand 5G security technologies.

Exercise 2

I. True or False

2.1 In CRAN architecture, each station is deployed independently in a server room, BBU and RRU are co-located, and power distribution and supply equipment and other supporting equipment are deployed independently. ( )
2.2 Channel coding is mainly used for error correction and error detection of digital signals. ( )
2.3 5G is enhanced in access authentication, and a unified access authentication framework is developed for 3GPP and non-3GPP accesses. ( )
2.4 Compared with the traditional 4G transport network, new technologies such as FlexE slicing and SRv6 are mainly introduced in the 5G transport network. ( )
2.5 SRv6 is essentially a protocol designed for forwarding IPv6 packets on the network based on the source routing concept. ( )
2.6 A core network can realize the control and forwarding of user Internet data. ( )
2.7 Only after the birth of 5G can a core network be CU-separated to achieve user plane sink. ( )

II. Choice Questions

2.8 (Multiple Choice) Which of the following fields in the SRH message header collectively determine IPv6 destination address information during SRv6 message forwarding? ( )

A. Next Header
B. Tag
C. Segments Left
D. Segment List

2.9 (Multiple Choice) Which of the following techniques can improve the efficiency of spectrum utilization? ( )

A. Uplink and downlink decoupling
B. Super Uplink
C. Massive MIMO
D. D-MIMO

2.10 (Multiple choice) Which of the following are the main application scenarios for 5G? ( )

A. eMBB
B. LPWA
C. URLLC
D. mMTC

2.11 (Single choice) Which of the following is used to secure user identification? ( )

A. SUPI encryption
B. Key length increase
C. User plane integrity protection
D. Slice protection

2.12 (Single choice) Which of the following descriptions is incorrect regarding components of FlexE technology architecture? ( )

A. FlexE Client
B. FlexE Server
C. FlexE Shim
D. FlexE Group

2.13 (Single choice) Which tunneling techniques can effectively reduce data forwarding latency in a network? ( )

A. MPLS-TP
B. MPLS-TE
C. FlexE crossover
D. SR-TE

2.14 (Single choice) Which of the following statements is correct regarding the location of control plane deployment? ( )

A. Generally deployed in central DC
B. Generally deployed in regional DC
C. Generally deployed at edge DC
D. Generally deployed in the user park

2.15 (Multiple choice) Which of the following can be solutions for a core network in order to respond to differentiated needs of toB customers? ( )

A. Data not leaving the park (UPF sinking)
B. Service isolation (5G slicing solution)
C. 5G local area network (5G LAN)
D. Service non-disruption (edge gateway high availability solution)

2.16 (Multiple choice) Which of the following network elements can be used by different slices in an end-to-end slicing scenario of deployment? ( )

A. AMF
B. SMF
C. PCF
D. UPF

III. Short Answer Questions

2.17 Briefly describe the basic principle of super uplink.
2.18 What are the scenarios where 5G LAN is applicable? In these scenarios, what problems does 5G LAN solve?
2.19 What are the advantages of FlexE over traditional network slicing technologies?

Chapter 3
5G Terminal Ecological Industry Chain

Currently, 5GtoB has been applied in various industries and promoted rapidly, but the inadequate capability of the terminals is affecting the development of the toB industry. For this reason, standards are needed to improve industry awareness of 5GtoB terminals and guide and help 5GtoB projects succeed.

This chapter focuses on the ecological chain and the industrial progress of 5G chips, modules, basic connection terminals, general-purpose terminals, and industry-specific terminals.

Learning Objectives

- To understand the classification of 5GtoB terminals.
- To understand the industrial progress of 5GtoB terminals.
- To understand the applications of 5GtoB terminals.

3.1 Overview of 5G Chips, Modules, and Terminals

The following briefly describes 5G chips, modules, and terminals.

Chip: It is the basis for the capability of 5GtoB terminals, which can implement the underlying communication protocol, sometimes referred to as MODEM or modem.

Module: Based on chips, a module encapsulates the Power Management Unit (PMU), wireless transceiver units, storage units, peripheral interfaces, etc. Modules can greatly reduce the difficulty and workload of terminal development.

Terminal: Based on modules, other functions required for terminals are integrated, such as Ethernet, Wi-Fi, and other interfaces, and a few terminals are developed directly based on chips.

According to application scenarios, 5GtoB terminals can be divided into three categories: basic connection, general-purpose, and industry-specific.

L. Song et al., *5GtoB From Theory to Practice*,
https://doi.org/10.1007/978-981-97-8453-0_3

Basic connection terminals: They provide basic 5G network access services, which are needed in most scenarios, such as Customer-Premises Equipment (CPE), routers, gateways, and data transmission terminals.

General-purpose terminals: General-purpose devices with integrated 5G connectivity can be applied in multiple industries, such as cameras, drones, Automatic Guided Vehicles (AGVs), vehicle-mounted terminals, etc.

Industry-specific terminals: Based on industry-specific devices, 5G connectivity is integrated through 5G modules and Dongle, such as coal mining machines, tunneling machines, and other large equipment in underground coal mines that integrate 5G capabilities.

3.2 5G Chips

The main role of 5G chips is to synthesize baseband signals or decode received baseband signals. 5G chips are core devices for developing upstream industries.

3.2.1 5G Chip Structure

5G chips can be divided into three categories: Application Processors (AP), Baseband Processors (BP), and RF chips, with BPs being the most challenging and important.

AP is responsible for executing the processing of the operating system, user interface, and application programs, such as CPU, GPU, and other processors.

BP handles signal communication between the terminal and the outside world.

The RF chip is responsible for converting RF signals to digital signals.

The 5G System on Chip (SoC) packages BP, CPU, Graphics Processing Unit (GPU), and Power Management Chip together in the structure shown in Fig. 3.1. 5G SoCs are mainly used on smartphones.

3.2.2 5G Chip Ecology

The 5GtoB chips based on eMBB capability can inherit and reuse the chip capability of toC to support the starting application of 5GtoB. Industry-specific capabilities, such as low-power consumption and high-precision positioning, need constant traction and promotion in conjunction with market demand.

In terms of 5G chip manufacturing, mainstream chips achieved mass production of the 5 nm chips in 2020 and will continue to evolve towards the 3 nm node chips.

At present, 5G chip design manufacturers include Huawei, Qualcomm, Samsung, ZTE, MediaTek, etc. Specific models of their designed BPs are shown in Table 3.1.

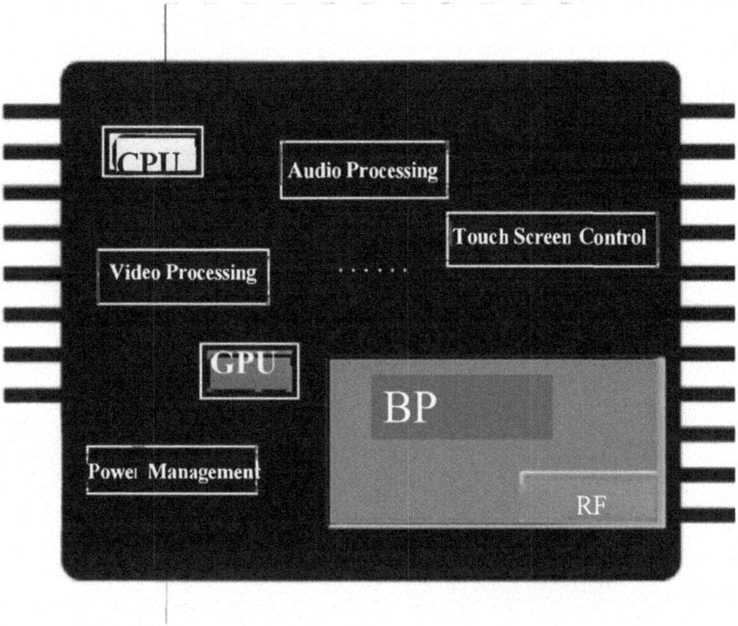

Fig. 3.1 5G SoC structure

Table 3.1 5G baseband processor

Chip specification	Chip model					
	Huawei Balong 5000	Qualcomm X55	Qualcomm X60	Samsung E5100	MediaTek Helio M70	Unisoc Ivy Spring V510
Manulaciuring process	7 nm	7 nm	5 nm	10 nm	7 nm	12 nm
Network	50/4G13G/20					
Networking	SA/NSA					

3.3 5G Module

The 5G module is the core component for connecting industry terminals to 5G networks. Industry users do not need to care too much about communication details when developing terminal products using 5G modules, thus reducing the threshold of terminal development.

Fig. 3.2 5G module structure

3.3.1 5G Module Structure

The structure of the 5G module is shown in Fig. 3.2. 5G module integrates various components, such as BP, RF chips, memory chips, etc., onto a circuit board and provides a standard interface to enable various IoT terminals to quickly realize communication functions by embedding the module.

3.3.2 5G Module Package Form

5G modules are used in a wide range of scenarios. Different scenarios have different requirements for terminals, and different terminal manufacturers have different packaging requirements for modules. Currently, the mainstream package forms of the 5G module are Land Grid Array (LGA) package and the M.2 package.

(1) LGA Package

LGA package adopts solder joints to mount on a circuit board, as shown in Fig. 3.3. It uses a metal contact-type package instead of the previous pin-like pins.

The advantage of the LGA package is that it is soldered reliably and can ensure product consistency.

Disadvantages of the LGA package: End manufacturers must redesign Printed Circuit Board (PCB). It also requires a high production process, and once the module is redesigned or malfunctions, the entire single board will be scrapped.

(2) M.2 Package

Fig. 3.3 LGA package

M.2 package is shown in Fig. 3.4. M.2 package adopts slot interface and follows the unified protocol standard. It has been widely used in PC products and is a more common package form in the industry.

Advantages of M.2 package: It is plug-and-play and compatible with modules from different manufacturers; the module and the board are separated, and the hardware revisions of the two are not affected by each other; it has better interface compatibility, and when a product is in upgrade, there is no need to revise the service board.

Disadvantages of the M.2 package: It is less vibration-resistant and has poor heat dissipation capability.

Fig. 3.4 M.2 package

3.3.3 5GtoB Module Ecology

5G modules from major chip manufacturers are currently available. There were about 27 5G modules in 2019 and 64 in 2020 in the market. 5G modules on the market all meet 5G common module standards proposed by the China Communications Standards Association (CCSA). 5G module manufacturers and models are shown in Table 3.2.

Table 3.2 5G Module Manufacturers and Models

Chip manufacturer	Module manufacturer	Model	Package type
Hisilicon	TD Tech Holding Limited	MH5000–31	LGA
	Quectel Wireless Solutions Co., Ltd.	RG801H	LGA
	Sichuan Ai-Link Technology Co Ltd.	A1-NR ICl/AI-Mtn	LGA, M.2
	China Mobile IOT Company Limited	M5	M.2
Qualcomm	Sunsea AIoT Technology Co., Ltd.	SIM82001SIM8300 LGA/S1M8200-M2/SIN 8300-M2	LGA, M.2
	Gosuncn technology group Co., Ltd.	GM800/GM801	LGA, M.2
	China Mobile IOT Company Limited	OueMa FO2X/F 03 X	LGA, M.2
	China United Network Communication Group Co., Ltd.	CU-QGIICU-QMI	LGA, M.2
	LongSung Technology (Shanghai) Co., Ltd.	EXGIO LGA/EX520 M.2	LGA, M•2
	ZTE Corporation	ZM9000	M.2
Unisoc	Quectel Wireless Solutions Co., Ltd.	RG500LI	**LGA**
	Fibocom Wireless Inc.	FG650	LGA
	Shenzhen Neoway Technology Co., Ltd.	N510M/N5106	LGA, M.2
MediaTek	Shanghai Mobiletek Communication Ltd.	T800	LGA
	Fibocom Wireless Inc.	FG3605G	LGA

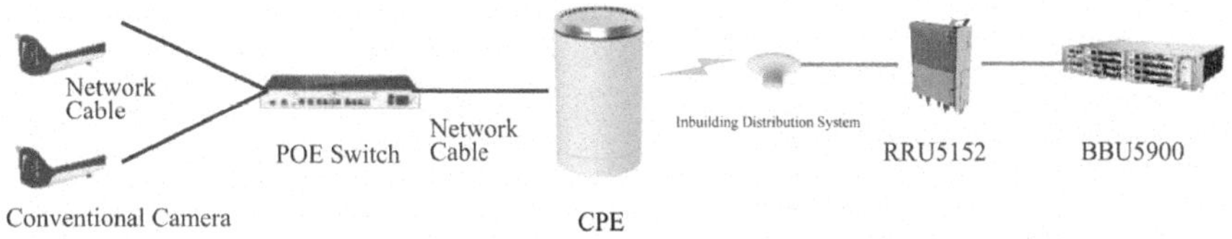

Fig. 3.5 CPE application example

3.4 5G Basic Connectivity Terminals

5G basic connection terminals provide basic 5G network access services for industry equipment, mainly including customer terminal equipment (CPE), industrial routers, industrial gateways, Dongle, and other terminals.

3.4.1 Customer Terminal Equipment (CPE)

The role of CPE is to perform secondary relaying (Wi-Fi or network port) of 5G signals for industry terminals to access 5G network.

An example of a CPE application is shown in Fig. 3.5. A traditional camera is connected to 5G CPE via a Power over Ethernet (POE) switch. This solution has two values.

(1) Equipment such as cameras, network cables, and POE switches can be fully utilized.
(2) Conventional devices can be quickly connected to 5G.

3.4.2 Dongle

The dongle is a 5G wireless device connected to other devices via USB, such as a 5G mobile network card, which enables a computer to communicate using a 5G network.

An example of a Dongle application is shown in Fig. 3.6. The AGV cart can be docked to 5G Dongle via USB. The driver of Dongle needs to be installed on the AGV cart, and Dongle is used as a network card for AGV. The dongle can assign an IP address through a 5G core network and communicate directly upward.

Value of the solution: Traditional AGV carts and other equipment can be fully utilized.

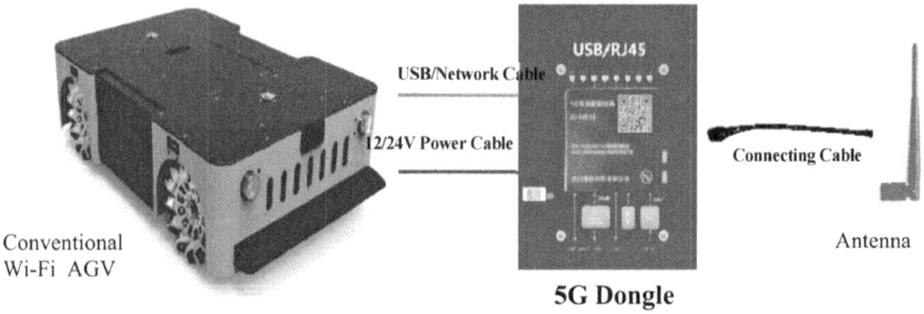

Fig. 3.6 Dongle application example

3.4.3 Industrial Router/Industrial Gateway

Industrial routers generally refer to industrial-grade wireless routers, which provide data transmission capability through public wireless networks and realize the routing function and data transmission security mechanism in data transmission. Although there are few types of ports in industrial routers, they have strong forwarding capabilities.

Industrial gateways enhance CPU processing power and memory capacity based on industrial routers to provide protocol conversion, data storage, data processing, and other edge computing capabilities. Industrial gateways have various ports (such as USB, RS232, etc.) and better support secondary development.

An example of a 5G industrial gateway/router application is shown in Fig. 3.7. It supports multiple access forms.

Serial server: Firstly, it converts current and voltage signals output from traditional meters into RS485/RS232 serial signals and then sends them to a 5G industrial gateway/router.

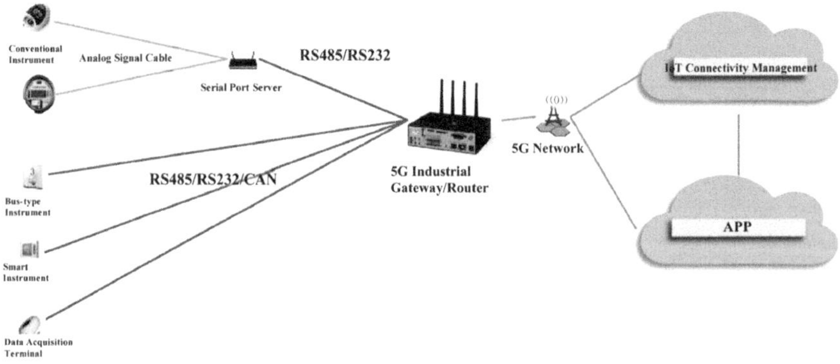

Fig. 3.7 5G industrial gateway/router application example

Other digital terminals: They are directly sent to the 5G industrial gateway/router through RS485/RS232/CAN industrial bus, etc.

3.4.4 Basic Connectivity Terminal Ecology

Suppliers, product types, product models, and 5G BP models of basic connection terminals are shown in Table 3.3.

Table 3.3 Suppliers, product types, product models, and 5G BP models of basic connection terminals

Supplier	Product type	Product model	*5G* baseband processor model
Askey Technology (JIANGSU) Ltd	CPE Pro	RTL0310 5G NR CPE	Qualcomm X55
TD Tech Holding Limited	CPE Pro	5G CPE Pro	Huawei Balong 5000
	Industrial CPE	5G CPE Ins	Huawei Baton 5000
Shenzhen Hongdian Technologies Corporation	Industrial CPE	Z 1	Huawei Baling 5000
	Industrial Gateway	Z2	Qualcomm X55
China United Network Communication Group Co., Ltd.	CPE Pro	50 CPE VNOO I	Qualcomm X55
	CPE Pro	5G CPE VN007	Unisoc Ivy Spring V510
	Industrial Gateway	5G industrial internet gateway	Huawei Balong 5000
China Mobile TOT Company Limited	CPE Pro	Forerunner P1/P2	Qualcomm X55/ Huawei Balong 5000
	Industrial Gateway	Onebox G5000/05100	Qualcomm × 55/ Huawei Balong 5000
Xiamen Four-Faith Communication Technology Co, Ltd	Industrial Gateway	F-NR100	Huawei Balong 5000
	Industrial CPE	F-NR200	Huawei Balong 5000
ZYXEL COMMUNICATIONS CORPORATION	CPE Pro	NR7101	Qualcomm X55
ZTE Corporation	CPE Pro	5G Outdoor CPE MC7010	Qualcom X55
Guangzhou Robustel Co., Ltd.	Industrial Gateway	R5020	Qualcom X55

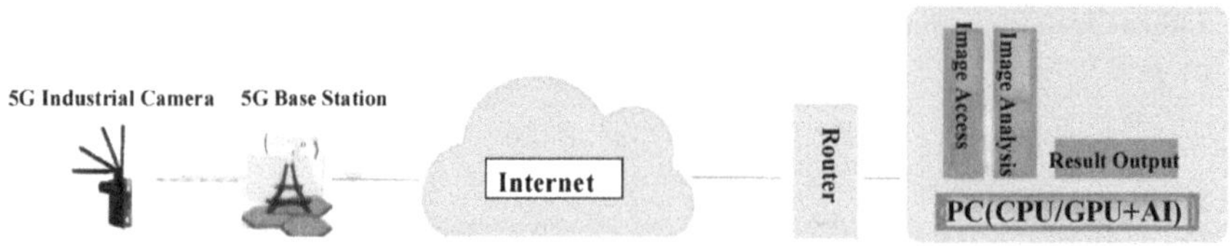

Fig. 3.8 5G industrial camera application example

3.5 5G General-Purpose Terminals

With the deepening of 5G in thousands of industries, the number of 5G general-purpose terminals is increasing and the application scenarios are widening. Developed devices include 5G cameras, 5G industrial cameras, 5G drones and so on.

5G general-purpose terminals can directly access the 5G network without basic connection terminals.

3.5.1 5G Camera

A 5G camera is a camera with a built-in 5G module integrated with innovative encoding mechanisms and transmission technology.

Value of 5G cameras: solve the problem of difficult wiring and temporary emergency deployment.

3.5.2 5G Industrial Camera

5G industrial camera refers to an industrial camera with a built-in 5G module.

Value of 5G industrial camera: improve detection efficiency and accuracy.

5G industrial camera application example: Fig. 3.8 shows that a 5G industrial camera pushes the real-time captured images to the operator's 5G network through the camera's built-in 5G module.

3.5.3 5G Unmanned Aerial Vehicle

5G Unmanned Aerial Vehicles are drones with built-in 5G modules or external 5G airborne cloud boxes.

Value of 5G UAV: meet the demand for UAV over-the-horizon control and video backhaul.

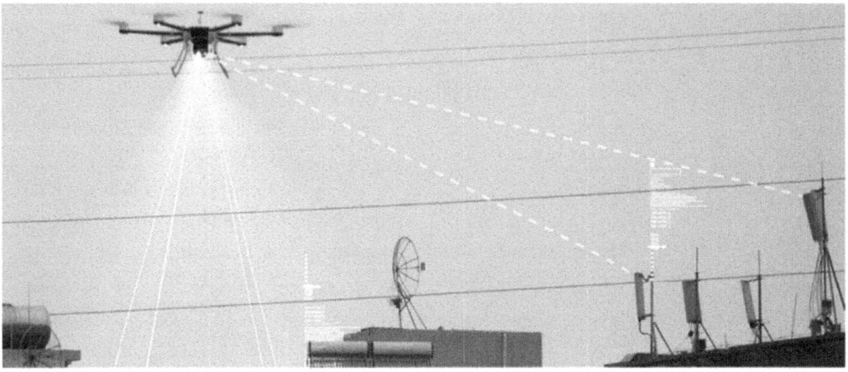

Fig. 3.9 5G UAV security inspection

5G UAV application example: 5G UAV security inspection is shown in Fig. 3.9. The ultra-low latency of 5G can provide stable tracking and accurate control of drones over the horizon, and the large bandwidth can guarantee that drones will backhaul HD video in real-time.

3.5.4 5G General-Purpose Terminal Ecology

Some suppliers and products of 5G UAVs, 5G cameras, and 5G industrial cameras are shown in Table 3.4.

Table 3.4 5G general-purpose terminal suppliers and products

Terminal	Supplier	Product
5G UAV	Tianyu Jingwei (Beijing) Technology Co., Ltd.	Tianyu cloud box
	Harwar International Aviation Technoogy (Shenzten) Co., Ltd.	Hawa UAV
	Qingdao Cloud Century Information Technology Co., Ltd.	Renwofei
	EHang Holdings Limited	Tianyu UAV solutions
	China United Network Communication Group Co., Ltd.	Rainbow No.1 onboard terminal
5G camera	Huawei Technologies Co., Ltd.	M6781-10-G740-W5
	Zhejiang Dahua Technology Co., Ltd.	HFS8841
5G industrial camera	Beijing Microview Technology Co., Ltd.	PG-A5010C-50

Table 3.5 Typical industry-specific 5G terminal

Industry	Typical scenario	Industry-specific equipment
Manufacturing	Process control and optimization	PLC controller
Mining	Video surveillance	Intrinsically safe 50 camera (downhole)
	Inspection	5G inspection robot (downhole)
	Underground coal mining operation	5G coal mining machine 50 drilling machine 5G support controller
Oil and gas	Video monitoring	Explosion-proof data terminal
Power	Differential protection	Timing CPE
Medical	Mobile ward round	5G medical cart
	Remote ultrasound/surgery	Remote ultrasound machine Remote surgical robot
Security	Smart policing AGV	50 law enforcement recorder
Port	Autonomous unmanned driving	Unmanned container truck
Live broadcasting	5G + 4 K live broadcasting	Live broadcasting backpack

3.6 5G Industry-Specific Terminals

5G industry-specific terminals are terminals that integrate 5G connectivity through 5G modules based on industry-specific devices. There is no clear demarcation line between 5G general-purpose terminals and 5G industry-specific terminals.

Currently, 5G applications have been integrated into various industries, such as manufacturing, mining, oil and gas, electric power, healthcare, security, ports, and live broadcasting. Typical industry-specific 5G terminals are shown in Table 3.5.

3.7 Summary

Industry terminals are key devices that enable a wide range of industries to use 5G, and they are an important entry point for the integration of 5G and industries. With the large-scale promotion of 5G network construction in China, the innovation space for industry application scenarios, such as 5G + smart agriculture, 5G + smart mines, 5G + smart ports, 5G + smart manufacturing, and 5G + smart cities, has increased.

With the traction of 5G Rel-16 technology standard, 5G terminals will continue to iterate, innovate, and continuously expand application scenarios. The strong demand for chips is also pushing back the acceleration of chip technology breakthroughs. Some chips are currently in 5 nm mass production and are evolving to the 3 nm node.

The huge changes in the external environment will also promote the upstream extension of China's 5G chip industry chain, thus accelerating the independent innovation of China's 5G chip technology.

This chapter firstly introduces the concept and the classification of 5GtoB terminals, including chips, modules, and terminals; secondly, it explains in detail 5GtoB chips, modules, basic connection terminals, general-purpose terminals, and industry-specific terminals.

With this chapter, readers should understand the ecology of 5GtoB terminals and know the industrial progress and applications of 5GtoB terminals.

Exercise 3

I. True or False

3.1 Size, power consumption, and packaging process of 5G chips have little impact on developing 5G in industry terminals. ()

3.2 5G module mode is better than 5G CPE mode in all aspects such as reliability and energy consumption, so using 5G module mode in all scenarios is recommended. ()

II. Choice Questions

3.3 (Single choice) Which of the following is the most appropriate way to quickly transform an existing network terminal into a 5G terminal in a space-constrained scenario? ()

A. 5G module
B. 5g CPE
C. 5G Dongle
D. 5G gateway

3.4 (Single choice) Which component in a PC does the 5G module function similarly in a terminal? ()

A. CPU
B. Hard disk
C. Keyboard and mouse
D. Network card

3.5 (Single choice) Which of the following chips functions as converting RF signals into digital signals? ()

A. AP chip
B. BP chip
C. RF chip
D. Baseband Processor

3.6 (Multiple choice) Which scenarios are suitable for using 5G cameras? ()

A. Difficult wiring scenarios
B. Mobile monitoring scenarios

C. Fixed monitoring scenarios
D. Temporary monitoring scenarios

3.7 (Multiple choice) In which aspects do industrial CPEs have advantages over consumer CPEs? ()

A. Low price
B. Dustproof and waterproof grade
C. Working temperature
D. Anti-vibration

3.8 (Multiple choice) Which of the following is the advantage of the M.2 package? ()

A. Reliable soldering
B. Good module compatibility with different manufacturers
C. Better interface compatibility
D. Ensuring product consistency

III. Short Answer Questions

3.9 What are the advantages and disadvantages of using the LGA package for 5G module?
3.10 What are 5G basic connection terminals?

Chapter 4
Basic Service Capabilities and Applications of 5GtoB

With the continuous development of 5G network construction, 5G applications in industry are gaining more maturity. 5G is a basic capability for the digital and intelligent society, and it has a wide variety of demands from all sorts of industries. But the basic services of different industries are similar, and so are the demands of the basic services for networks.

This chapter will introduce the basic service scenarios in different industries and the basic network capability requirements for each business scenario.

Learning Objectives

- To understand 5G basic service capabilities.
- To master 5G basic service applications.

4.1 Understanding Industrial Requirements in 5GtoB

To promote 5G applications, it can only close the loop in the business of industry applications with the ability to replicate at a large scale. To make 5G applications replicable, the division of industrial applications and scenarios is needed. First, application scenarios of different industries need to open to extract replicable scenario-based applications, which can later be summarized as network requirements based on scenario applications, namely, "network atomic operation capability". For example, although the scenarios of ports, iron and steel manufacturing and mines are different, they all use remote control at a horizontal level. The remote control requirements for networks are mainly uplink wide bandwidth and low latency. For this example, the scenario application is remote control, and the "network atomic operation capability" is uplinking to wide bandwidth and low latency. Based on abundant 5G industry practices, Huawei has refined multiple 5G industry horizontal applications, such as video surveillance, machine vision, indoor positioning, data collection, cluster scheduling,

L. Song et al., *5GtoB From Theory to Practice*,
https://doi.org/10.1007/978-981-97-8453-0_4

controller-to-controller management, and remote control. The "network atomic operation capability" extracted from the horizontal application requirements include wide bandwidth, low latency, high reliability, high precision timing, network positioning, and minimal operation and maintenance. With "network atomic operation capability" and scenario-based applications, 5G product solutions can have greater potential to replicate and popularize. 5G-related applications will be integrated into enterprises and become part of their business processes.

4.2 5GtoB Basic Service Capability

Basic 5GtoB service capabilities will be introduced, including end-to-end wide bandwidth, latency, high precision timing and positioning, high reliability, and minimal operation and maintenance.

4.2.1 End-to-End Wide Bandwidth

While 4G networks primarily serve the needs of people, the vision of 5G is to achieve connectivity of everything. In 5G industry applications, most of the services have a greater demand for uplink bandwidth, so the requirements for 5G networks are to support wide uplink bandwidth in addition to wide downlink bandwidth. In 5G network architecture, the wireless network, as the nearest node to users, represents the comprehensive capability of the network to a certain extent. According to the 3GPP protocol, 5G supports a maximum uplink rate at 10 Gbps. To achieve this, key technologies were introduced into 5G, including super uplink, UL and DL decoupling, as well as carrier aggregation.

Super UL (Super UL)

5G industry applications mainly focus on uplink services. The current commercial 5G network mainly applies time division duplex, which means that terminals cannot always send uplink data. This mechanism limits the uplink rate to a certain extent. Super UL technology greatly increases the uplink spectrum resources available to 5G users by introducing new spectrum resources and sending uplink data on different spectrums in a time-sharing manner. See Sect. 2.2.1 of this book for details of Super UL technology.

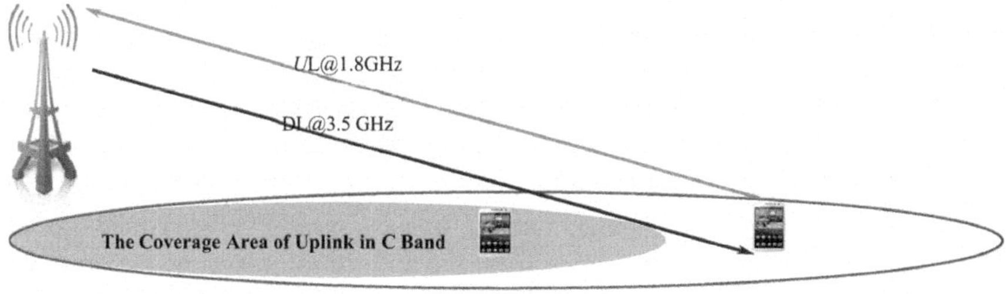

Fig. 4.1 UL and DL decoupling

UL and DL Decoupling

Limited uplink coverage has always been a bottleneck in mobile communications. UL and DL decoupling technologies define a new way of pairing spectrum, as shown in Fig. 4.1. Downlink data is transmitted in the high-frequency band: when the uplink signal is strong, the terminal transmits data in the high-frequency band, for example, utilizing the 3.5 GHz frequency for uplink transmission; when the uplink signal is weak, the terminal transmits data in the low-frequency band, for example, utilizing 1.8 GHz for uplink.

Carrier Aggregation

By aggregating multiple contiguous or non-contiguous carriers into a larger bandwidth, carrier aggregation technology allows the terminal to occupy more spectrum resources, thereby increasing uplink and downlink rates. As shown in Fig. 4.2, the terminal aggregates 3.5 GHz and 1.8 GHz spectrums for transmission.

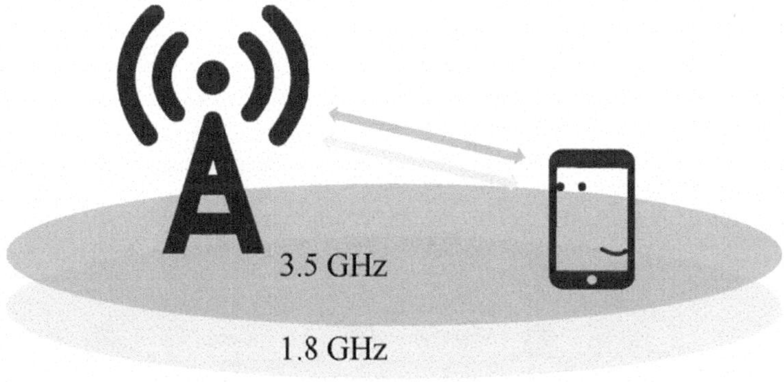

Fig. 4.2 Carrier aggregation

4.2.2 *End-to-End Delay*

Requirements for latency in different industries and scenarios vary widely. Latency requirements for typical 5G service scenarios are shown in Table 4.1.

From the 5G network topology perspective, latency mainly comes from the processing of the network equipment and the transmission of network medium, as shown in Fig. 4.3. End-to-end latency from the application server to the terminal needs to be considered from four aspects: architecture, air interface, equipment, and service, in which architecture and air interface are the keys.

Low-Latency Technologies for Air Interface

Low-latency technology for air interface reduces air interface latency by optimizing the time domain and scheduling process from the dimension of base station scheduling.

1. **Licensing-Free Scheduling**

Traditional scheduling and licensing-free scheduling are shown in Fig. 4.4. In traditional scheduling in mobile communications, when a terminal has a service demand, it needs to send a request first. The base station allocates resources, and the terminal sends data based on scheduling. From resource requests to data sending, a certain latency was generated. In contrast, 5G introduces licensing-free scheduling. The base station reserves resources in advance so the terminal can use them without request when data transmission is needed, thereby reducing latency.

2. **Symbol-Level Scheduling**

Table 4.1 Latency requirements for typical 5G service scenarios

Scenario	Application	End to end latency (ms)
Automatic driving	Queue control	< 3
	Remote driving	10–30
Virtual reality	VR multiplayer games	10–20
Intelligent manufacturing	Remote control	50
Unmanned aerial vehicle (UAV)	Operating UAV remotely	10–30

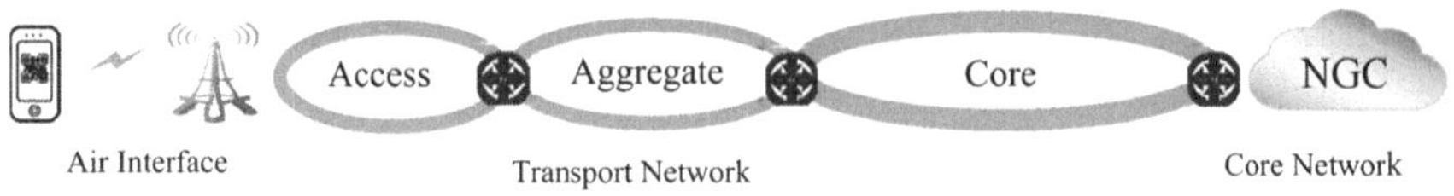

Fig. 4.3 5G network topology

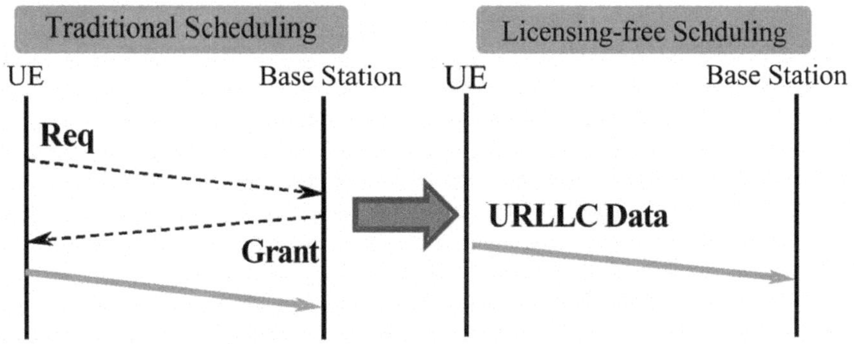

Fig. 4.4 Traditional scheduling and licensing-free scheduling

The cycle of traditional scheduling is scheduled in time slots, with one slot consisting of 14 symbols. When 5G subcarrier spacing is 30 kHz, the time length of 1 slot is 0.5 ms. 5G introduces symbol-level scheduling, which reduces the scheduling cycle and air interface latency, as shown in Fig. 4.5.

3. **Embed Air Interface (EAI)**

In a certain scheduling cycle, when eMBB service has already allocated all the resources of air interface, but there is a request for URLLC service, the base station can "punch" part of eMBB air interface resources to URLLC, which can also be regarded as URLLC services preempting eMBB resources. In this way, URLLC services can be satisfied as soon as possible, thus reducing the latency of URLLC services, as shown in Fig. 4.6.

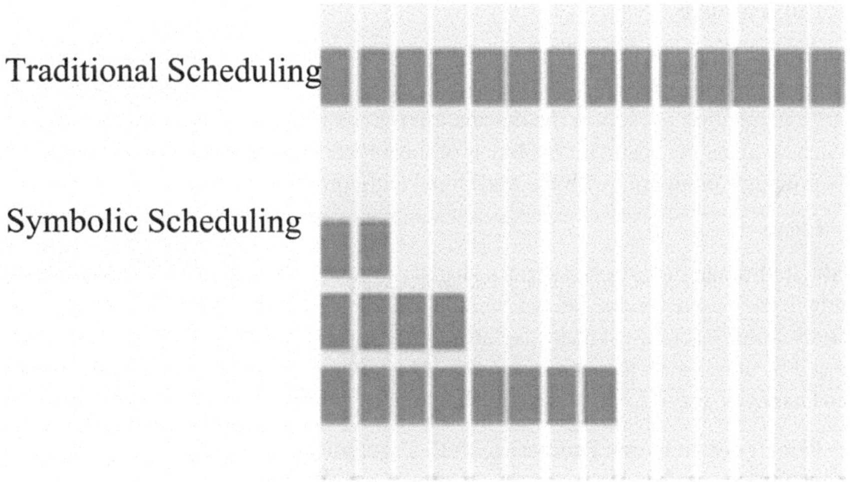

Fig. 4.5 Symbol-level scheduling

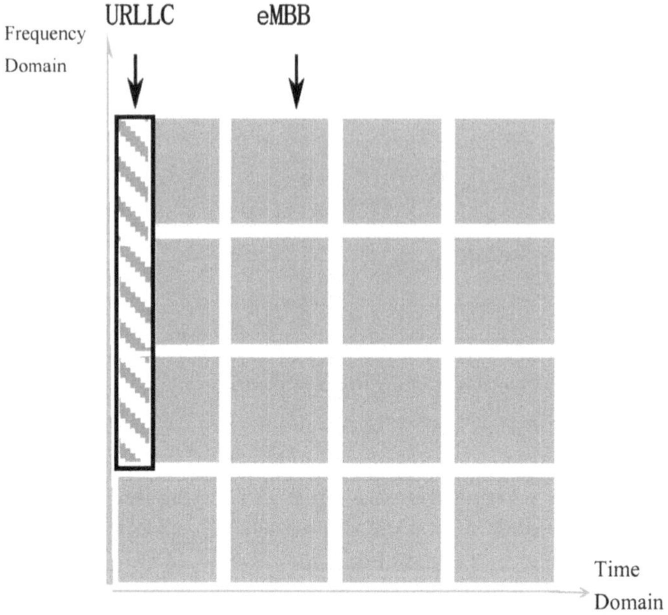

Fig. 4.6 EAI

Low-Latency Technologies for Transport Networks

Similar to how wireless networks reduce latency, transport networks also have the "preemption" mechanism for high-priority service. In addition, 5G transport networks also introduce some crucial technologies to reduce latency.

1. **Cut-Through**

Traditional data forwarding verifies and forwards data after the port has acquired a complete data packet, so some latency is introduced, while cut-through technology is to start sending packets to the destination port as soon as the network forwarding device acquires the destination address of the packet, which has higher data rate and lower latency compared with the traditional technology.

2. **FlexE**

In the traditional model, each service shares the bandwidth of the MAC layer, which leads to increased service latency once congestion occurs. Flexible Ethernet technology (FlexE) achieves service isolation between sub-MACs. It reduces the latency caused by congestion through independent bandwidth guarantees, thus ensuring the low latency of service perception, as shown in Fig. 4.7.

3. **Priority of Network Processing (NP) Perception**

Traditional processors do not perceive service priorities when forwarding. Introducing the dedicated low-latency NP means opening up a processing channel for

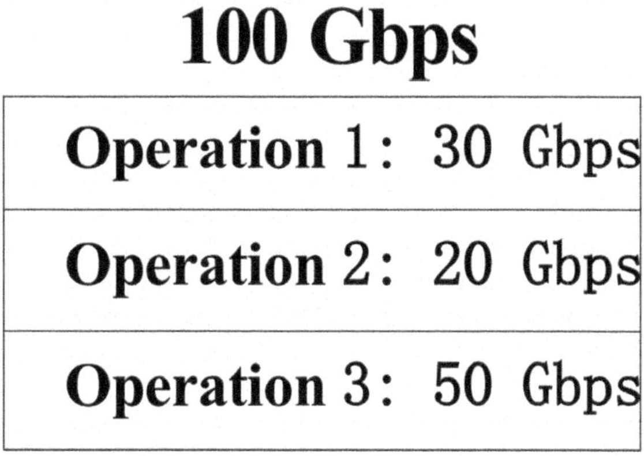

Fig. 4.7 FlexE

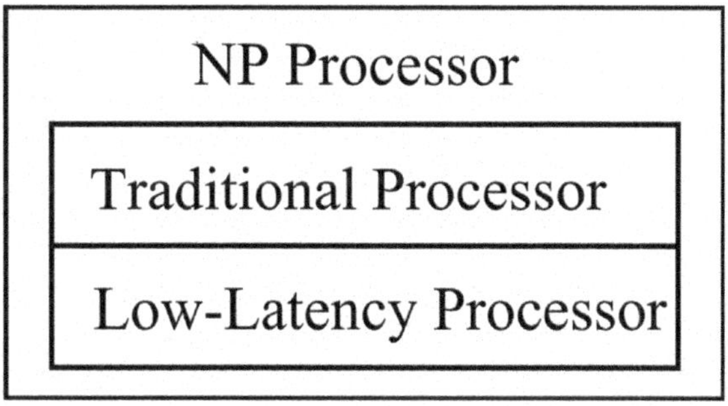

Fig. 4.8 Priority of NP perception

low-latency services, which can reduce the processing time of low-latency data at the network nodes, thereby reducing the overall latency. Priority of NP perception is shown in Fig. 4.8.

Core Network Low-Latency Technologies

The main solution for the core network to reduce latency is to sink the user plane gateway, and the location of the sinking can be deployed according to service requirements, as shown in Fig. 4.9. If the service does not have a high requirement for latency, the user plane gateway corresponding to the service can be deployed to the core data center. If the service has a high requirement for latency, the user plane gateway corresponding to the service can be deployed to the edge data center.

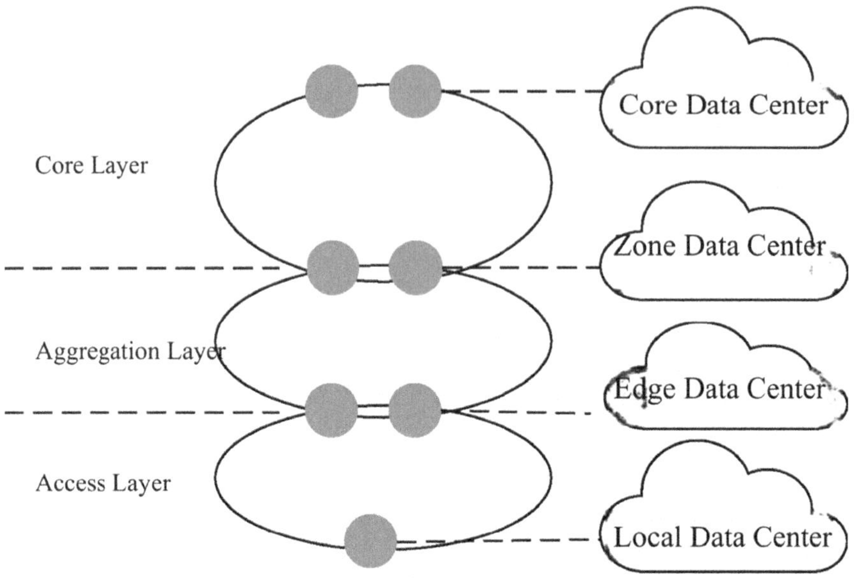

Fig. 4.9 Location of core network gateway deployment

4.2.3 High Precision Timing

5G timing refers to the technology of sending synchronized time to the terminal through a 5G wireless base station. Accurate timing is the precondition for many high-precision science and technology applications. An accurate clock can be used for distance measurement, altitude measurement, synchronization, and speed measurement. 5G network has introduced network timing at the beginning of the protocol design. Take the example of a lower computer (such as PLC/SCM) obtaining 5G base station timing; the overall process is divided into three steps: ① A base station receives satellite signals first; ② The base station sends time messages to the CPE device periodically; ③ CPE device forwards time information to a lower computer. The 5G network timing system is shown in Fig. 4.10.

The time information sent by the base station is broadcast to CPE in the form of periodic system messages, and the system messages carrying reference time is system message 9. CPE receives and parses the system messages. By combining time advance information sent by the base station, IRIG-B time code is generated, which is the abbreviation of InterRange Instrumentation Group-B. It is a widely used form of transmission of information about time at present. CPE sends IRIG-B information to the slave computerat a fixed frequency and cycle so that the lower computer can synchronize time with the network, as shown in Fig. 4.11.

The clock signal terminal gets from the base station is based on the relative accuracy of the clock source. There are two main clock source solutions for base stations: satellite and network clock solutions.

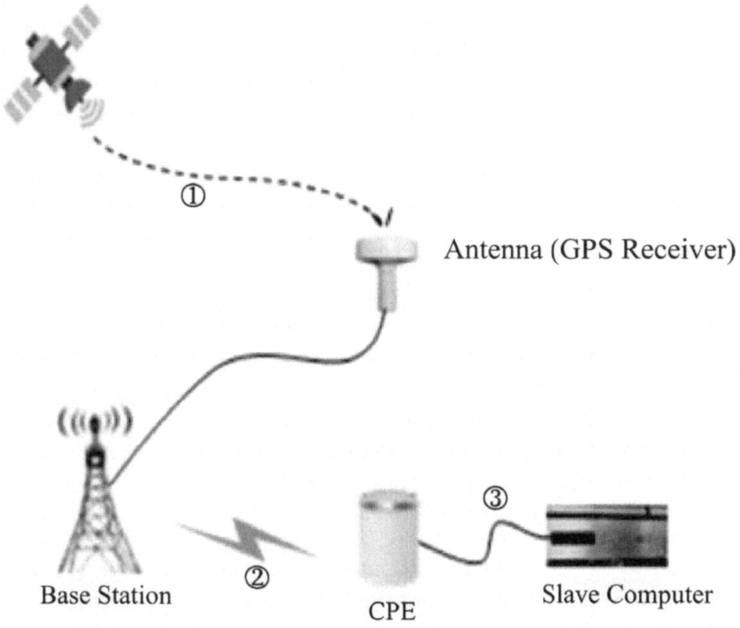

Fig. 4.10 5G network timing system

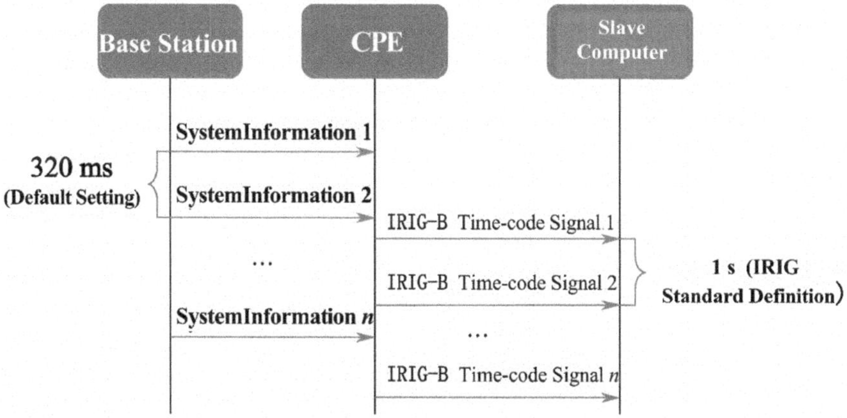

Fig. 4.11 Timing process in 5G base station

Satellite Clock Solution

Global Navigation Satellite System (GNSS) provides clock signals for base stations. Each GNSS satellite has an atomic clock, enabling satellite signals to contain precise

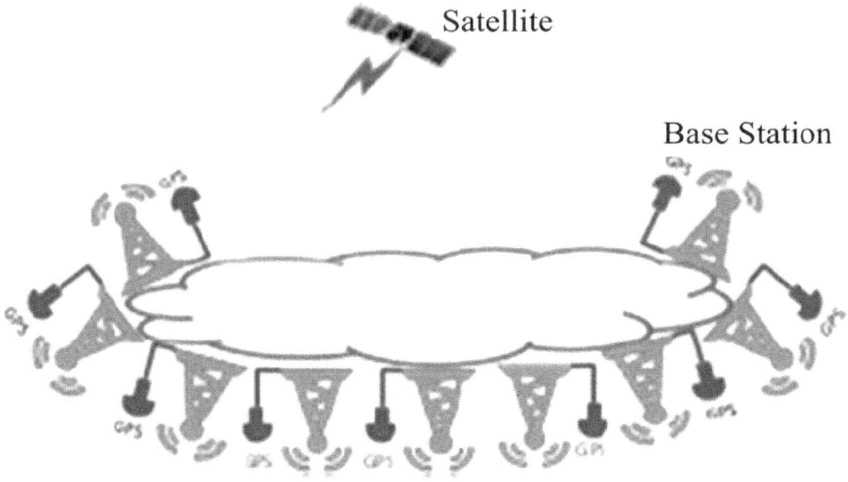

Fig. 4.12 Satellite clock solution

data. A base station receives satellite time through GNSS timing module and synchronizes its equipment with the satellite atomic clock to obtain a clock, as shown in Fig. 4.12.

GPS is Global Navigation Satellite System (GNSS) by the US, which is also the earliest GNSS in the world. BeiDou Global Navigation Satellite System (BDT) is a GNSS developed and built completely by China, with global coverage capability. GNSS also includes GLONASS by Russia and Galileo by Europe. GNSS provides time information as the upper-level clock, and the lower-level user terminal receives the time information and adjusts the local clock to keep the time difference within a certain range. To improve timing accuracy, the user terminal must accurately calculate the latency of time information in the propagation link. Satellite timing, such as GPS and Beidou, can determine the electromagnetic wave transmission latency through the distance between the user terminal position and the satellite so as to eliminate most of the errors. GNSS usually requires the number of satellites to be searched for at the receiving end. Usually, a minimum of four satellite signals are demanded. However, satellite signals can be easily blocked by something tall and big, such as buildings and mountains, affecting the quality of satellite signal reception.

Network Clock Solution

When receiving GNSS satellite signals (such as indoor scenarios) is difficult, IEEE 1588v2 synchronization technology can be used to supplement or replace the base station clock source. IEEE 1588v2 is one of the network clock solutions, with the basic idea of providing clock signals by the clock server. The clock signal will be sent to the base station through the bearer network, as shown in Fig. 4.13.

Fig. 4.13 Network clock solution

The source of 5G timing system clock uses a double backup of GPS/BeiDou satellite. Double-system backup of satellite and IEEE 1588v2 greatly improves security and accuracy. Besides, thanks to wide coverage and synchronization of whole network time, 5G can support more diversified scenarios.

4.2.4 High-Accuracy Positioning

The indoor applications of Location Based Service (LBS) have been integrated into various toB industries, such as automatic tractor-trailers used in manufacturing and self-driving vehicles to everything.

Figure 4.14 shows that the 5G positioning standard is divided into two versions, NSA and SA, with NSA architecture using the original LTE positioning specification and SA architecture using the NR positioning specification. The first protocol of SA-based NR positioning standard Rel-16 was released in 2020. 3GPP protocol group will continue to update NR positioning specifications and techniques in subsequent versions.

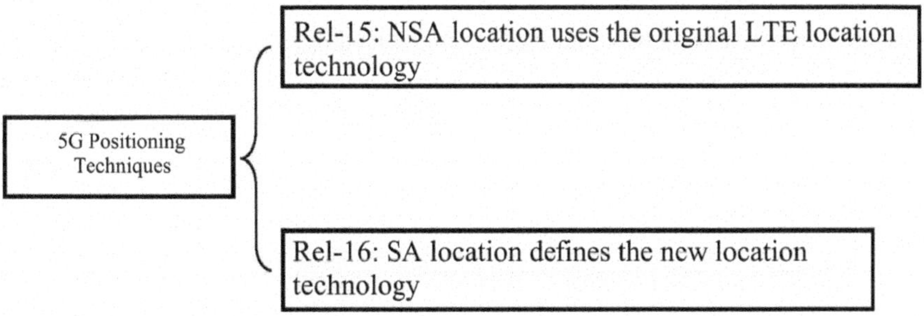

Fig. 4.14 NR positioning specifications and positioning techniques

Table 4.2 5G positioning accuracy

Position level	Position accuracy		Position reliability (%)	Position latency
	Horizontal accuracy (m)	Vertical accuracy (m)		
1	10	3	95	1 s
2	3	3	99	1 s
3	1	2	99	1 s
4	1	2	99.9	15 ms
5	0.3	2	99	1 s
6	0.3	2	99.9	10 ms
7	0.2	0.2	99	1 s

3GPP protocol specification provides a detailed definition of the level, accuracy, reliability, and latency for 5G network positioning, as shown in Table 4.2.

5G network positioning technology mainly uses wireless positioning technology, where the network measures characteristic parameters of the received radio waves and uses the measured wireless signal data to estimate the geographical location of the mobile terminal using specific algorithms. Wireless positioning signal measurements mainly include Time Of Arrival (TOA) measurements, Arrival Of Angle (AOA) measurements, and signal field strength measurements, with field strength measurements mainly used indoors.

TOA Measurement

Distance is measured by measuring the arrival time of the signal, as shown in Fig. 4.15. *A*, *B*, and *C* are nodes with known positions, and point *P* is the location of the terminal. The distances R_1, R_2, and R_3 from *P* to *A*, *B*, and *C* can be calculated according to the time difference sent from *P* to *A*, *B*, and *C*. Then, the three nodes A, B, and C are used as the center of the circle, and R1, R2, and R3 are used as the radius to draw three circles. The intersection of the three circles is the position of point *P*. This method is based on the fact that the position of the measurement node is known, and the signal received by UE must be propagated in a straight line. In reality, the signal detected by UE is likely to be reflected from multiple paths, resulting in a large measurement error in this way.

AOA Measurement

AOA measurement is a technology that uses directional rays of arrival angles for intersection positioning. The technology sets up directional antennas or arrays of antennas at more than two location points. First, obtain angle information of radio signals transmitted by terminals, then intersect directional rays of terminals estimated

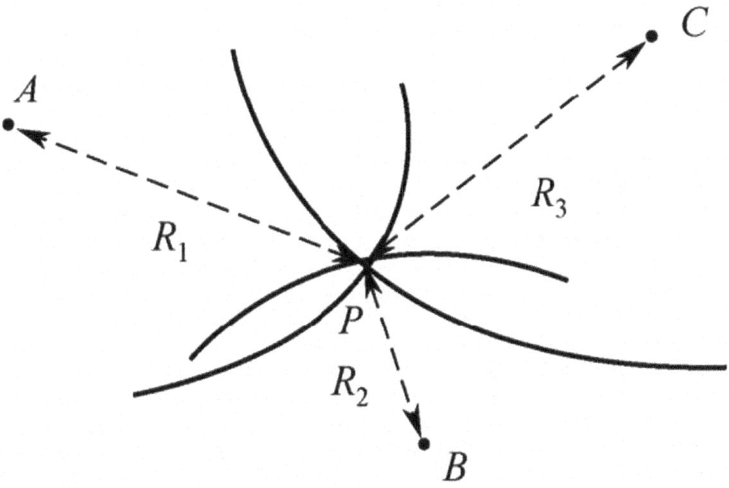

Fig. 4.15 TOA positioning

by each array antenna to obtain intersection points, which are the estimated locations of terminals, as shown in Fig. 4.16.

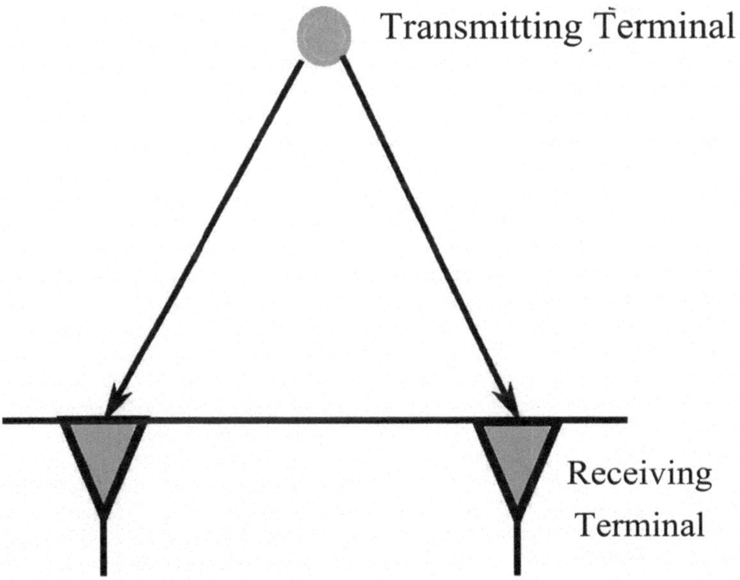

Fig. 4.16 AOA positioning technology

Signal Field Strength Measurement and Positioning (Fingerprint Database Positioning)

The positioning method for signal field strength measurement mainly uses the field strength of different locations as fingerprint features, such as wireless signal strength, geomagnetic strength, etc. Fingerprint database positioning generally includes two steps: offline training of fingerprint collection and online positioning by fingerprint matching. Establishing a fingerprint feature database requires a large amount of data collection and measurement and has high requirements for the measurement accuracy and stability of the field strength. Meanwhile, the database also needs to be updated quickly. When the location of a user needs to be determined, the field strength information of the user's location is obtained first, and then the fingerprint database needs to conduct a backward search to match the user's location. As shown in Fig. 4.17, an area is divided into 12 grids, and each grid is numbered. The signal characteristics of each grid are tested and listed in the database. When the signal of a user's location matches grid number 4, it means that the user is in grid number 4. Since the accuracy of the field strength measurement and positioning is related to the accuracy of the fingerprint database, a lot of testing is needed in the early stage. The fingerprint database has to be updated in time once the environment changes, therefore, this method is suitable for the positioning of indoor micro-stations.

The above positioning technologies are applicable to both 5G and other wireless communication methods, but 5G has certain enhancements in positioning capability compared to previous cellular mobile networks due to higher site density. From a global perspective, the frequency used in 5G is generally higher than that in 4G, which means that the site density of 5G is higher than that of 4G when covering the same area, and the high-density base stations are conducive to enhancing UE positioning accuracy. Meanwhile, 5G supports Massive MIMO. The more antenna arrays there are, the more accurate the measured signal will be, thus the higher the accuracy of 5G positioning. 5G not only supports the positioning of a single device but enables more location-based push services in the future, such as electronic fence, location interaction and location tracking. The typical positioning accuracy of 5G can reach the decimeter level indoors. The accuracy can reach the meter level outdoors with the combination of other technologies. Therefore, the real-time positioning and

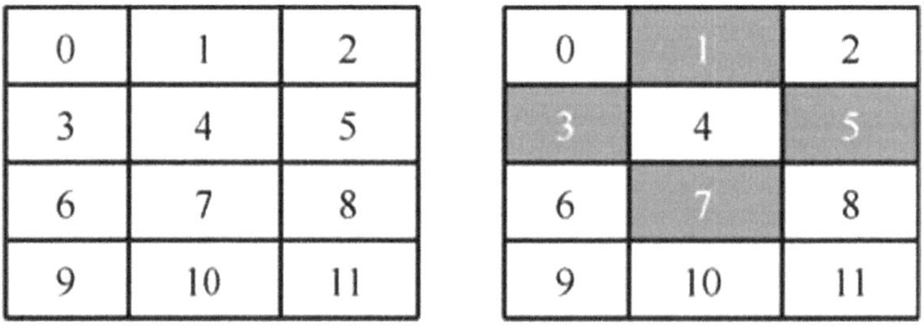

Fig. 4.17 Field strength positioning

interaction latency have fully met the application requirements. Some of the key possible application scenarios include:

In the outdoors, mobile broadband positioning (eMMB) can achieve 10 m level positioning accuracy. Integrating satellite positioning systems will further improve positioning accuracy to 10 m level and below. As a complement to GNSS, it can cover most outdoor positioning needs.

Indoors, 3GPP focuses on positioning and industrial application scenarios, with horizontal positioning accuracy at the meter level or even the decimeter level and vertical positioning accuracy at the meter level. High-precision positioning technology enables 5G to meet the positioning needs of most indoor scenarios.

4.2.5 High Reliability

With the advancement of 5G industry applications, scenarios enabled by 5G are getting more diversified. The industry's demand for 5G networks has been reflected not only in network bandwidth but in the high reliability of 5G network performance. Reliability requirements for typical application scenarios in the 5G era are shown in Table 4.3. Reliability refers to network equipment availability and data connection reliability. Network equipment availability refers to the ratio of network equipment available, taking 99.999% (5 of 9) as an example, 86,400 s per day, that is, $86{,}400 \times (1 - 99.999\%) = 0.86$ (s). Therefore, the time of network equipment failure cannot exceed 0.86 s per day (network equipment availability). Data connection reliability refers to the success rate of 99.999% of packets sent to the destination, which means that only one packet is allowed to be lost in 100,000 packets in the specified time.

Network equipment availability improvement mainly eliminates the impact of a single point of failure of equipment, with the active standby mode the most effective solution for connection reliability. 5G network equipment includes CPE, wireless base station, transport network, and core network. CPE on the terminal side can use industrial-level equipment and have mutual backup between the active and the standby. The base station can use dual base station backup. Baseband single board,

Table 4.3 Typical application scenarios for reliability requirements

Application scenario	Reliability (%)
Cooperative control of vehicle	99.9999
Remote driving	99.9999
VR remote shopping	99.9
Augmented reality	99.9
High-voltage grid communication	99.9999
Real-time industrial motion control	99.9999
Monitoring	99.9
Remote surgery	99.9999

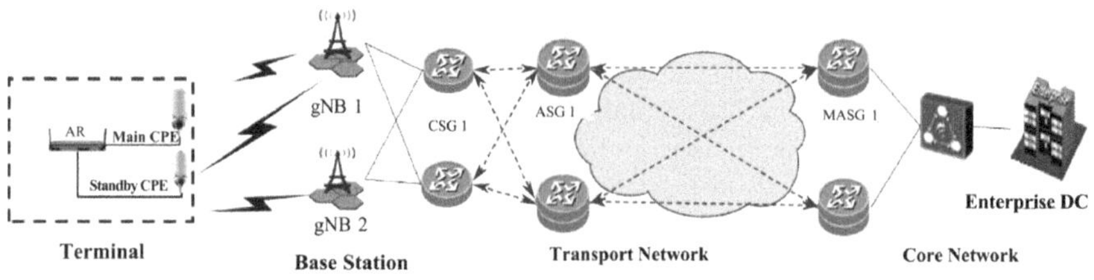

Fig. 4.18 Active-standby mode

master control single board, and other base station hardware can also be used for backup.

Wireless coverage can use dual carriers. Transport networks can use multiple links to achieve mutual backup. Core networks can be organized into a resource pool model, and enterprise parks can also introduce lightweight core networks for mutual backup with public networks, as shown in Fig. 4.18.

Data connection reliability can shorten impact time and scope of failure. Dual-link selection technology can effectively improve the reliability of data connection. When the sender sends data through the main link, it simultaneously copies the message to other links and sends it out. The receiver recombines the packets of the two links. Even if one link is interrupted, its copied link message can still be sent to the other end, preventing service interruption or packet loss due to poor link transmission quality.

In addition, compared to wired network coverage, wireless network coverage has lower signal transmission reliability. To improve the reliability of wireless network links, 5G introduces technologies to improve the reliability of data transmission and reception in wireless networks, such as the repetitive transmission mechanism, which increases the reliability of data transmission through repetitive transmission in the time domain, air domain, and frequency domain. In addition, narrow beams can be reduced to reduce interference between users in different cells, as shown in Fig. 4.19. In terms of modulation and coding, 5G also introduces multiple Modulation and Coding Scheme (MCS) to suit different service requirements and complex wireless environments.

In 5G industry applications, there are also some mobile scenarios, such as unmanned driving, drone inspection, etc. These mobile scenarios are sensitive to service interruptions caused by handover, so 5G introduces a zero-interruption technology for handover in 3GPP Rel-16—Dual Active Protocol Stack (DAPS) handover, as shown in Fig. 4.20. UE differs from the original handover process of "disconnect first, connect later". During the handover, the terminal first establishes a connection with the target base station, holds it for a period of time, performs the transmission, and then releases the connection with the source base station. The original data transmission is not interrupted after the source base station issues the handover command; the original data transmission is not interrupted. UE initiates access to the target side,

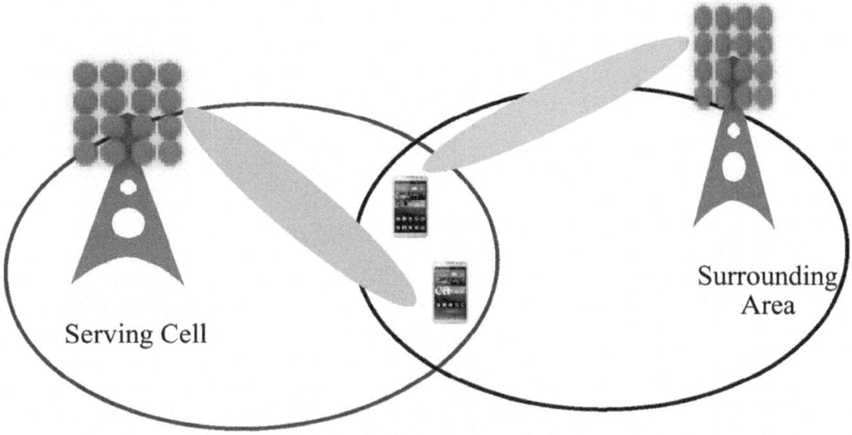

Fig. 4.19 Reliability enhancement technologies for wireless network links

and the original data is forwarded to the target station. The target side starts to send data only after UE synchronizes with the target station.

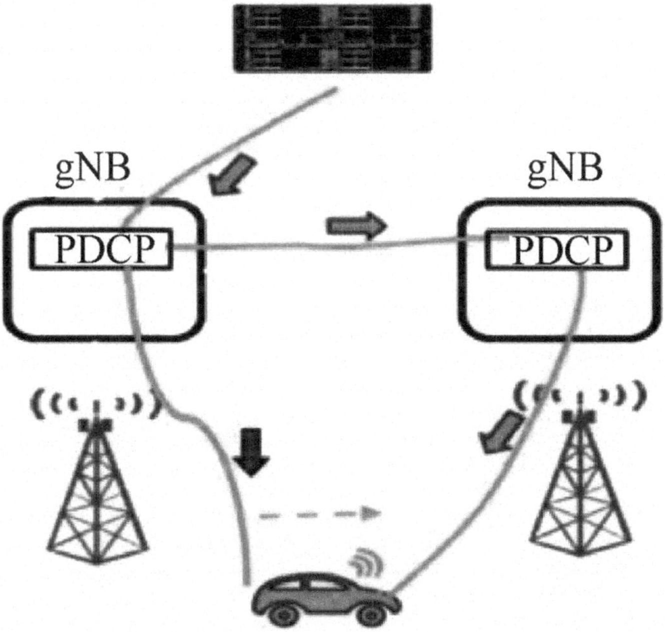

Fig. 4.20 Handover of dual activation protocol stack

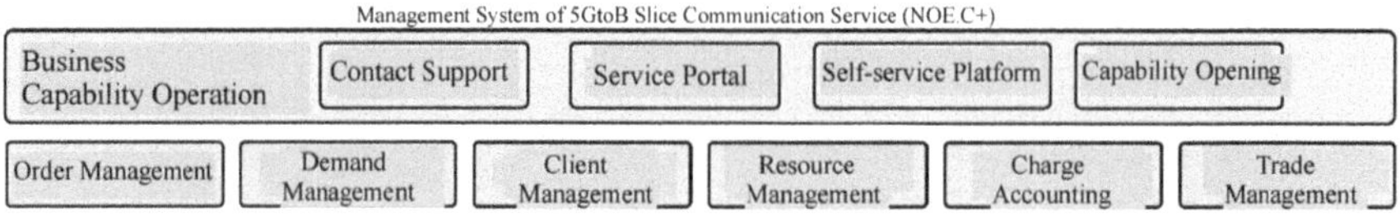

Fig. 4.21 Functional modules related to sliced communication service management system

4.2.6 Minimal Operation and Maintenance

From the operator's perspective, there are many challenges in current 5GtoB scenarios, such as the analysis of customer requirements and network design plans in the pre-sales stage and how to give recommended network plans in combination with service level agreement (SLA) specifications. The mid-sales stage faces problems of slow service activation, non-centralized planning, and low capacity reuse. In the after-sales stage, challenges are how to achieve user SLA management, how to discover network failures or quality problems, how to quickly identify problems and locate boundaries.

In addition, operators and enterprises need to make 5GtoB services visible and manageable, from demand matching to activation and operation and maintenance. Huawei Network Operation Engine (NOE) is an enabling platform for 5GtoB network operation and maintenance, which is used for service activation, SLA guarantee, capability openness, and network model precipitation of 5GtoB cloud network/private line/park and other scenarios. The system mainly includes the following modules: NOE.C+ (sliced communication service management system), NOE.N+ (network operation enablement platform), and NOE.M+ (park network self-management system).

NOE.C+

NOE.C+ provides a one-stop service, including slicing ordering/billing, resource evaluation, order submission, and order status query. Relevant functional modules are shown in Fig. 4.21.

NOE.N+

NOE.N+ supports one-stop service provision for operators and tenant-level SLA monitoring, with relevant functional modules shown in Fig. 4.22.

(1) Slice Solution Design: Compared with toC business, there are more differences between toB business. NOE.N+ improves the efficiency of slice solution design by providing auxiliary operation services.

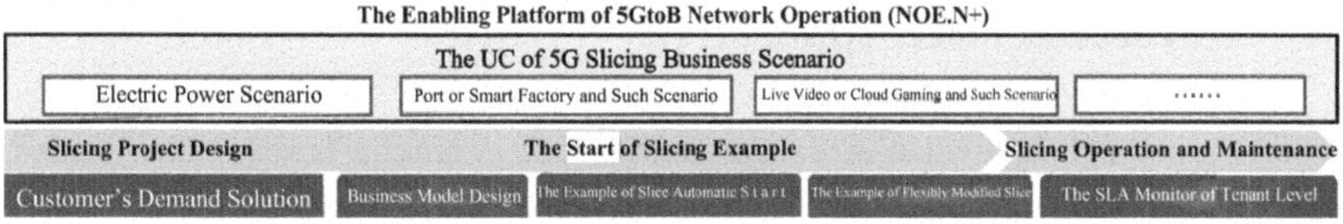

Fig. 4.22 Functional modules related to network operation enablement platform

(2) Slice Instance Activation: The activation of slice instances mainly involves three modules: service model design, automatic activation of slice instances, and flexible modification of slice instances, of which the service model design module describes the service model as slice instances and the resource model as cross-domain network resources; the automatic activation of slice instances module mainly realizes automatic activation of slice networks; the flexible modification of slice instances module can quickly realize on-demand modification, activation/deactivation and expansion of slices through graphical operations to meet the needs of toB service development.
(3) Sliced Operation and Maintenance: Based on sliced instances, tenant-level SLA monitoring is provided and activated to tenants for self-service use on demand. On the basis of slice topology, performance information and real-time hierarchical alarm information are superimposed.

NOE.M+

NOE.M+: In response to the self-management needs of 5GtoB park network enterprises, 5GtoB park network self-management system (NOE.M+) for operators and enterprise park 5G private network is a lightweight self-management platform. Relevant functional modules are shown in Fig. 4.23.

5GtoB park network self-management system has the characteristic of visualization of slice resources and terminal information.

(1) Slicing Resources Visualization: Visualized slicing quality level, alarm distribution, key SLA contracted value and current value.
(2) Terminal Information Visualization: Users can view terminal connection status, basic information, and access information and perceive the terminal, network, and service problems in the park in real-time.

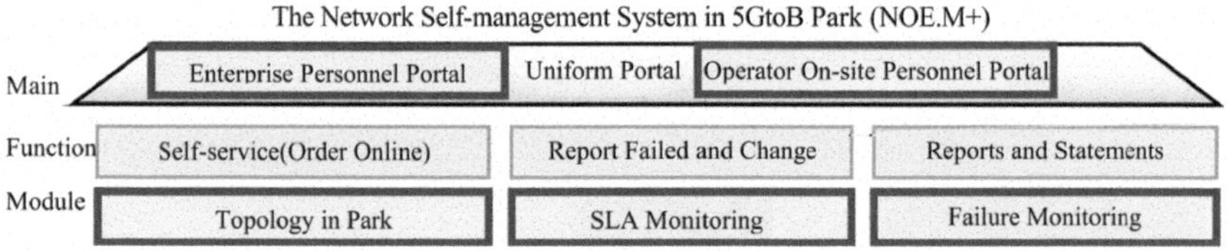

Fig. 4.23 Functional modules related to campus network self-management system

4.3 5G Basic Service Applications

5G has been widely applied in vertical industries such as mining, ports, steel, logistics, power, public security, and media. However, applications vary greatly from industry to industry. In order to facilitate the extraction of vertical industry requirements for 5G and correspond to the modelling and mapping of the network, mainstream industry applications are classified into three common applications: Industrial eMBB, Cloud Vision, and Wireless Automation, as shown in Fig. 4.24. In addition, the maturity of 5G applications in industry and future market size is highly relevant to technical solutions of 5G networks. Of the three main application scenarios defined by 5G, large bandwidth and large connectivity technologies have been defined in 3GPP protocols, and technical protocols related to low latency are still being refined. In terms of implementation difficulty, among the three common applications, technologies for industry production broadband services and cloud vision are more mature with rich applications, while wireless industrial automation services require higher latency and network reliability, which are also the most difficult in terms of technical solutions. Some of the application scenarios will be introduced.

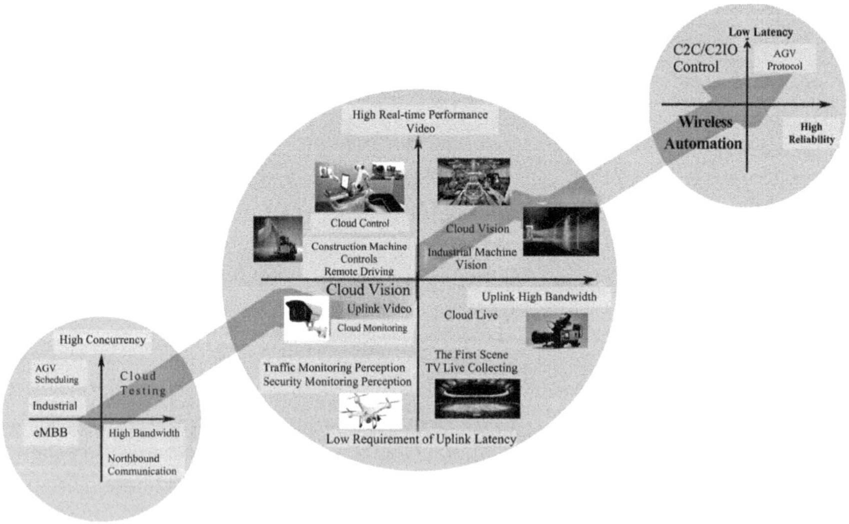

Fig. 4.24 Mainstream industry application types

4.3.1 *Uplink Video Cloud Monitoring Technology and Applications*

The number of Chinese annual shipment of network cameras has exceeded 100 million, of which demanded by the government, public utilities, energy, manufacturing industry account for about 40%. Traditional video surveillance industry has pain points in wiring difficulties, dispersed geographical locations for video surveillance and high cost. Mobile surveillance scenarios, such as police cars, police drones and other scenarios cannot use wired connection. In addition, in scenarios needing temporary emergency deployments, such as large-scale gatherings, concerts, and sports events, the deployment wired connection is slow and easy to cause resource waste. With the enrichment of business scenarios, the video surveillance industry is also developing towards HD and intelligence. In addition to presenting video, intelligent video surveillance can not only present video but also make use of AI to intelligently analyze video content and achieve functions such as behavior analysis, anomaly detection, identification detection, and statistics.

The demand for video surveillance on the network is mainly reflected in uplink bandwidth. The uplink bandwidth rate depends on the bit rate of the video, video encoding format, resolution, etc. Take 1080P resolution for example, the uplink rate requirements are shown in Table 4.4.

Since the 5G wireless network has the characteristics of large bandwidth, wide coverage, and large connection, it can better match the demand for video surveillance. Therefore, 5G-based video surveillance applications are becoming more widespread. A typical 5G video surveillance networking scenario is shown in Fig. 4.25. Cameras with 5G modules can be directly connected to 5G base stations, and cameras without 5G modules can first connect to a router and then to a 5G CPE before connecting to the 5G network. The video surveillance platform can be flexibly deployed based on enterprise needs in enterprise park MECs, offload to park video platforms, or be deployed outside the enterprise park as needed.

In video surveillance applications for enterprise parks, a 5G-based park video surveillance system is shown in Fig. 4.26. 5G enables flexible access to video surveillance, allowing on-demand deployment of video surveillance nodes, including drones, park robots, etc. Applications, such as HD video and finer visual recognition, are supported.

Table 4.4 Rate requirements for video

Resolution ratio	Typical code rate	Condition (Times code rate)	Speed rate requirement (Mbps)
1080P	3.5 Mbps	1.4	4.9
		1.7	6
		1.9	6.7

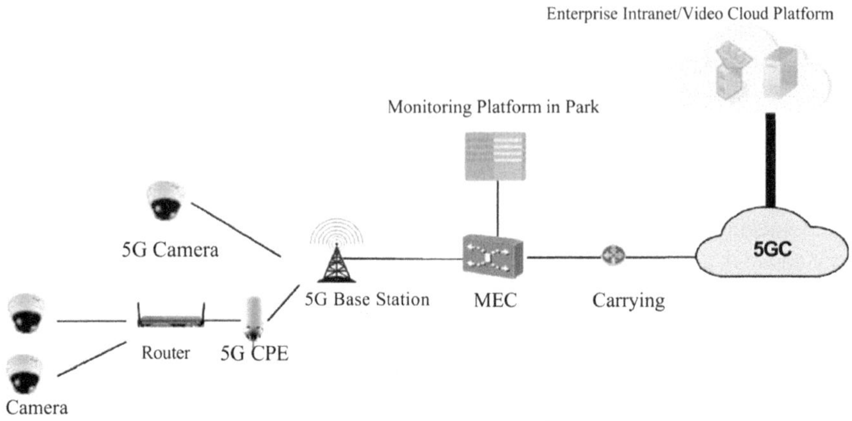

Fig. 4.25 Typical networking scenario for 5G video surveillance

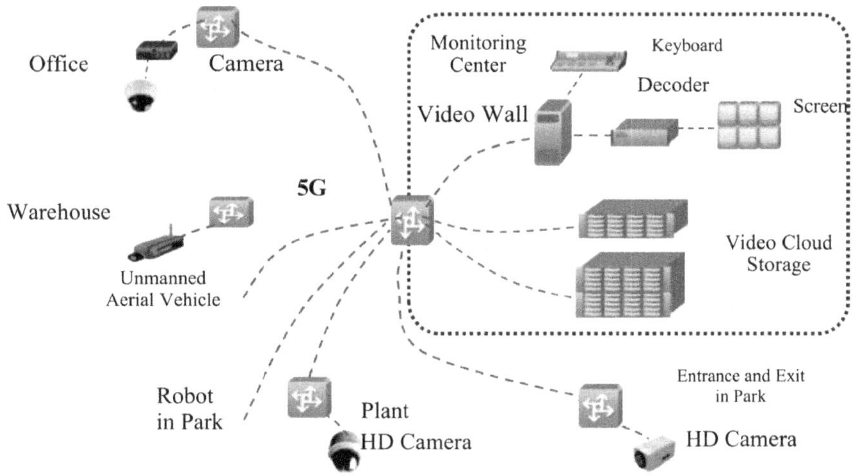

Fig. 4.26 5G park video surveillance system

4.3.2 Remote Control Technology and Applications

In industrial manufacturing environments, there are often scenarios where the environment is harsh or the working environment is dangerous, for example, hazardous locations that are toxic, flammable, and explosive and workplaces with geological hazards. For example, mining, steel making or excessively narrow workspaces are difficult to construct and pose a threat to the personal safety of the operator. In addition, the frequent construction safety accidents are constantly troubling construction companies. For these pain points, more companies are considering using remote control for operations. The remote control process usually involves two roles: the

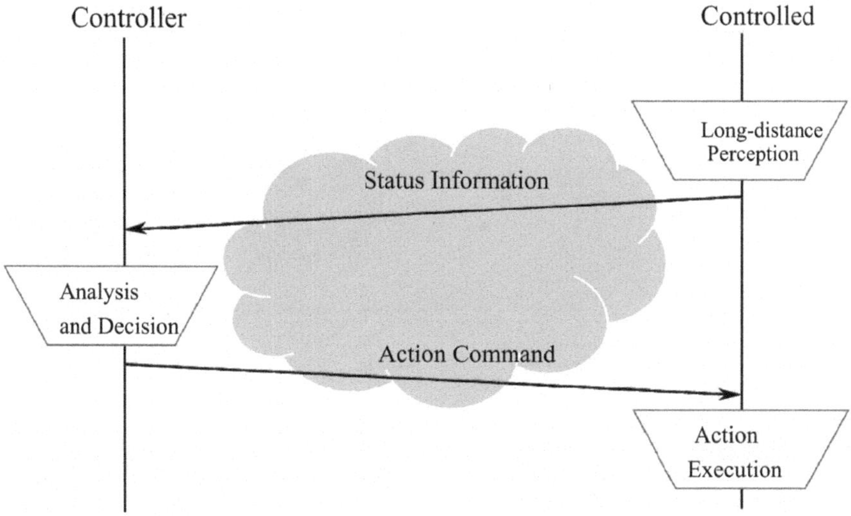

Fig. 4.27 Process of remote control

controller and the controlled, as shown in Fig. 4.27. The control needs to send status information to the controller through the communication network. The controller first analyzes and makes decisions based on the received status information (such as location, sound, environment, temperature, and humidity) and then sends corresponding action instructions to the controller via the network. The controller executes the corresponding action according to the action instructions received to complete the remote control.

In a simple remote control system, video information is usually included in the status information of the controlled, and remote control requires real-time video backhaul to the operation site. In a complex remote control system, a large amount of sensor information from IoT terminals is also included. In remote control, the stability of the transmission of downlink control instructions is very important. If a certain control instruction is lost in transmission, a control error will lead to abnormal equipment operation, or even a serious safety accident will occur.

4.3.3 Machine Vision Technology and Applications

Machine vision refers to the process of machine judgment and recognition of an image or video. With the development of technology, the viewing object of images is shifting from the human eye to machines. The application of machine vision combined with AI technology is becoming more widespread. Currently, the four more mature types of machine vision applications include positioning, detection, measurement, and recognition, as shown in Fig. 4.28.

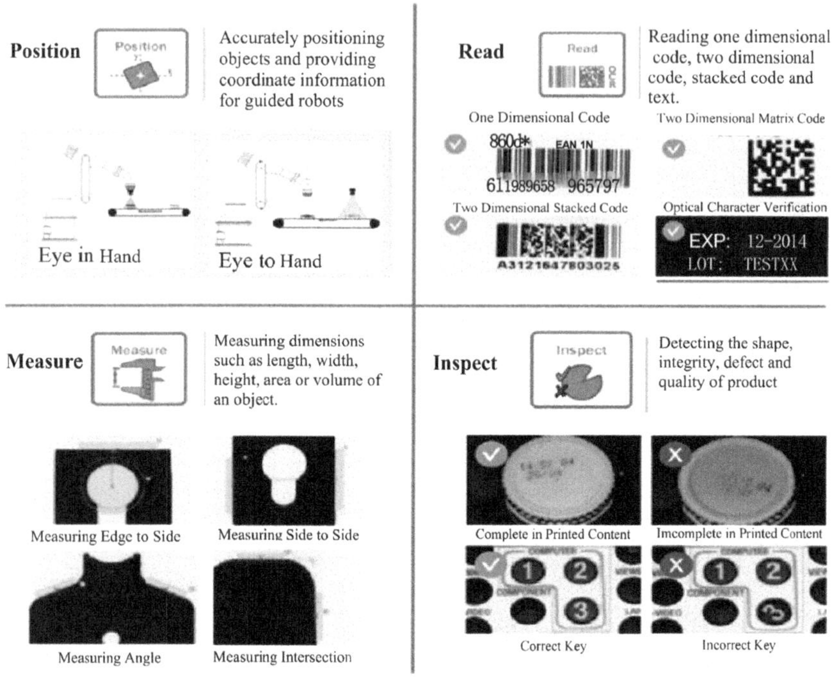

Fig. 4.28 Four types of machine vision applications

5G machine vision is widely used. For example, in road and bridge inspection, the traditional inspection is mainly carried out manually, which has low efficiency and high labor costs. In a 5G-based road and bridge visual recognition system, real-time recognition of surface damage can be achieved quickly and accurately through video recognition. The road and bridge monitoring vehicle continuously transmits 4K video back to the visual recognition platform in real-time through 5G, and the platform automatically identifies surface damage such as subsidence, potholes, cracks, mesh cracks, breakage, and congestion, enabling real-time inspection round the clock. Intelligent and accurate analysis of problem sections can be achieved, as well as closed-loop management of automatic dispatching and maintenance and acceptance, as shown in Fig. 4.29.

5G+ intelligent video surveillance inspection solutions for roads and bridge can not only address safety management problems but enhance inspection efficiency.

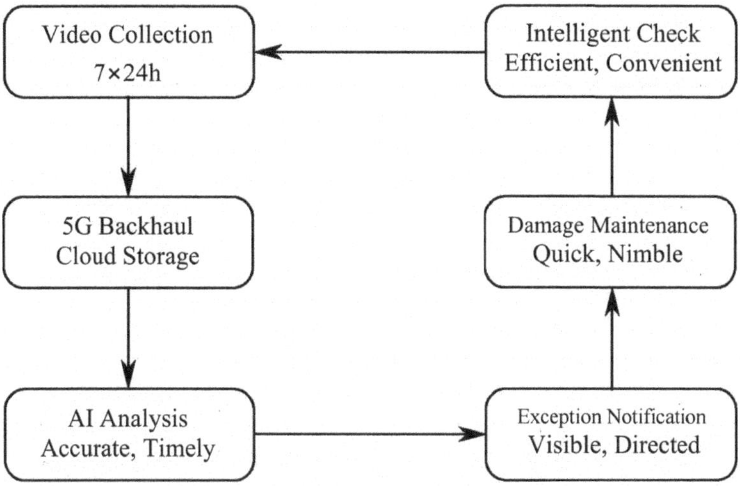

Fig. 4.29 Intelligent inspection system for road and bridges based on 5G+ machine vision

4.3.4 Indoor Positioning Technology and Applications

Outdoor positioning is usually based on global satellite navigation systems but cannot meet the needs of indoor location services (such as real-time tracking of work-in-progress, asset location, and digital management of people and property). Commonly used indoor positioning technologies include Wi-Fi, Bluetooth, Ultra Wide Band (UWB), vision, optical, laser, and electromagnetic. The current fragmented indoor positioning market and the high cost of proprietary positioning solutions are the main factors limiting the application. Relying on a 5G network and integrating indoor proprietary positioning, a network can achieve multiple uses and multiple functions. The integration of indoor positioning technologies into the large 5G industry has become an important initiative to promote the widespread application of positioning services. The combination of 5G and other positioning technologies, as well as the comparison of advantages and disadvantages and application scenarios, are shown in Table 4.5. The 5G hybrid positioning approach is important to achieve high-accuracy positioning anytime and anywhere.

Take the positioning of indoor Automatic Guided Vehicle (AGV) as an example. AGV has now become an important piece of equipment for flexible production lines, flexible assembly lines, and warehouse logistics automation systems and is also a core component of intelligent and unmanned factories. The current positioning and navigation methods of AGVs mainly include fixed path guidance, semi-free path guidance, and free path guidance. The fixed path guides AGV to move along the pre-planned track by laying out markers along the path, including magnetic tape, magnetic strips, etc. This method is less flexible because tapes need to be relayed to change the driving track of the AGV. Semi-free path guidance constructs a map by setting markers in the path so that AGV can self-plan the path based on the

Table 4.5 Combined indoor positioning technology solutions, comparison of advantages and disadvantages, and application scenarios

Integrated positioning scheme	Advantage	Disadvantage	Application scenario
5G + UWB/Bluetooth	High positioning accuracy, enables positioning at the decimeter level	Higher cost	Fields for high precision positioning requirement, such as manufacturing
5G + Bluetooth	Low cost, meter level positioning accuracy	Average positioning accuracy	Hospitals, Shopping malls, Parking lots
5G + Wi-Fi	Low cost, meter level positioning accuracy	Average positioning accuracy	Shopping malls, Transportation hubs
5G + Laser/Vision	High cost, centimeter level positioning accuracy	Immature technology	Intelligent manufacturing, Unmanned driving

map during movement, such as laying QR codes on the driving track, but the QR code is easy to wear out; the laser positioning device is more costly and has higher requirements for the environment (such as light and visibility). Free path guidance achieves simultaneous localization and map construction (SLAM) mainly through laser. This method doesn't need to lay markers on the driving track. AGV uses laser scanning to model the work scene and construct a map, as well as self-planning the moving track. But the laser SLAM method is more costly.

5G cloud-based AGV vision navigation solution is shown in Fig. 4.30. The integrated 5G module and camera first scan the surrounding environment, then data is transmitted through the 5G network to the machine vision recognition system deployed on MEC. The location information then comes out and is sent to the control and dispatch center of AGV.

Compared with traditional AGVs, the unified dispatching and coordination capability of 5G cloud-based AGVs is enhanced, thanks to large connectivity, low latency

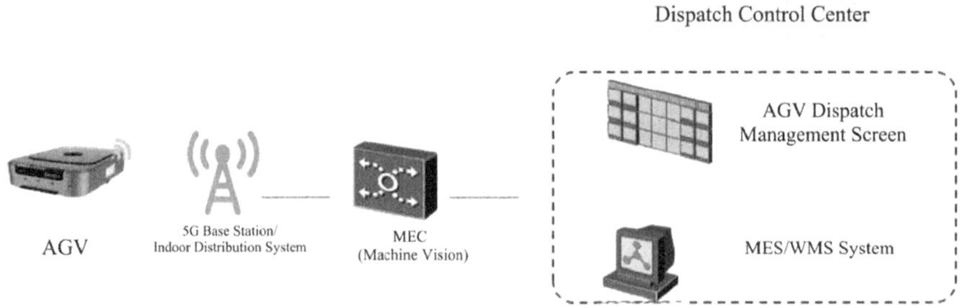

Fig. 4.30 5G cloud-based visual navigation solution for AGVs

of 5G, and the improvement of AGV control capabilities. Meanwhile, centralized control can also reduce the cost of individual AGVs.

4.3.5 Other Technologies and Applications

1. **Data Collection/Loading Techniques and Applications**

With the advancement of digital transformation of enterprise business, new challenges are posed for the collection and loading of production data. The data output and manual entry capabilities of the original platform are no longer sufficient to meet the needs of enterprises operating under digitalization. To build data-perception capabilities, enterprises need to adopt modern methods to collect and acquire data and reduce manual entry. Take the industry as an example; data sources mainly come from various components, and data types include structured and unstructured data. Moreover, the data is massive and real-time in transmission, and the decision-making issued by the system is also real-time after AI processing. The combination of 5G + cloud computing + big data + artificial intelligence is the technical basis for digitalization.

2. **Cluster Scheduling Technology and Applications**

The large-scale application of cluster scheduling technology is mainly concentrated in public security and security industries. Take the police cluster business as an example, the police cluster business mainly includes cluster command and dispatching, police unmanned aerial vehicles, vehicle scheduling, video surveillance, and police dispatching. In emergency response, it is necessary to simultaneously dispatch voice, view on-site video, hold video conferences and query associated data. If the degree of data sharing and business linkage is low, it will lead to inefficient emergency response and delay of opportunity. 5G is a three-dimensional coverage network, and a 5G-based cluster scheduling system can not only command large-scale dispatching on the ground but also support the dispatching of unmanned aerial vehicles, increase collaboration of various service streams, and achieve the interconnection between data.

3. **IoT Sensor Technology and Applications**

IoT sensor technologies mainly enable short-range communication, such as Radio Frequency Identification (RFID) and Near Field Communication (NFC). With the rapid development of IoT, services have put forward higher requirements for wireless communication technology. Narrow Band Internet of Things (NB-IoT) technology and Long-Range Radio (LoRa) technology, which are specifically designed for low bandwidth, low power consumption, long-range and large-connection IoT applications, are becoming more extensively used. The development of digitalization will lead to an increase in the number of IoT terminal connections and data types, but NB-IoT and LoRa are not as good as 5G networks in terms of linking numbers,

Fig. 4.31 C2C Control

mobility, and bandwidth. In addition, in the age of intelligence, 5G will empower every terminal with "AI" capabilities so that every terminal can think.

4. **C2C Control Technology and Applications**

The full name of C2C is Control to Control, which usually refers to the communication between master and slave programmable logic controllers (PLCs). Taking advantage of the large connection and the low latency of the 5G network, communication between master–slave PLCs can be carried out via the 5G network, as shown in Fig. 4.31.

4.4 Summary

This chapter introduces the industrial demand for 5GtoB. Unlike 5G in the personal field, requirements for customization will be a norm state as 5G advances in industry applications. Through the analysis of segmented application scenarios, it is found that they have certain commonalities—the demand for key network capabilities is convergent, which is also a characteristic of 5G in industry applications. Meanwhile, this chapter introduces 5G video surveillance, remote control, machine vision, indoor positioning, data collection/loading, cluster scheduling, IoT sensing, and C2C control application scenarios from an application perspective.

By studying this chapter, readers should understand 5G network capabilities and applications, be able to grasp indicators of 5G network capabilities and have a clear understanding of the role 5G plays in applications.

Exercise 4

I. True or False

4.1. Carrier aggregation technology is primarily designed to reduce the latency of the network. ()

II. Choice Questions

4.2. (Multiple choice) When the latency from the application server to the terminal is high, in what way can it be optimized? ()

A. Network Architecture

B. Air Interface
C. Equipment
D. Service.

4.3. (Single choice) In which scenarios can the positioning method based on field strength measurements be used? ()

A. Along the high-speed railway
B. Urban roads
C. Indoor shopping malls
D. Unmanned aerial vehicle positioning.

4.4. Multiple choice) Which of the following are short-range communication IoT sensor technologies? ()

A. Rfid
B. Nfc
C. NB-IoT
D. LoRa.

4.5. (Multiple choice) The main application scenarios for machine vision are____ ___ ()

A. Positioning
B. Testing
C. Measurement
D. Identification.

III. Short Answer Questions

4.6. Briefly describe the principle and purpose of uplink and downlink decoupling.

Chapter 5
Integrated and Innovative Applications of 5G+ New Technologies

Nowadays, human society is facing a new round of technological revolution and industrial transformation and is about to evolve into an intelligent society with the perception of everything, interconnection of everything, and intelligence of everything. Emerging technologies based on 5G, the Internet of Things (IoT), cloud computing, big data, and artificial intelligence (AI) will become important engines to promote industrial transformation and economic development.

This chapter will explore the role of ICT technologies in the era of digital economy, introduce emerging technologies such as the Internet of Things, cloud computing, big data, and artificial intelligence, and discuss how to realize the integrated empowerment and innovation of 5G and these emerging technologies.

Learning Objectives

- To understand the development trend of ICT technology.
- To master the characteristics and current situation of new technologies.
- To master the integrated and innovative application of 5G+ new technologies.

5.1 Digital Economy Driven by the Integration of ICT Technology

5.1.1 History of ICT

IT originally refers to Internet Technology. Before the integration of IT and CT, IT was not Information Technology, which is why the IT industry was called the Internet industry. In the beginning, IT and CT were two distinct technologies, and the IT industry could be said to represent the computer industry, and some software and hardware companies involved in computers were also called IT companies. CT refers

L. Song et al., *5GtoB From Theory to Practice*,
https://doi.org/10.1007/978-981-97-8453-0_5

to Communication Technology. The earliest CT industry was also known as Telecommunication. The earliest communications were telegraph, telephone, and other technologies, so they were also known as telecommunication technologies. Communication enterprises are divided into operators, communication manufacturing, communication service support and communication construction units. Famous operators in China are China Mobile, China Unicom, China Telecom, and China Broadnet. Huawei is a worldwide, well-known communication manufacturing enterprise.

The integration of the IT industry and the CT industry started with the invasion of the IT industry into the CT industry. IP technology is the technology of the IT industry. After IP defeated Asynchronous Transfer Mode (ATM) technology, the ICT industry began to integrate greatly. Software and hardware companies in the IT industry began to march into the CT industry on a large scale, and CT companies also began to develop IT technology. As a result of this two-way integration, barriers between the IT industry and the CT industry became less and less obvious, and a new industry, ICT, was formed.

The United Nations released the fourth edition of the International Standard Industrial Classification (ISIC) on August 11, 2008, and the Organization for Economic Co-operation and Development (OECD) gave the definition of ICT in 2017 with reference to this classification: ICT refers to products and services that mainly complete information processing and communication by electronic means, or make them have information processing and communication functions, which includes ICT manufacturing, ICT trade and ICT service.

Economically, the future development of the ICT industry will facilitate the transformation of the traditional economy into a digital economy. In recent years, digital infrastructure has become an important support for developing the national economy. The future of ICT will reflect the deep integration of the digital economy with traditional power grids, highway networks and railway networks, forming an intelligent integrated information infrastructure with the interconnection of everything, ubiquitous perception, and space integration, which will greatly enhance the networking, digitalization, intelligent level of economic activities and operational efficiency, and the formation of "digital technology-economy paradigm" has been accelerated. The integration and penetration of the digital economy with various industries and fields has given a strong impetus to the technological progress of traditional industries and triggered a new industrial revolution. Digital knowledge and information have become the "multiplier" productivity that drives the amplification of all productivity, becoming the leading force in reshaping the economic structure and improving productivity.

5.1.2 Acceleration of the ICT Industry in Transforming Towards Ecology and Intelligence

According to research data from the China Academy of Information and Communications Technology, the scale of China's digital economy industry reached 22.6 trillion yuan in 2016, accounting for 30.3% of GDP, and has become a core component of China's economic growth. In particular, along with the combination of cloud computing, big data, artificial intelligence, and other technologies with vertical industries, various ecological members of the ICT industry accelerate to help different types of traditional enterprises achieve "Internet+" transformation and upgrading. Under the trend of ecological development of the ICT industry, IT hardware and software equipment, operators, service providers, solution vendors and enterprise users together form an essential part of the cloud ecology, cooperating at different levels to achieve mutual benefits and win–win results, while continuously strengthening the development of the ICT ecology. In addition to the development trend of ecology, more and more enterprises have begun to focus on using emerging technology such as artificial intelligence, big data and 5G to accelerate the digital transformation of traditional industries. Customers' business in key industries such as finance, telecommunications and manufacturing began to expand from the processing of unstructured data such as video, images, and voice to intelligent technologies such as image identification and semantic context analysis. At the same time, the integration of various emerging technologies has begun to accelerate, such as combining artificial intelligence and deep learning technologies to make unstructured data applications "smarter" through the training and learning of neural networks, and intelligence has become another mainstream trend in the development of the ICT industry. It can be said that, from 2018 to 2020, with the continuous improvement of data infrastructure and the awakening of enterprises' awareness of "informatization-internetization-intelligence", ecology and intelligence will be the two major directions for the accelerated development of the ICT industry.

5.1.3 ICT Driven Digital Economy

1. **The Shift of the Digital Economy from Consumer Internet to Industrial Internet**

Against the backdrop of the transformation of old and new driving forces, digitization is accelerating its penetration into all fields and regions. And how to achieve digital transformation with the help of cloud computing, big data, artificial intelligence, and other emerging technology applications has attracted attention of the whole society. Correspondingly, as a service provider of enterprise digital transformation, ICT enterprises are also entering the stage of accelerating ecological and intelligent transformation.

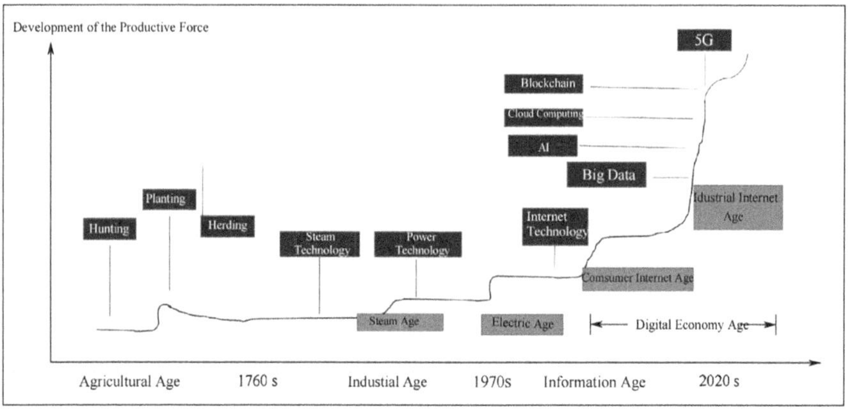

Fig. 5.1 The world entering a new era of digital economy

Productivity development has gone through the primitive era, the steam era, and the electrical era. It is now transforming from the era of consumer Internet to industrial Internet. Figure 5.1 shows key technologies in the industrial transformation in each period, and 5G, as an opportunity for a new round of transformation, will push the digital economy to a new height.

(1) From the agricultural age to the industrial age to the information age, throughout the history of world development, each technological revolution and industrial transformation has promoted a great leap in productivity and great progress in human civilization. Technological forces have continuously promoted human beings to create a new world, and human beings are standing at the forefront of a new era. The digital economy, as a new form of economic and social development in the information age, makes it easier to achieve economies of scale and scope and has increasingly become a new driving momentum for global economic development. The digital economy is a new economic form where digital knowledge and information are taken as key production factors, digital technology innovation as the core driving force, and modern information networks as the important carrier. It deeply integrates digital technology with the real economy to continuously improve the digitalization and intelligence of traditional industries and accelerate the reconstruction of economic development and government governance models.

(2) 5G + Cloud + AI has become an important engine to drive the development of the digital economy. The reliable network of 5G, the massive computing power of cloud computing, and the application intelligence of AI are collaborating with each other, penetrating various industries, creating new business experiences, industry applications and industrial layouts. From digital government to smart cities, from industrial automation to agricultural smart management, the integration and innovation of 5G + Cloud + AI will open up new development space

Fig. 5.2 Industrial integration in digital economy

for thousands of industries and inject new momentum into the transformation of government and enterprises and industrial upgrading.

Currently, the digital economy is changing from consumer Internet to industrial Internet. With the continuous deepening of the global digital economy development, digital development has entered a new stage of kinetic energy transformation. The focus of the digital economy has shifted from consumer Internet to industrial Internet, and the digital economy is entering a new era. Figure 5.2 demonstrates that ICT technologies will reshape economic development and political governance models, digitize industries such as intelligent manufacturing, smart agriculture, and smart services, industrialize digital industries such as basic networks, electronic components, and software and services, and realize digital governance in urban operations, urban management, and urban services.

2. **Terminal Management Cloud: A New Network Architecture and Service Model**

In the next twenty to thirty years, human society will evolve into an intelligent society. Figure 5.3 shows the three characteristics of a smart society: perception of everything, interconnection of everything, and intelligence of everything. In an intelligent society, everything can be perceived. The physical world can be perceived and transformed into digital signals, and multiple sensory channels (temperature, space, touch, smell, hearing, vision) can be utilized to achieve situational awareness, interaction, and immersive user experience. The network connects everything, connecting all data online—wide, broad, multiple, and deep connections from different areas such as cities, mountains, and space—to achieve intelligence. Applications based on big data and artificial intelligence will make everything intelligent, and Digital Twin will gradually spread across individuals, families, industries, and cities to meet the needs of a better physical world, while a second life of digital existence will emerge to make the spiritual world richer. Thanks to advanced ICT technologies, these three characteristics will be realized. At the same time, ICT infrastructure will become the cornerstone of the intelligent world.

3. **5G Assisting Global Perception**

Fig. 5.3 ICT infrastructure becoming the cornerstone of an intelligent world

5G, as the vital link between the physical and digital world, will bring a new global perception with its capacity and speed advantages. The physical world is fully connected and unified through a variety of sensor devices and high-speed networks, mapping out a twin digital world. Figure 5.4 shows the real-time experience of VR under the high-speed network in the future.

5G brings a new opportunity and a new era for the development of smart cities. Every person, every object, and every organization in the city will become an intelligent part. 5G provides connectivity for urban intelligence anywhere and anytime; people, objects, and organizations will connect and exchange data and needs in real-time in the digital twin city. The seamless integration and exchange between the digital twin city and the physical city will drive each intelligent agent in the city to connect and become a distributed super brain.

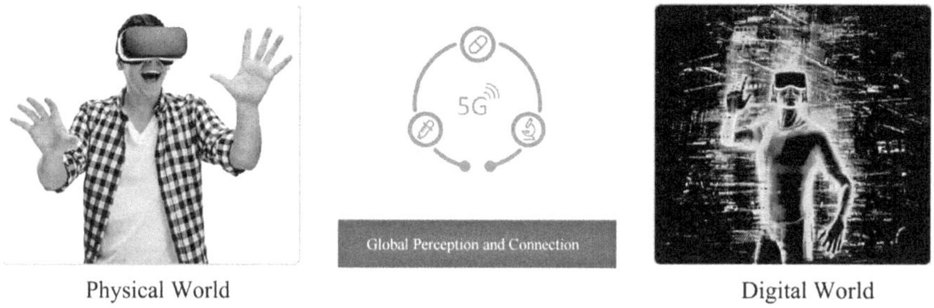

Fig. 5.4 5G assisting global perception

5.1.4 Trends of Talent Demand in Emerging Fields of ICT

ICT industry development has the typical characteristics of advanced driving backward progress. That is, based on the initial achievements of the test site, the progress and upgrading of the entire industry will be driven by the ICT industry. Therefore, the successful application of emerging technologies determines the potential and growth rate of the ICT industry. Emerging technologies often come with higher technical thresholds and more emerging application fields, which put forward higher requirements for talents' technical ability, application threshold, and comprehensive ability. Talents in emerging fields have become the key to maintaining rapid growth in the ICT industry. Take artificial intelligence as an example. Artificial intelligence not only technically requires practitioners to have a deeper understanding of obscure machine learning, algorithmic models, etc., but also builds a higher learning threshold in advanced mathematics, framework applications, etc. In addition, the application of artificial intelligence is often a pioneering field, which requires a comprehensive review and reconstruction of business processes and reconstructed. Thus, practitioners should have an in-depth understanding of industry user processes and cutting-edge guided innovation. The reserve of industry knowledge will further highlight their value. The talent demand gap in the emerging field represented by artificial intelligence will continue to enlarge. With the increasing attention of industrial policies, funds, and industry attention towards the emerging fields, the pattern of industry talent demand is undergoing positive changes. According to CCW (China Computerworld) Research, from the total scale of personnel, the overall talent demand in the ICT field was about 7.65 million in 2017, reaching 12.46 million in 2020. The annual growth rate is close to 20.8%.

With the increasing extension of ICT in the whole industry, the application of emerging technologies in different scenarios starts to accelerate, which further puts forward particular requirements for talents in emerging field applications. Figure 5.5 comprehensively compares application scenarios and talents in different industries. The figure covers four fields: cloud computing, big data, Internet of Things, and artificial intelligence. It also shows the core competitiveness needed to enter each field. The current popular fields urgently need talents whose number will exceed 2 million. The trend of talent demand will be closely combined with the characteristics of the field, with more specialization and scenario specificity.

5.1.5 ICT Integration Trends

In China, the ICT industry is currently in a period of high-quality development, mutual demands for cutting-edge technologies, a focus on expanding industry capabilities, and shoring up deficiencies in core competencies. Against this backdrop, innovation in China's ICT industry is poised to exhibit ten major trends: data

Areas	Capability Requirement	Technical Direction	Talents Gap in 2020	Talents Trend
Cloud Computing	Technical Capabilities: Core Cloud Platform, Monitoring Tools and Configuration Management System, Cloud Security, Container Technology Comprehensive Capabilities: Emphasis on Cooperation, Focusing on Communication, Cooperation, Accountability, etc	Such Languages as Java/Linux/Puppet Such Platforms as OpenStack/Apache/ Zabbix	2.1 million	The cloud computing industry is accelerating the process of industrial applications with the cloud, and there is an urgent need for consulting business on the cloud and the pressure on the demand of talent is prominent.
Big Data	Technical Skills: Business Modeling, Data Mining Algorithms, Database Architecture, Visualization Technology, Data Security. Comprehensive Abilities: Emphasizing Familiarity with Business, Communication, Mathematics, Modeling, etc.	Such Languages as Java/Python/SQL Such Platforms as Caffe/Hadoop/Spark/ Storm	2.57 million	The demand for big data will shift toward a greater understanding of the data and the business logic behind the data, and data scientists will replace the data as the core competency of the industry.
Internet of Things	Technical Skills: IoT System Design, IoT Hardware Development, Embedded Development, Communication Protocol. Comprehensive Abilities: Emphasizing Integration Skills, Teamwork,Cooperation, Communication, etc.	Such Protocols as ZigBee /Wi-Fi/RFID/Bluetooth/ NB-IoT Such Architectures as C++/Linux/CAN	2.11 million	IoT is accelerating its development towards segmentation, specialization and integration, and the requirements for the comprehensive cognitive ability of the industry chain will be more prominent.
Artificial Intelligence	Technical Skills: Machine Learning, AI Algorithms, Deep Learning, Image Processing. General Skills: Emphasis on Foresight, Active Learning, Math Skills	Such Languages as Python/Java/C++ Such Platforms as Caffe/TurnsorFlow/ SciKit-learn	2.26 million	AI will drive industrial model innovation and accelerate the transformation to scenario-based application practices.

Fig. 5.5 Four emerging areas of ICT diverge

value release, edge empowerment, Blue Ocean of 5G, extensive security, integration of "AI+", blockchain consensus, multi-cloud management, core software, native penetration, superlative user experiences.

1. Data Value Release. The steady emergence of essential resource data is entering a phase of releasing its inherent value. The application scenarios for data are becoming increasingly diverse, with continuous integration of data resources. Simultaneously, the transformation of data resources into the value of data assets has reached a critical juncture.
2. Edge Empowerment. With the proliferation of intelligent applications and data volume, network bandwidth, and computing throughput have become performance bottlenecks of computing. Meanwhile, the demand for real-time processing of massive amounts of "small data" generated by terminal devices has surged, driving edge computing to become an important computing platform for the implementation of technology in the intelligent era, as well as key support for agile connectivity, real-time business, and privacy protection in the digital transformation of the industry.
3. Blue Ocean of 5G. With the deepening commercial use of 5G, the blue ocean market is getting into gear. Although VR/AR and Ultra HD (ultra-high definition) are the few 5G applications that can be foreseen at present, the industrial Internet, vehicle networking, telemedicine and other fields will drive the outbreak of 5G applications in the future.
4. Extensive Security. With the acceleration of the digitalization process in various industries, the demand for network security in manufacturing, education, energy, and other industries continues to increase. Network security continues

to expand to emerging fields like artificial intelligence, cloud computing, big data, the Internet of things, mobile Internet. Network security is not only a problem of informatization but also involves national security, social security, personal security, and other aspects which cannot be ignored.

5. Integration of "AI+". As the core technology in the era of intelligent economy, the development of AI continues to integrate. On the one hand, it refers to the integration of technology, accelerating the deep integration and innovation of emerging technologies such as big data, cloud computing and the Internet of Things; on the other hand, it refers to the integration of applications in industries from key breakthrough to balanced distribution, and more and more market opportunities arise in small scenes and traditional scenes.
6. Blockchain Consensus. Under the guidance of national policies, blockchain technology has returned to the mainstream of empowering the real economy. The decline of cryptocurrency speculation and the launch of digital currencies from the People's Bank of China have formed a sharp contrast, and various industrial entities have formed a new consensus. Government, industry, academia, research institutions, and users have focused on empowering the real economy with blockchain technology—moving away from the superficial appearance of "currency" and returning to the essence of "chain".
7. Multi-cloud Management. As the main deployment mode of cloud computing, the private cloud has poor flexibility and high costs, the public cloud has information leakage vulnerabilities and complex migration processes, and the hybrid cloud faces difficulties in application compatibility, adaptation connectivity, and maintenance. However, Multi-cloud Management can effectively solve the above problems and is becoming the trend. By 2023, more than 90% of large and medium cloud enterprises have adopted multi-cloud management.
8. Core Software. Software technology is becoming the core competitiveness of the new round of IT transformation. On the technology side, software has become the key to the development of new technologies such as AI, cloud computing and big data; on the enterprise side, the infrastructure construction in the digitalization process tends to be more stable. Software is becoming the new driver to enhance the value of IT and to complete digital transformation and innovation; on the industry side, more and more traditional enterprises are spinning off new companies focused on software and services, while software talents are also flowing to traditional enterprises to become an important force supporting the digital transformation of traditional industries.
9. Native Penetration. In the new era of the digital economy, demand is changing at an unprecedented speed, and it is urgent to build enterprise digital architecture with a "native" mindset that includes cloud-native, data native, and intelligence-native principles to break through resource and business boundaries, eliminate information isolated island, and flexibly adapt to ongoing shifts. This approach enables organizations to transform their structures, expand resources, innovate business models, and break traditional industry boundaries to achieve digital transformation and innovation-driven growth.

10. Superlative User Experience. With the continuous emergence of new technologies and applications, "experience" is moving from consumers' personal needs to the industry market, driving the development of product technologies that prioritize human–machine interaction, greatly enhancing user experience, and forming experiential business models that become the new competitiveness of enterprises. Related product technologies and business models are forming a new excellent user experience.

5.1.6 Future Prospects of ICT Industry

Currently, global ICT innovation is more susceptible to the influence of the macro environment, which includes international trade, policy environment, and unexpected epidemics. Countries are paying more attention to the role of industrial policy on innovation and are committed to introducing relevant outline plans from multiple perspectives like financial, market, technology, and talent.

In terms of products and services, high-end technology products better reflect a country's ability to commercialize its R&D achievements in the international market. The competition for ICT product innovation in many countries tends to be high-end products, and the competition in the upstream of the industry chain will be more intense. In terms of mode, artificial intelligence, blockchain, and other technologies have achieved wide applications with the help of open source, reducing the cost of entry and avoiding technology monopoly. At the same time, leaders in high-tech are more inclined to protect the market with intellectual property rights and promote standards to unify the industry in the direction of their own expectations so as to grasp the market discourse. Open source and standards play an increasingly significant role in the ICT industry.

From the perspective of the internal structure of the ICT industry, the global ICT manufacturing industry is widely distributed in China, the US and South Korea. In the short-term direction, many economies may prioritize the completeness of innovation factors for the purpose of ensuring the local security of the industrial chain and delay or slow down the pace of de-industrialization to enhance the commercial transformation capabilities. However, in the long term, the return of manufacturing and industrial diversification will lead to a wider distribution of economic activities and provide more opportunities for new entrants. The ICT industry, as a new industrial element, will be integrated into the industrial manufacturing sector, promoting a new industrial revolution marked by intelligence. Ultimately, based on the innovation and promotion of industrial technology, information technology and intelligent technology, intelligent manufacturing will achieve a leap forward.

In the next 20 years, it can be predicted that the ICT industry will accelerate its transformation into an ecosystem, and China's rapid digitalization will bring enormous market opportunities. Every participant in the ICT ecology cannot remain unaffected. The only constant is change. There will be no steady company or position here, and only by constantly maintaining a humble mind and thirst for exploration can development be kept. In the era of ICT technology integration driving the digital economy, ICT industry members work together to serve the new development of the ICT industry, provide new momentum of the ICT industry, and prosper the new ecosystem of the ICT industry.

5.2 Characteristics and Current Situation of New Technologies

With the development of technology in recent years, the number of global Internet users has increased rapidly, leading to the emergence of numerous new technologies, including the Internet of Things, cloud computing, and big data. This section provides a brief introduction to these technologies.

5.2.1 Internet of Things (IoT)

Overview of Internet of Things

The Internet of Things (IoT) refers to a network that connects any object or process that needs to be monitored, connected, or interacted with using various information sensors, radio frequency identification technologies, global positioning systems, infrared sensors, laser scanners, and other devices and technologies. It collects sound, light, heat, electricity, mechanics, chemistry, biology, location, and other necessary information in real-time and achieves ubiquitous connection between objects and people through different network access methods. This allows for intelligent perception, identification, and management of objects and processes. The Internet of Things is an information carrier based on the Internet, traditional telecommunication networks, and other related technologies. It enables ordinary physical objects that can be independently addressed to form an interconnected network.

IoT is an important part of the new generation of information technology, which is a network of connected objects, as shown in Fig. 5.6. The core and foundation of IoT is still the Internet. It is an extended and expanded network based on the Internet, with its user terminals extending and expanding to exchange and communicate information between objects. Therefore, IoT is a network that connects any object with the Internet for information exchange and communication through information-sensing devices, such as radio frequency identification, infrared sensors, global positioning systems, laser scanners, and other devices. It carries out information exchange and communication according to a predetermined agreement to achieve intelligent identification, positioning, tracking, monitoring, and management of the objects.

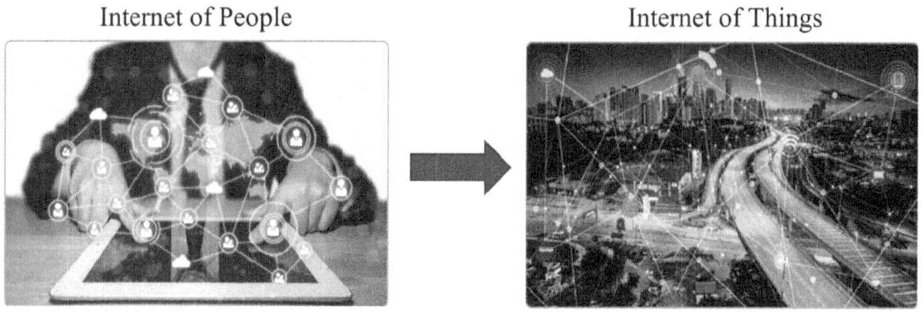

Fig. 5.6 Classification of networks

Application of Technology Architecture for Internet of Things

The architecture of IoT is generally divided into three or four layers, as shown in Fig. 5.7. The three-layer architecture from bottom to top is divided into the perception layer (terminal), network layer (including cloud, management, and edge), and application layer (application); the four-layer architecture from bottom to top includes perception device layer (or perception layer), network connection layer (or network layer), platform and tool layer, and application service layer. The difference between the three-layer and the four-layer architectures is that the four-layer split the three-layer "application layer" into "platform tool layer" and "application service layer" to make more detailed distinctions for software applications. From the end to the industry application, it passes through the following three layers: edge, management, and cloud. The signal is transmitted from the terminal by the edge, accessed, and transmitted to the network through management, then the cloud solves the problems of data storage, retrieval, usage, and business planning, and finally, the industry application presents the data to users and carries out interaction.

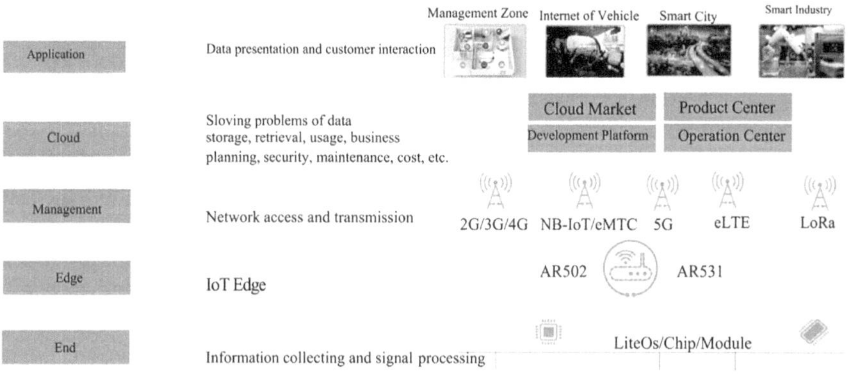

Fig. 5.7 Architecture of Internet of Things

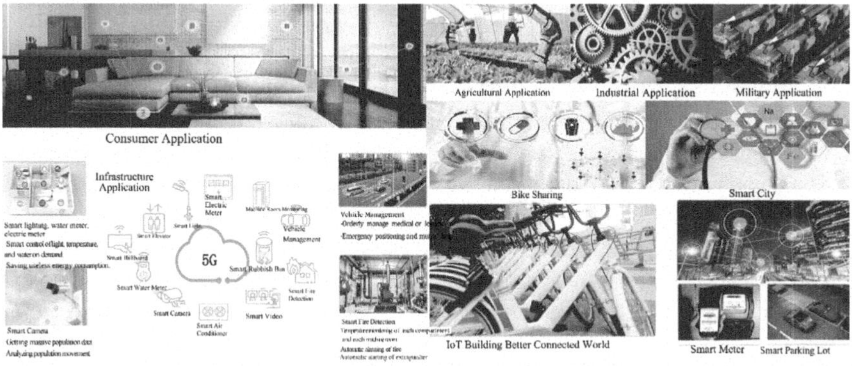

Fig. 5.8 Application of the Internet of Things

Applications of Internet of Things

The applications of IoT involve various aspects, as shown in Fig. 5.8. Applications in industries, agriculture, environment, transportation, logistics, security, and other fields effectively promote the intelligent development of related industries, enabling more reasonable use and allocation of limited resources, thus improving the efficiency and benefits of industries.

(1) Consumer Applications: Vehicles, home automation, connected wearable devices, health monitoring devices, remote monitoring devices, intelligent parking, electronic road charging systems, etc.
(2) Industrial Applications: Robotics and software-defined production processes.
(3) Agricultural Applications: Collecting data on temperature, precipitation, humidity, wind speed, plant diseases and pests, and soil composition, together with analyses and applications.
(4) Infrastructure Applications: Monitoring and controlling various types of infrastructure facilities, such as railways, bridges, wind farms, waste management, etc.
(5) Military Applications: Reconnaissance and surveillance of combat-related targets, including sensors, vehicles, robots, weapons, wearable smart products, and the use of related intelligent technologies on the battlefield.

5.2.2 Cloud Computing

Definition of Cloud Computing

In the early days of the Internet era, network service providers had to purchase dedicated servers and hire full-time operations and maintenance staff. However, with the diversification of network services, it is difficult for companies to develop websites or software that can meet user needs independently. Meanwhile, the number of global Internet users has been increasing daily in recent years. According to Internet World Statistics (IWS), global Internet users reached 5.17 billion in 2021, accounting for 65.6% of the total population. Such a large number of users brings new challenges to the Internet infrastructure. Thus, cloud computing was born.

The concept of Cloud Computing was first proposed by Eric Schmidt, a former CEO of Google, in 2006, to allow users to use computing and storage resources as conveniently as using water, electricity, and coal while hiding complex technologies behind the scenes, freeing users from time and space constraints, as well as professional training, thus effectively improving productivity and reducing production costs.

The US National Institute of Standards and Technology (NIST) defines cloud computing as a model that provides convenient and on-demand access from shared configurable computing resources (such as networks, servers, storage, applications, and services) anywhere and anytime. The resources can be provisioned and released quickly, minimizing the effort of managing resources and interacting with service providers.

Wikipedia defines cloud computing as a computing model that provides dynamically scalable and virtualized resources in a service way through the Internet.

Another more common definition of cloud computing is that it is a network of hundreds of thousands or even millions of inexpensive servers that provide users with the computer services they need, which is a new development in recent years in the field of computer science in Distributed Computing, Parallel Computing, and Grid Computing.

Key Features of Cloud Computing

Cloud computing has five key characteristics:

(1) On-demand Self-service (ODS). Since a customer's demand for IT resources is random, the dynamic allocation of IT resources can be realized by technical means, and virtual resources can be allocated according to the customer's demand so that efficient use of resources can be realized.
(2) Ubiquitous Network Access (UNA). In order to guarantee the quality of service, service providers need to guarantee round-the-clock and ubiquitous network access services, which can help users enhance flexibility and efficiency of work.
(3) Location Independent Resource Pooling (LIRP). Service providers build a "pool" of IT resources (network, servers, storage, applications, services, etc.) and allocate resources from the "pool" on demand.
(4) Rapid Elastic. Through technical means to achieve rapid allocation and release of resources, efficiency, and service quality can be improved.
(5) Pay Per Use (PPA). The billing standard of cloud services generally adopts granular measurement and is based on the number of resources used and the duration of use, which effectively improves operational efficiency.

Common Classifications of Cloud Computing

According to service level, the classification of cloud computing, as shown in Fig. 5.9, can be divided into Infrastructure as a Service (IaaS), Platform as a Service (PaaS), and Software as a Service (SaaS). The following are brief descriptions of each of them.

(1) IaaS: It mainly provides/leases basic IT services such as computing, storage, network, etc. Users can manage the virtual resources they have applied for through the operating system and install the software they need by themselves. IaaS gives customers greater freedom, but it also requires professional management, such as elastic cloud servers.
(2) PaaS: It mainly provides application operation, development environment and application development components. Cloud service providers offer customers programming languages, libraries, services, and development tools to create and develop applications and deploy them on related infrastructure, such as database services.
(3) SaaS: It mainly provides software-related functions through a Web interface, and cloud service providers provide direct client applications, such as Office 365 and Office Shimo.

The following is a classification of cloud computing based on how the service is deployed and to whom it is provided.

(1) Public Clouds. "Public" indicates that this type of cloud service is not owned by individual users. It is a type of service that provides computing resources to the

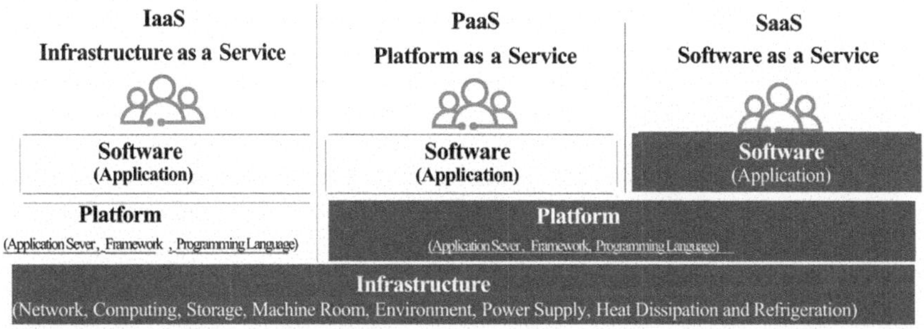

Category	Content	Business Opportunity
SaaS	● Internet Web 2.0 Application (MM) ● Enterprise Application (ERP/CRM, etc.)	● Wal-Mart of Application and Software
PaaS	● Providing environment for application's operation and development ● Providing application development components (E-mail,Message,Charging,Paying)	● Tapping the potential of web-based resources ● Control of Network Platform
IaaS	● Provision/Rental of basic IT service, such as computing, storage, network, DNS, etc.	● Low-cost elastic IT resource services ● Application and web hosting

Fig. 5.9 Classification of cloud computing

public. The resources, such as applications and storage resources, are generally provided by IDC service providers or third parties. The advantages are the lower cost and the better scalability. The disadvantages are the lack of control over resources in the cloud, weak data security, and network performance and matching problems.

(2) Private Clouds. Private clouds are extensions and optimizations of traditional enterprise data centers that provide storage capacity and processing power for various functions and are built for individual use, so these data, security, and quality of service are better guaranteed than public clouds. Users can build the basic settings of the cloud and deploy applications on the basic settings.

(3) Hybrid Clouds. In a hybrid cloud model, the cloud platform is a combination of two different models of cloud platforms (private or public). These platforms remain separate entities, using standardized or proprietary technologies to achieve binding and enabling data and application migration between them (for example, the balance between different cloud platforms). Organizations can deploy minor applications and data to the public cloud, taking advantage of the scalability and cost benefits of the public cloud while keeping critical applications and data in the private cloud, resulting in higher security.

Virtualization Technologies

Virtualization has a broad meaning. Technology that abstracts any form of resource into another form can be called virtualization. Virtualization is a logical representation of resources that is not bound by physical conditions. Virtualization technology is the foundation of cloud computing. In simple terms, virtualization makes it possible to run multiple virtual machines on a single physical server at the same time. Virtual machines share the CPU, memory, and I/O hardware resources of the physical machine, but virtual machines are logically isolated from each other.

Common virtualization applications are:

(1) Memory Virtualization (Page File). Virtual memory technology under the Windows system, often called page files, uses hard disk space to simulate memory space in the case of insufficient memory, so as to achieve the purpose of expanding memory.

(2) Disk Virtualization (RAID Volume). Redundant Array of Independent Disks (RAID), referred to as disk arrays, solves the problem of insufficient disk capacity and uses virtualization technology to combine multiple disks into a disk array group, so as to improve read and write efficiency and reduce data redundancy.

(3) Virtual Local Area Network (VLAN). Virtual Local Area Network (VLAN) is the most commonly used LAN virtualization technology. Since the number of mainframes on the Internet is rapidly increasing due to limited IP addresses, the use of VLAN technology can realize multiple mainframes sharing a public IP address.

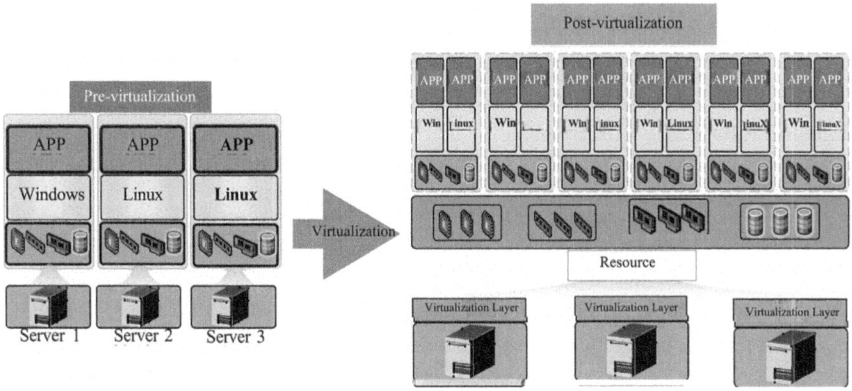

Fig. 5.10 Virtualization schematic

Virtualization creates a layer of isolation that separates the hardware from upper-layer applications, allowing multiple logical applications to run on a single hardware, as shown in Fig. 5.10.

FusionSphere is Huawei's self-developed cloud operating system that combines a virtualization platform and cloud management to enable the virtualization of x86 servers. More precisely, it includes the following three aspects:

(1) Computation (Virtualization of CPU/Memory). The virtualization of CPU resources enables multiple virtual machines to share CPU resources by time-sharing and null-sharing and intercepts and simulates the execution of sensitive instructions of virtual machines; the virtualization of memory resources is packaged into multiple virtual memories for virtual machines, and physical and virtual memories are mapped (mapping table) through Virtual Machine Monitor (VMM), but not necessarily one-to-one, and supports memory over-allocation (memory multiplexing technology). The physical memory and virtual memory are mapped (mapping table) through a Virtual Machine Monitor (VMM), but not necessarily one by one, and memory over-allocation (memory multiplexing technology) is supported.
(2) Storage (VIMS File System). Virtual Image Management System (VIMS) is a high-performance clustered file system that breaks through the limitations of a single storage system and is designed, built, and optimized for virtual server environments, allowing multiple virtual machines to share access to a consolidated, clustered pool of storage, thereby significantly improving resource utilization. As the foundation for virtualization across multiple storage servers, VIMS enables various services, such as storage thermal migration, dynamic resource scheduling for storage, and high availability.
(3) Network (Distributed Virtual Switch). A distributed virtual switch is a virtual network management method that manages virtual switches (software-based virtual switches or intelligent NIC virtual switches) on multiple mainframes, including the management of physical ports of mainframes and virtual ports

of virtual machines. A distributed virtual machine switch ensures consistent network configuration when virtual machines migrate between mainframes.

Characteristics of Virtualization

Virtualization has four characteristics, as shown in Fig. 5.11.

(1) Partitioning. Partitioning refers to dividing multiple virtual machines on a server, each of which can run a separate identical or different operating system simultaneously, enabling users to run multiple applications on a single server. Each operating system can only see the "virtual hardware" (virtual NIC, CPU, memory, etc.) provided by the virtualization layer, and logically, it can be considered that each operating system runs on its dedicated server.
(2) Isolation. Virtual machines are isolated from each other. The crash or failure of one virtual machine, such as an operating system failure, application crash, or driver failure, will not affect other virtual machines on the same server; viruses, worms, and other problems in one virtual machine are isolated from each other, just like each virtual machine is located on a separate physical machine. Resource control can be carried out to provide performance isolation, and it is also possible to run multiple loads, applications, and operating systems on top of a single machine simultaneously without conflicts between applications.
(3) Encapsulation. Encapsulation means storing the entire virtual machine information (hardware configuration, BIOS configuration, memory state, disk state, and CPU state) in a small group of files independent of the physical hardware so that users only need to copy a few specific files, save and move the virtual machine whenever and wherever needed.
(4) Standalone. Since the virtual machine runs on top of the virtual layer, only the virtual hardware provided by the virtual layer is visible. This virtual hardware

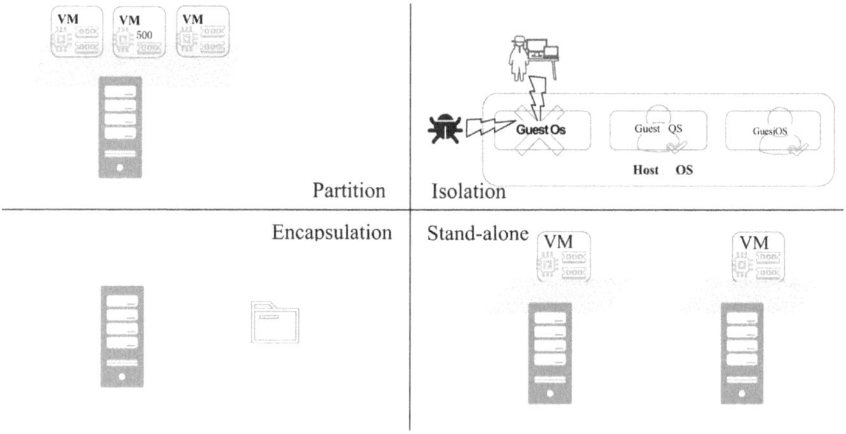

Fig. 5.11 Four main features of virtualization

does not have to take into account the physical server, so the virtual machine can run on any x86 server (IBM/DELL/HP, etc.) without any modification, thus avoiding constraints of the hardware on the operating system and those of the operating system on the application.

Cloud Computing in the 5G Era

1. **New Development Opportunities for Cloud Computing in the 5G Era**

First, in the 5G era, cloud services will be comprehensively upgraded. The popularity of cloud computing in the 4G era provides many enterprise users with the convenience brought by the cloud, but for individual users, there are not many opportunities to contact or use the cloud. The advent of the 5G era and the improvement of computer performance will enable more cloud services to be upgraded, directly affecting people's food, clothing, housing, and transportation. 5G will be deeply integrated with the Internet of Things, the Internet of Vehicles, Smart Cities, Industrial Internet, Smart Healthcare, and other scenarios, thus truly entering the era of smart life.

Second, the advent of the 5G era will drive the comprehensive upgrade of cloud vendors. In the 5G era, the speed and the scale of network construction are bound to be greatly enhanced, which will then drive the comprehensive development of cloud infrastructure. Cloud service providers need to upgrade from network architecture, infrastructure, service model, and operation system and accelerate cloud solutions for vertical industries to keep pace with the development of the cloud computing era.

Third, in the 5G era, cloud computing will shift from the center to the edge. With the further popularization of the Internet, more and more devices will be connected to the network, and users' demand for data will increase greatly. If the user-side data needs to be obtained from the data center every time, it will greatly affect the use experience of 5G network. With edge computing technology, users only need to transmit data from the closest edge data center to further reduce network latency and meet the delivery demand of 5G real-time response service in the future. At the same time, with edge computing, the integration of industrial ecology can be further accelerated, new business scenarios can be created, and cloud service models for vertical industries can be explored.

2. **Applications in the 5G Era Based on Terminal Management Cloud Collaboration**

It is believed that applications in the 5G era will be mainly based on terminal-management-cloud collaboration in mobile scenarios, as shown in Fig. 5.12, which is defined as Cloud X service.

5G brings a new eMBB management that is closer to users with edge computing, which will hopefully change the entire business chain.

The ubiquitous 5G connectivity and edge cloud on the terminal side can put the original computing, storage, and even rendering capabilities on the client side in the cloud, while the terminal will become more "simplified", lower in cost, and more

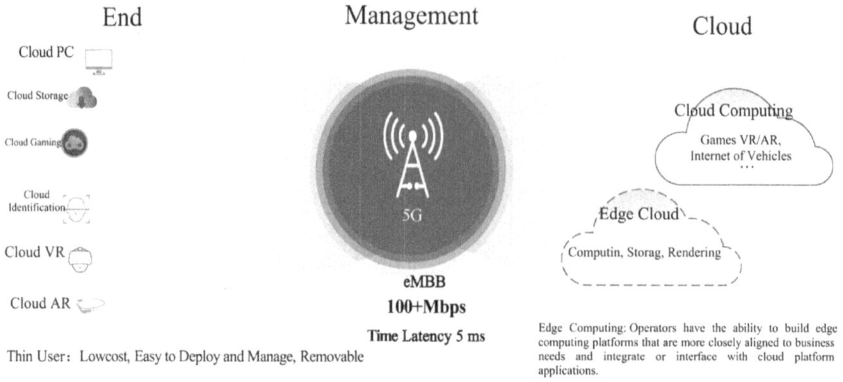

Fig. 5.12 Terminal management cloud collaboration

mobile. More importantly, it can lower the threshold of service deployment and popularization and enhance the vitality of the service.

The cloud side integrates many services, thus relying more on the management, edge computing and future network slicing capabilities, further highlighting the importance of network capabilities and strengthening the operator's control.

It is known that in the 4G era, network capacity is measured in terms of video quality and operators are managed, while in the 5G era, network capacity will be measured in terms of Cloud X, which also gives operators the opportunity to reinvent their business model and network values.

3. **Application Cases—5G + Cloud AR/VR**

Huawei Wireless X Labs research shows that Cloud VR, rendered through the cloud, will be the future development trend of VR. In the local VR mode, VR terminals need to be connected to local servers via cables, which has a poor user experience and high cost, while Cloud VR realizes wireless terminals and completes image rendering through cloud servers, greatly reducing terminal costs and improving user experience.

Cloud VR brings higher requirements for mobile networks, mainly the two key requirements: bandwidth and latency. For example, the entry-level VR experience requires 125Mbps bandwidth and 5 ms latency, while the ultimate experience requires 3.2 Gbps bandwidth and 2 ms low latency. Only 5G networks can meet the demand for ultimate VR experience, as shown in Fig. 5.13. At present, VR application scenarios are mainly videos and games, and they will be expanded to more application scenarios in the future.

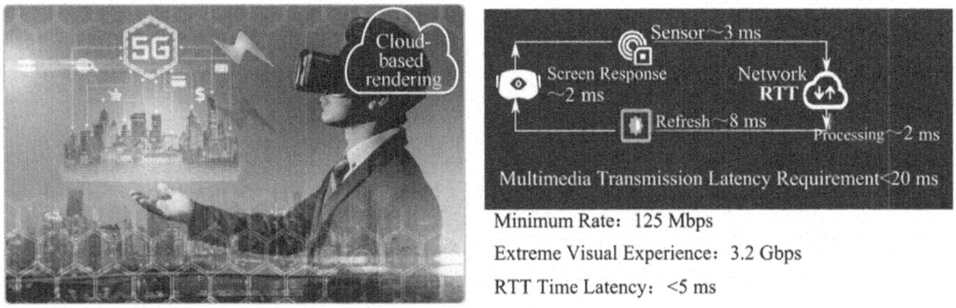

Fig. 5.13 Cloud VR application scenario

5.2.3 Big Data

Overview of Big Data

Big Data is a collection of massive, complex, and numerous types of data. By the end of 2020, there have been 50 billion networked devices around the world that are continuously generating data, and the interconnected network is changing business models and people's lifestyles. The development of new industries such as the Internet of Things, Industrial Internet, Telematics, and Artificial Intelligence all need big data to provide basic support, and all walks of life are increasingly relying on big data to carry out their work. Human daily life has become inseparable from data. According to the latest data provided by Baidu Index and Google Trends, since the concept of big data attracted attention in 2012, the search index of big data has increased explosively and continues to attract attention. With the advent of the 5G era, the volume of data has reached an unprecedented scale, as shown in Fig. 5.14. According to the white paper "Data Age 2025" released by International Data Corporation (IDC) in April 2017, it is predicted that the volume of data generation will continue to maintain high growth in the future, and by 2025, the global data volume will reach 163 ZB. Moreover, the speed of data growth will outpace technological evolution, and the world will enter an era of data-driven research applications. As an emerging and potentially valuable asset, big data is profoundly changing the world's economic, interest, and security patterns.

There is no universally accepted definition of Big Data, and various definitions of Big Data are given from the perspective of characteristics. In the book *The Age of Big Data*, written by Viktor Mayer-Schönberger and Kenneth Cukier, it is proposed that Big Data has "4V" characteristics, which are Variety, Volume, Velocity, and Value.

The contrast between traditional and big data analysis is shown in Fig. 5.15. Compared with traditional data analysis, there are four main changes in big data analysis.

1. **Change of Data Format**

Compared with traditional data analysis, there are changes in data formats for big data analysis as follows:

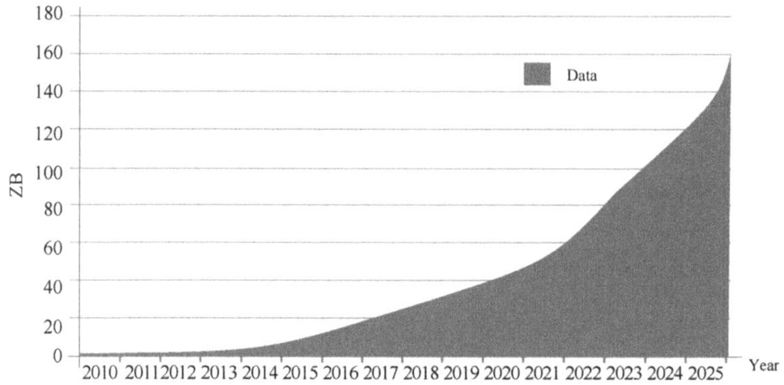

Fig. 5.14 Data scale forecast chart of data age 2025

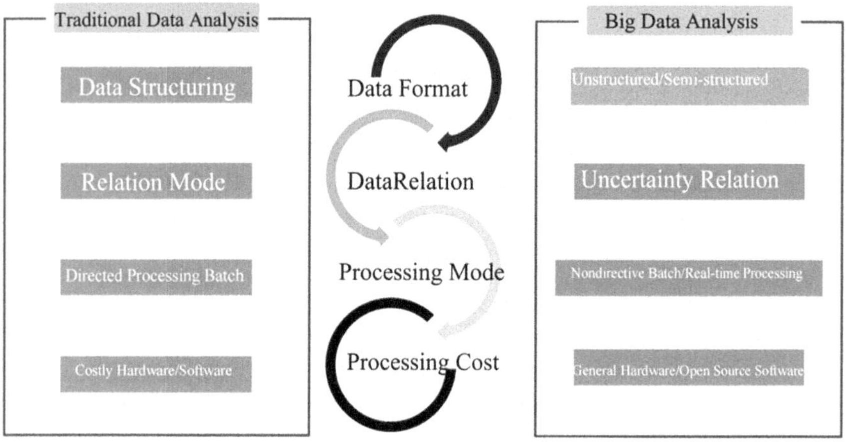

Fig. 5.15 Comparison between traditional data analysis and big data analysis

(1) Structured Data. Traditional data analysis is usually geared towards structured data and has small data sets. Most data warehouses have an elaborate Extract-Transform-Load (ETL) process and database constraints, meaning the data loaded into the data warehouse is easily parsed and clean-formatted.
(2) Unstructured Data or Semi-structured Data. The biggest advantage of Big Data is the ability to capture unstructured data beyond traditional data. This means a wider range of data formats and more challenging ways to analyze them.

2. **Changes in Data Relationships**

Compared with traditional data analysis, there are changes in data relationships in big data analysis as follows:

(1) Relational Model. The traditional analysis is built from the relational database, and the relationship between subjects is established within the system, and the data is analyzed on a based basis.
(2) No Deterministic Relationships. In the form of unstructured data (video, audio, etc.) targeted by big data analysis, it is difficult to establish a correct relationship between all information, so it is difficult to have definite relationships between the data.

3. **Changes in Processing Methods**

Compared with traditional data analysis, big data analysis has changes in processing methods as follows:

(1) Targeted Batch Processing. Traditional data analysis is a targeted batch processing. Extraction, transformation, and loading work are completed before required insights are obtained.
(2) Non-targeted Batch or Real-time Processing. Big data analysis is the non-targeted batch or real-time analytical processing, using support from software meaningful for data.

4. **Changes in Processing Costs**

Compared with traditional data analysis, there are changes in processing costs for big data analysis as follows:

(1) Expensive Hardware/Software. Traditional data analysis systems usually rely on expensive hardware and software, such as Massively Parallel Processing (MPP) systems or Symmetrical Multi-Processing (SMP) systems, which are relatively costly.
(2) Generic Hardware/Open source software. Big data analysis systems can be implemented based on general-purpose hardware and open-source software such as Hadoop.

Key Technologies of Big Data

With the vigorous development in the Big Data era, multiple heterogeneous, massive aggregation and fast-growing data have penetrated into all walks of life, changing the way people think about data and pushing forward the reform of the information society. However, at the same time, the low-value density of big data makes it necessary to support a variety of technologies to fully explore the valuable part of the data that can be used. Therefore, a variety of big data technologies have emerged. The so-called big data technologies refer to the related technologies associated with the collection, storage, analysis, and application of big data. They are a series of processing and analysis technologies that use non-traditional tools to process large amounts of structured, semi-structured and unstructured data to obtain analysis and prediction results. As shown in Fig. 5.16, the key technology architecture of big data can be divided into six aspects: data visualization, data analysis, data processing, data

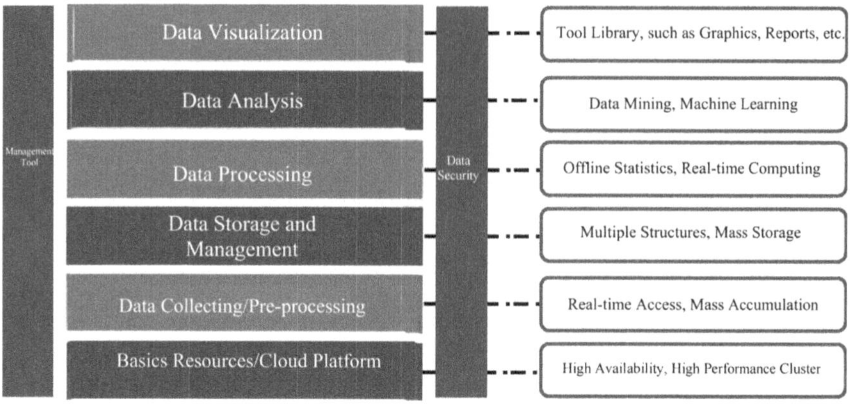

Fig. 5.16 Key technology architecture diagram of big data

storage and management, data acquisition/pre-processing, and underlying resources/cloud platforms. Next, there will be an introduction to key big data technologies around these aspects.

1. **Data Visualization**

Data visualization is one of the key technologies of big data, which provides an intuitive and visual way to show the practice of big data. Data visualization technology is still evolving, and many cross-domain emerging technologies, such as graphic image processing technology and computer vision technology, are used to enhance the visualization of big data. To summarize, the data visualization process mainly includes four steps: data collection, data processing, data analysis, and data visualization display. To meet the demands of big data visualization, there are many targeted tools, such as Power BI, Tableau, QlikView, Apache Superset, Apache Zeppelin, etc.

2. **Data Analysis**

Data analysis refers to extracting useful information from the collected data by applying appropriate methods to analyze it. By analyzing a large amount of seemingly irregular data in a centralized manner, the inherent patterns are extracted to assist people in making judgments and predictions. In the field of Statistics, data analysis can be divided into three categories: descriptive statistical analysis, exploratory data analysis, and validation data analysis. Among them, descriptive statistical analysis provides a statistical description of the data set as a whole and is a kind of primary analysis; exploratory data analysis uses several keywords to describe the overall data set, focusing on the discovery of new features in the data set; and validation data analysis is to analyze whether the conditions of the model and hypothesis testing are met, which is a supplement to the traditional statistical hypothesis testing means. As shown in Fig. 5.17, a typical data analysis usually contains data collection steps, pre-processing, data mining, and results presentation.

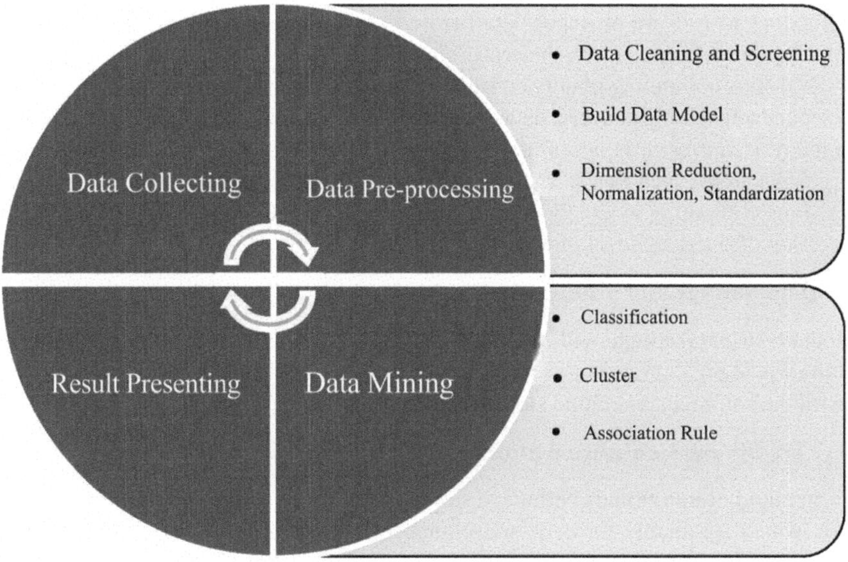

Fig. 5.17 Several steps of data analysis

3. **Data Processing**

The data that has been collected and stored in various storage devices needs further processing. As shown in Fig. 5.18, according to the timeliness of data storage, big data processing can be broadly divided into three categories: offline statistics, near-line processing, and real-time computing. Since near-line processing mainly considers data from minutes or hours ago, it is in between offline statistics and real-time computing, and its boundary is rather blurred, so here what is merely briefly introduced are the two major types of data processing methods: offline statistics and real-time computing.

Offline statistics mainly considers the data processing cycle with the delay unit of days and does not have a strong need for timeliness. Offline statistics uses various

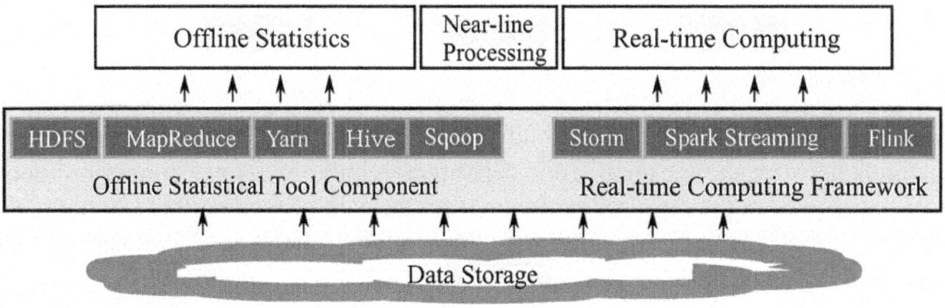

Fig. 5.18 Data processing

techniques to prepare all input data before the processing starts and to obtain the calculated results as soon as the processing is completed.

Real-time computing mainly considers the data processing cycle with a delay unit of seconds. In the big data era, data demand is not just limited to analysis and mining. In many scenarios, it is necessary to capture data generated in real-time and make targeted actions, such as advertising push and product recommendations. Therefore, real-time computing is widely used in such time-sensitive fields, which is the future direction of big data development.

4. **Data Storage and Management**

Methods of data storage and data management can directly affect the performance of the whole big data system to a certain extent. Next, traditional centralized storage, distributed storage, and cloud storage will be introduced.

(1) **Traditional Centralized Storage**

The method of traditional centralized storage is relatively simple and stable in structure, which is suitable for early traditional Internet data. However, with the rapid development of 5G, big data, and other technologies, data types and architectures have changed dramatically, and traditional centralized data storage systems are difficult to meet the needs of Internet development in terms of capacity size, expansion speed, read and write speed, and data backup, etc. Therefore, new data storage technologies have emerged one after another, mainly including direct-attached storage, network-attached storage, and storage area networks.

(2) **Distributed Storage**

The most intuitive difference between distributed storage and traditional centralized storage is the distribution of storage devices in terms of geographic location. Distributed storage distributes data on multiple storage nodes, and each node is interconnected and managed through the network, solving the limitations of local storage systems in terms of storage capacity and maintenance costs. A typical distributed storage architecture is shown in Fig. 5.19. The current mainstream distributed storage system architecture is a master/slave architecture, which usually includes a master node, multiple data nodes, and various clients consisting of big data applications or terminal users. Through distributed storage of data, big data can be divided into small data, reducing the size of data to a granularity that can be handled by a single node.

(3) **Cloud Storage**

With the rapid development of cloud computing, cloud storage has become an emerging network storage method as a derivative technology. Cloud storage abstracts storage space from the physical hardware layer through virtualization technology to build out a cloud environment. The resources to be stored are archived, organized, and distributed on demand in the cloud environment by management tools. According to the openness of the cloud, cloud storage can be divided into three categories: public cloud storage, private cloud storage, and hybrid cloud storage, and according to the

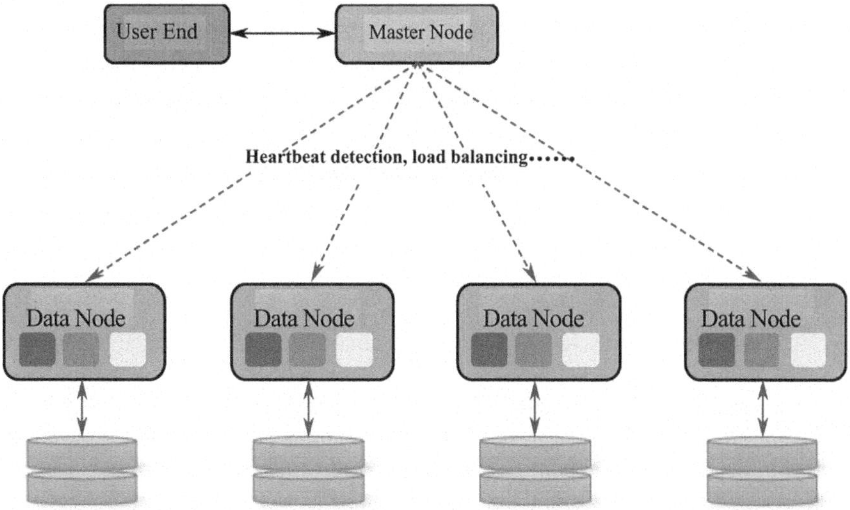

Fig. 5.19 A typical distributed storage architecture

format of storage objects, it can be divided into three categories: block storage, object storage, and file storage.

5. **Data Acquisition/Pre-processing**

Data acquisition and data pre-processing is the preliminary processing of data while acquiring data, so as to improve the quality of the collected data and provide high-quality data input for subsequent processing of big data.

(1) **Data Acquisition**

Data acquisition, also known as data collection, refers to the various types of structured, semi-structured and unstructured massive data obtained through data sources such as RFID radio frequency data, sensor data, social network interaction data, and mobile Internet data, which are fundamental to Big Data services. Big data collection methods can be classified as system log collection methods, network data collection methods (crawler technology) and other data collection methods (through specific interfaces and devices), among which system log collection methods are the most traditional way of big data collection.

(2) **Data Pre-processing**

To effectively analyze the massive data collected, further pre-processing of data, such as data cleaning, data integration, and data transformation, needs to be done to improve the overall data quality. Due to the low-value characteristic of big data, the pre-processing of data becomes crucial. As shown in Fig. 5.20, the pre-processing work mainly includes four aspects: data cleaning, data integration, data transformation and data reduction.

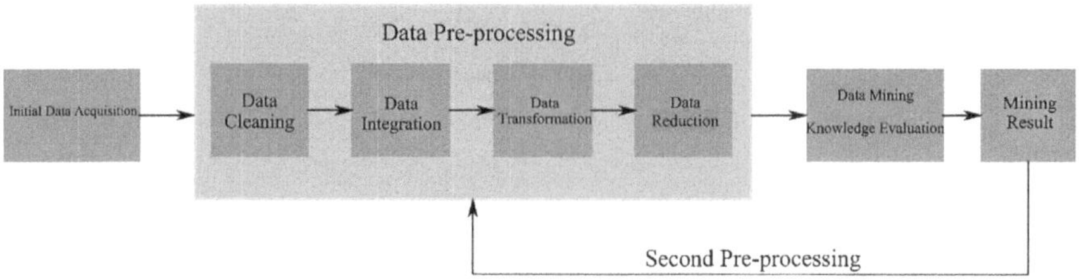

Fig. 5.20 Data pre-processing steps

(1) Data Cleaning: Data screening focuses on abnormal data cleaning, missing data processing, noisy data removal, consistency checking, and other work.
(2) Data Integration: Data integration refers to the combination and storage of data from different application systems and data sources in different forms to establish a data warehouse. The purpose is to solve problems, such as inconsistent multiple data formats and data redundancy, so as to improve the efficiency of subsequent data analysis.
(3) Data Transformation: Data transformation means using linear or nonlinear mathematical transformation methods, such as smoothing aggregation and normalization, to reduce the dimensionality and compress the data, eliminate unnecessary features, and transform the data into a form suitable for data mining.
(4) Data Reduction: Data reduction is used to discover features related to the target and to streamline the data volume to the maximum extent without affecting the integrity of the original data. In general, data reduction can be divided into feature reduction, sample reduction and eigenvalue reduction.

6. **Basic Resources/Cloud Platform**

Due to the complexity, diversity, massive volume, and high speed of big data, extremely high requirements are put forward for big data processing platforms. Currently, big data processing is usually done with the help of distributed cloud platforms. The mainstream distributed computing cloud platforms include Hadoop and Spark.

(1) **Hadoop**

Hadoop is a distributed system infrastructure architecture developed by the Apache Software Foundation that allows users to process massive amounts of data larger than 1 TB without understanding the underlying details of distribution. Core components of Hadoop include the Hadoop Distributed File System (HDFS) and distributed computing framework MapReduce. HDFS is highly fault-tolerant and provides high-throughput access to data on low-cost hardware, and MapReduce provides efficient distributed computing for massive amounts of data. Such an architecture is designed to achieve a high degree of cooperation between computation and storage and has become the de facto standard for Big Data technologies.

(2) **Spark**

Spark is an open-source big data processing engine released by AMP Labs at the University of California, Berkeley, in 2010. It is compatible with the Hadoop ecosystem. Due to its fast processing power and advanced design concept, Spark became the top project of Apache eight months after its launch, and components, such as Spark SQL, Spark Streaming, MLlib and GraphX, were quickly launched around Spark. Currently, Spark has become one of the mainstream data analysis, data streaming processing, and machine learning platforms.

Integration and Empowerment Between 5G and Big Data

The practical application of big data needs to be realized on the basis of the popularization of 5G, and it is difficult for the existing communication transmission technology to transfer the huge data volume to meet its performance requirements. 5G is far ahead of 3G and 4G in terms of performance and other aspects, which promotes big data from theory to practice. It can provide users with a high-quality experience at a reasonable cost. In the mMTC scenario, 5G communication technology provides massive connection support, focusing on solving the problem that traditional mobile communication cannot support the development of IoT well, realizing the intelligent interconnection of massive terminal devices. However, it also causes the explosive growth of terminal data, which puts forward higher requirements for real-time data processing, reliability, security, and other capabilities and further promotes the development of big data technology. The integration of 5G and big data will burst out more wonderful sparks and bring more convenience to our lives.

5.2.4 *Artificial Intelligence*

Overview of Artificial Intelligence

1. **Basic Concept of Artificial Intelligence**

Artificial Intelligence (AI) is a new technical science that studies and develops theoretical methods and application systems for simulating, extending, and expanding human intelligence. The concept of artificial intelligence was first introduced by John McCarthy at Dartmouth Conference in 1956 and was then defined as "the science and engineering of making intelligent machines". Its purpose was to enable machines to think like humans and have intelligence. Today, the meaning of AI has been greatly expanded, and as shown in Fig. 5.21, AI has evolved into an interdisciplinary that integrates Computer Science, Mathematics, Brain Science, Cognitive Science, Psychology, Linguistics, Logic, Philosophy, and Social Science. Artificial intelligence, like other intelligence, needs to be accomplished through constant repetition of phenomenal observation, data collection, information extraction, and knowledge

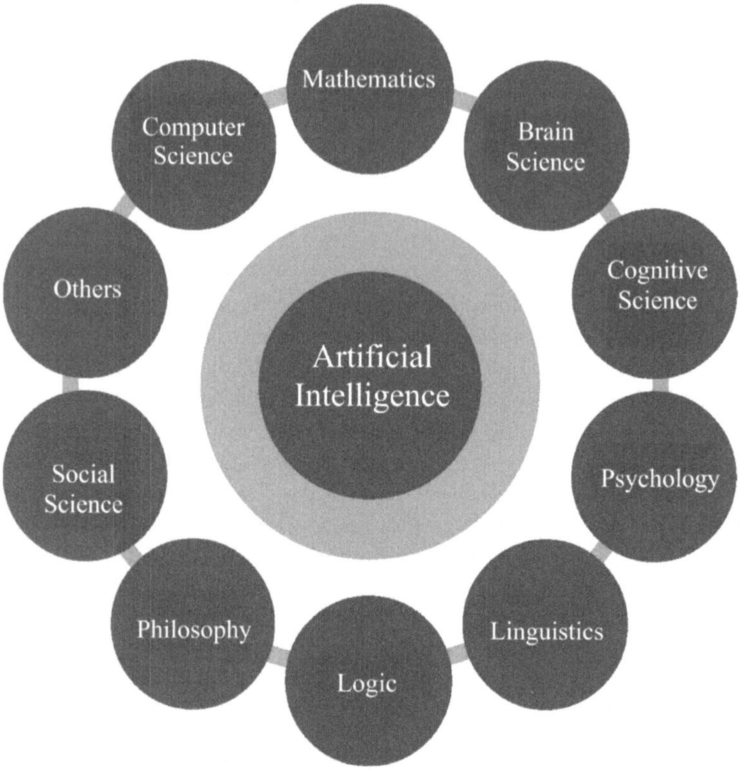

Fig. 5.21 Artificial intelligence: cross-integration of multiple disciplines

refinement, which eventually equips machines with certain thinking and cognitive abilities.

Data, algorithms, arithmetic power, and scenarios are the four elements of AI systems. Specifically, AI is based on data and combined with human-designed algorithmic logic. It sets computer programmatic algorithms into hardware (arithmetic power) and repeats training with massive data volume so that the machine can obtain a certain degree of intelligence and be applied to specific scenarios. Currently, artificial intelligence has been able to use its powerful computing and storage capabilities to do some search-oriented and compute-intensive tasks (such as robots and chess). However, performance at higher levels of cognition is still far inferior to human intelligence, especially in language, thinking and culture.

2. **Classification of Artificial Intelligence**

There is no internationally recognized standard for the classification of artificial intelligence so far. If divided according to strength and ability, intelligent machines are now commonly classified into two categories: Artificial Narrow Intelligence (ANI) and Artificial General Intelligence (AGI). ANI is good at completing one-sided tasks, but it is only good at specific tasks and is incompetent at other tasks. For example,

Google's AlphaGo only uses its powerful computing power to win as quickly as possible, and it is unlikely to imitate the way chess masters think. Artificial Narrow Intelligence seeks to quickly accomplish special tasks as perfectly as possible. AGI is a machine made with human-like level intelligence. It possesses consciousness and can think like a human, solve problems, think abstractly, understand complex ideas, and learn quickly, and can perform multiple tasks (or even any task), which is the highest ideal of artificial intelligence. However, there are no suitable examples yet because the technology is still under development. Also, it is not clear whether strong AI has moral standards similar to human beings, so moral and ethical issues it may bring deserve deeper investigation.

In addition, artificial intelligence can also be classified based on its technology. It can be classified into traditional machine learning, deep learning, computer vision, natural language processing, and recommendation systems. Among them, traditional machine learning can be further divided into supervised learning, unsupervised learning, reinforcement learning, and other learning.

3. **The Relationship Between Artificial Intelligence, Machine Learning, and Deep Learning**

As shown in Fig. 5.22, Machine Learning belongs to a branch of artificial intelligence that mainly designs algorithms that enable computers to learn autonomously, and these algorithms are classified as supervised learning algorithms and unsupervised learning algorithms. Deep Learning is a sub-domain of machine learning that is usually used to enhance feature representation and train deep neural networks through traditional machine learning methods. In recent years, Deep Learning has received a lot of attention from academia and industry, and its research has progressed rapidly. Some novel neural network structures have been designed and show excellent performance so more and more people are studying and discussing it as an independent learning method for artificial intelligence. The motivation for Deep Learning mainly lies in building and imitating the neural networks of the human brain to analyze and interpret data, such as images, sounds, and texts, by imitating the human brain.

Machine Learning and Deep Learning

1. **Machine Learning**

Machine Learning (including the branch of deep learning) is the science of "learning algorithms". The so-called "learning" means that for a certain type of task, T, and performance measurement, P, if the performance of a computer program measured by P on T improves itself with experience E, then this computer program is learning from experience E. As shown in Fig. 5.23, the basic process of machine learning includes data collection, data cleaning, feature extraction and selection, model training, model evaluation and testing, and model deployment and integration.

(1) Data Collection and Data Screening. Data collection can be done through database pulling, web crawlers, and API acquisition, etc. Different Machine

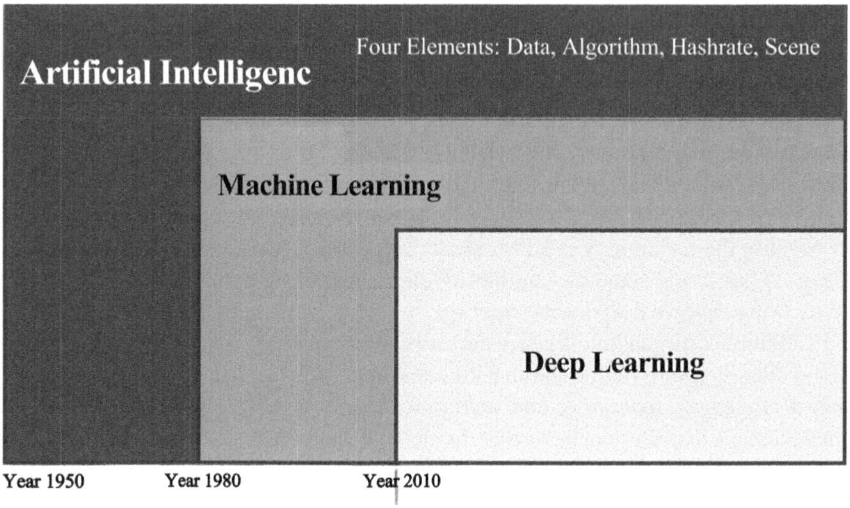

Fig. 5.22 Differences and connections between artificial intelligence, machine learning, and deep learning

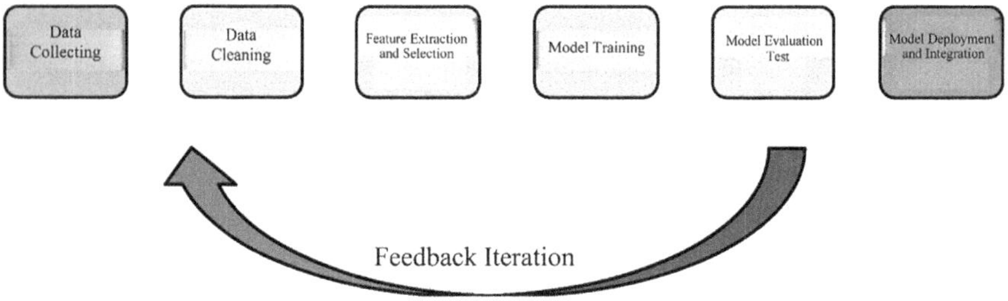

Fig. 5.23 Basic process of machine learning

Learning methods have their own advantages and disadvantages in different application scenarios, but all of them have data greediness, no matter which algorithm can achieve better results by providing more data. Therefore, collecting as much valid data as possible is a fundamental and important step in the Machine Learning process. However, the original data is often heterogeneous and noisy and cannot be used directly, so a lot of data screening is usually required before feature engineering, and special attention should be paid to handling invalid values, missing values, and outliers.

(2) Feature Extraction and Selection. Feature extraction and selection is the process of converting data attributes into data features, the attributes representing all dimensions of the data. Although all attributes of the original data are learned during data modeling, it is not always possible to find the potential trend of

the data, so it is necessary to pre-process the original data with feature engineering first to reduce the interference of the algorithmic model by noise, so as to find out the better trend. Commonly used methods for feature engineering include discrete variable processing, timestamp processing, binning/partitioning, feature scaling, feature selection, and feature extraction.

(3) Model Training. During model training, attention should be paid to selecting appropriate initialization methods, preventing overfitting (increasing the samples and decreasing the features), selecting appropriate activation functions and parameter tuning.

(4) Model Evaluation Testing. In model evaluation testing, the performance of the model is usually evaluated by cross-validation. Cross-validation aims to repeatedly slice the data into training and testing sets, then train the model and evaluate its predictive ability. In repeatedly slicing, samples from different test sets and training sets intertwine.

(5) Model Deployment and Integration. Trained models are stored and put into use.

2. **Deep Learning**

Deep Learning is a kind of Machine Learning and an advanced study of Machine Learning, which simulates the hierarchical structure model of the human brain to extract the high-level features of input data and solve some complex pattern recognition problems. The comparison between Deep Learning and Machine Learning is shown in Table 5.1. Deep Learning originates from the study of the artificial neural network, and its core idea is to transform the feature representation of the original data into the new feature space by multi-layer feature transformation so that the extracted high-level features can better represent the inherent patterns of the input data. Common, typical neural network structures include traditional fully connected Deep Neural Networks (DNN), Convolutional Neural Networks (CNN), and Recurrent Neural Networks (RNN). DNN can be interpreted as a multilayer neural network with many hidden layers, and the internal layers of a DNN can be divided into an input layer, hidden layers, and an output layer. CNN is a feedforward neural network with convolution algorithm and a deep structure composed of a five-layer structure, including input, convolution, activation, pooling, and full connection. RNN is a recurrent neural network that takes sequence data as input, recurses in the direction of sequence evolution, and all nodes are connected in a chain.

Application Direction of Artificial Intelligence

After understanding basic concepts and key technologies of artificial intelligence, focus is then put on various scenarios of AI applications. Although AI technologies are widely used in all aspects of life, they are mainly focused on several aspects, such as computer vision, speech recognition, and natural language processing. This chapter will briefly introduce several key applications of AI, such as computer vision, speech processing and natural language processing.

Table 5.1 Comparison of machine learning and deep learning

Machine learning	Deep learning
Fewer requirements for computer hardware	Performs a large number of matrix operations, which can be optimized with GPU
Suitable for small data volume training, difficult to improve performance if data volume increases	High-dimensional weighting parameters for high performance with large amounts of training data
Need to decompose problem layer by layer	"End-to-End" learning
Manual feature selection	Self-learning with algorithms for automatic feature extraction

1. **Computer Vision**

Computer vision is a field of artificial intelligence (AI). The research topics mainly include image classification, target detection, image segmentation, target tracking, text recognition, and face recognition. By using technologies related to the field of computer vision, computers and systems obtain meaningful information from digital images, videos and other visual inputs and act or make suggestions based on this information. If artificial intelligence enables computers to think, computer vision enables them to see, observe and understand. The working principle of computer vision is basically the same with that of human vision, but the training methods are different. Human vision has a huge advantage. That is, it can be trained through previous experience to distinguish objects, how far they are, whether they are moving, and whether there is anything wrong with the image. Computer vision, on the other hand, performs these functions by training machines. Compared with the human eye, it must be done with cameras, data, and algorithms rather than the retina, optic nerve, and visual cortex in a shorter time. In the field of computer vision, the accuracy of computer recognition of objects has improved with the development of new hardware and algorithms. In less than a decade, systems have surpassed 99% accuracy in recognizing images, making them more accurate than humans in responding quickly to visual input, and the speed at which systems can recognize visual information continues to accelerate.

Computer vision techniques tend to imitate the way the human brain works to process visual information, that is, through pattern recognition. Computers are trained to process images on large amounts of visual data, to label objects on the image, and to find patterns in the objects. For example, to train a computer to recognize patterns in "flowers", a computer is provided with a million images of flowers, and then the computer analyzes these images, recognizes patterns that are similar to all the "flowers", and at the end of this process, creates a model of "flowers". Thus, each time images are provided to the computer, it will be able to accurately detect whether a particular image is a flower or not. Deep Learning (DL), especially convolutional neural network (CNN) techniques, are mainly used in the implementation of this process of learning patterns.

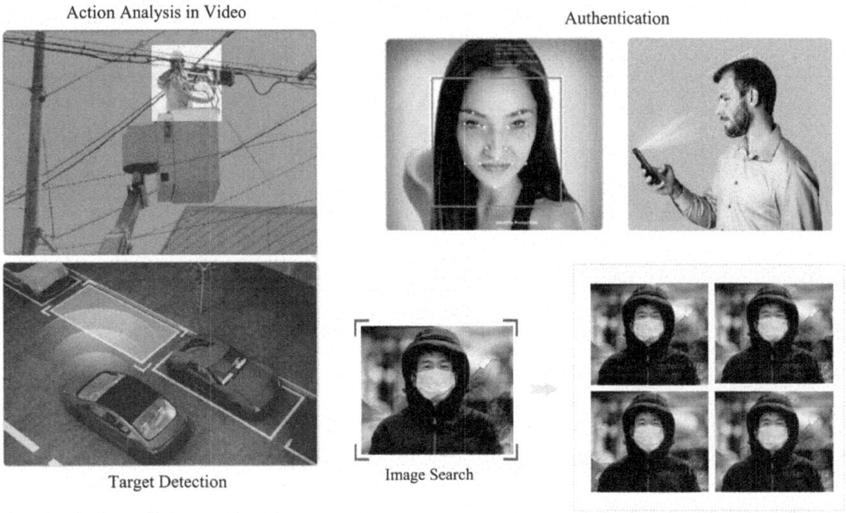

Fig. 5.24 Computer vision application scenarios

Computer vision application scenarios are shown in Fig. 5.24. Currently, computer vision is widely used in various industries, for example, tracking customer behavior by computer vision in retail, identity verification in online payment scenarios, monitoring crop diseases in agriculture, segmenting medical images in healthcare, recognizing and classifying objects in autonomous driving, and so on. With the development of artificial intelligence technology and other technologies, computer vision technology will be widely used in more fields.

2. **Voice Processing**

Speech processing is the process of interpreting, understanding, and processing speech signals. It refers specifically to processing human speech by computer systems, such as speech recognition software or speech-to-text programs. Speech processing aims to model and process the speech signal so that it can produce or synthesize natural language or recognize human languages.

Since the aim of speech processing is digital signal, to obtain a digital signal, the first step is to use a microphone or other device to collect and sample the signal, which is then quantified and coded by an analog-to-digital conversion device; secondly, the speech signal is standardized so that its values fall in the same range and the frame selection is performed; finally, after endpoint detection to accurately determine the range of signal processing, a simple high-pass filter is used to remove some noise, to obtain desired speech digital signal. After acquiring the digital signal, it is ready to do the related work in speech understanding and recognition. Speech understanding refers to using artificial intelligence technologies, such as the expression and organization of knowledge for automatic sentence recognition and semantic understanding. Speech recognition refers to the automatic recognition of phonemes, syllables, or

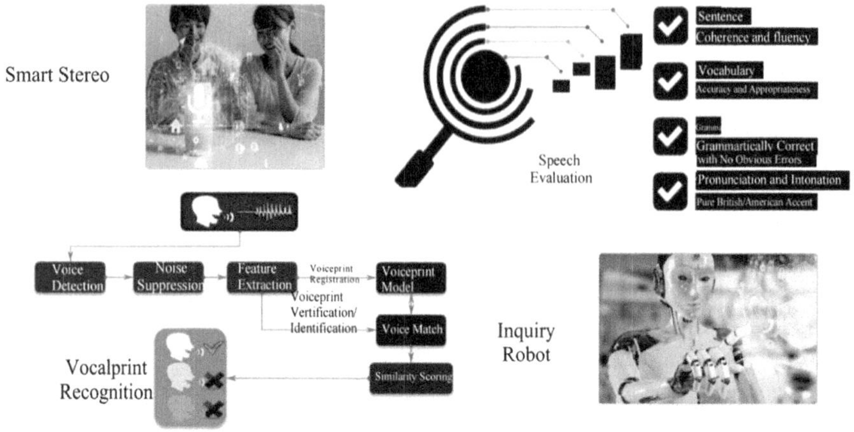

Fig. 5.25 Speech recognition application scenarios

words of the speech signal with computers. The main difference between the two is the degree of full utilization of grammatical and semantic knowledge.

Speech processing is widely used in interactive speech systems, virtual assistants, and robots, as shown in Fig. 5.25, such as smart speakers, voiceprint recognition, spoken language assessment, and diagnostic robots. With the development of artificial intelligence technology and other technologies, speech processing will be promising in other fields as well.

3. **Natural Language Processing**

Natural Language Processing (NLP) is a branch of computer science and, more specifically, a branch of artificial intelligence. Its purpose is to build machines that can understand and respond to text or speech data in the same way as humans do, so that computers can understand text and spoken language in the same way humans do.

Human language is full of ambiguity that makes it exceptionally difficult to write software to accurately determine the intended meaning of text or speech data. Irregularities in human language, such as synonyms, homophones, sarcasm, idioms, metaphors, exceptions to grammar and usage, and variations in sentence structure, which humans have had to spend years learning, are all well solved by NLP algorithms. NLP algorithms combine computational linguistics with statistics, machine learning, and deep learning to enable computers to process human language in the form of text or phonetic data and "understand" all meanings, including the intentions and emotions of a speaker or author. NLP-driven computer programs can perform tasks such as translating text from one language to another, responding to spoken commands, and rapidly summarizing large amounts of text in real-time conditions. Figure 5.26 shows several application scenarios for natural language processing. The rapid development of NLP has already profoundly impacted society, and the results achieved by NLP have far exceeded expectations. In the foreseeable future, NLP

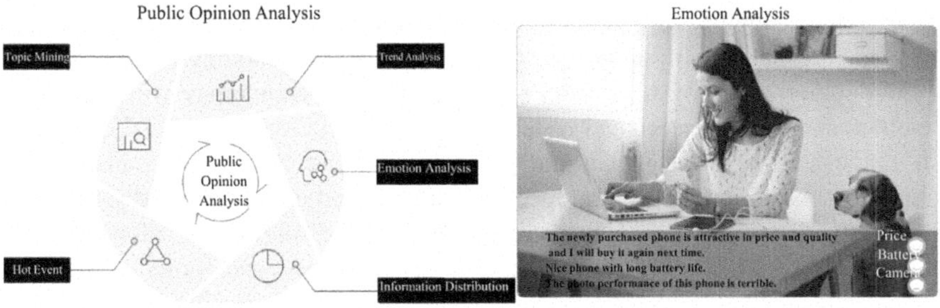

Fig. 5.26 Application scenarios for natural language processing

will dramatically change many fields of industrial and consumer experience and will become a necessity in our intelligent lives.

Integration and Empowerment Between 5G and AI

5G and AI are two of the most disruptive technologies in the world. Although each technology brings changes and new experiences to industries, the integration of 5G and AI will bring real transformation (as shown in Fig. 5.27). In fact, this intersection of technologies is the foundation for realizing the vision of an intelligent wireless edge network in which devices, edge clouds, and 5G work together to create a connected infrastructure where intelligent devices and services are ubiquitous. This section focuses on how 5G and AI technologies are integrated and empowered, that is, how 5G technology makes AI-driven experience better and how AI technology empowers 5G to bring better communication services.

1. **5G Empowering AI**

5G has the characteristics of high bandwidth, large connection, and low latency, which can provide a higher transmission rate and comprehensively improve the

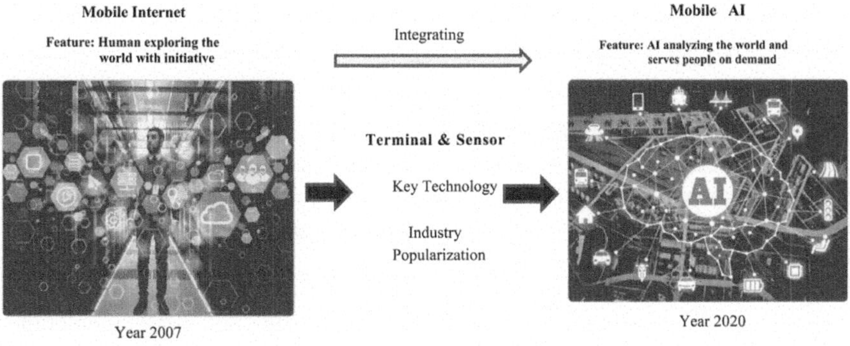

Fig. 5.27 Integration and empowerment between 5G and AI

capacity and reliability of the mobile networks, providing a strong communication infrastructure platform to support the development of AI technology. Firstly, in terms of the popularization of AI technology, the use of 5G technology will accelerate the efficient communication and collaborative work between terminal AI and cloud AI, which will make the whole unified operation more efficient, thus making AI technology more popularly developed in terminals and clouds. Secondly, in terms of data, the use of 5G technology will create an era of connectivity of everything, which will bring explosive growth in the volumes, types, and forms of data and guarantee the collection of massive high-quality data for AI training and modeling. Finally, regarding usage scenarios, the high speed and reliability of 5G technology will enrich usage scenarios of AI technology, thus bringing more possibilities of using AI technology, such as wireless federal learning in edge scenarios.

2. **AI Empowering 5G**

The application of AI to 5G networks and devices will lead to more efficient wireless communications, longer battery lifespan, and an improved user experience. As a powerful, intelligent tool, AI is suitable for solving important and challenging wireless problems in 5G networks that are difficult to solve with traditional methods. Much research and practice has been done in the communication industry around how AI can make 5G networks better. Practice shows that AI will have a powerful impact on several key areas of 5G network management, such as enhancing the quality of service of 5G networks, simplifying the deployment of 5G networks, reducing the complexity of 5G network operations and maintenance, and improving network efficiency and network security of 5G networks. In addition, AI technologies can be used to improve the terminal-to-terminal systems in 5G networks. For example, AI technologies improve the user experience of devices by enhancing the perception of radio, improving system performance, and bringing a safer wireless network.

Overall, the mutual integration and empowerment of AI and 5G technologies bring the effect of 5G and AI and the multi-effect of 5G and AI. It is believed that with the development of 5G technology and AI technology, the in-depth integration of the two technologies will bring more surprises to the communication industry.

5.3 Summary

This chapter explores the role and functions of ICT technologies in the era of digital economy, introduces basic concepts, principles, and applications of some emerging technologies, including Internet of Things, cloud computing, big data, and artificial intelligence, and discusses how to realize the integration of 5G and new technologies for empowerment, integration, and innovation.

By learning this chapter, readers shall understand the development of the integration of ICT technologies, understand the characteristics and status of emerging technologies, such as the Internet of Things, cloud computing, big data, and artificial intelligence, and master how to realize the innovative integration and application of 5G and new technologies.

Exercise 5

I. Blank Filling

5.1. The application of technology architecture of IoT are ________, ________, and ________.

5.2. The four characteristics of cloud computing virtualization technology are ________, ________, ________, and ________.

II. Choice Questions

5.3. (Single choice) Which of the following is not a type of cloud computing service? ()

A. Infrastructure as a Service (IaaS)
B. Platform as a Service (PaaS)
C. Software as a Service (SaaS)
D. System as a Service (SaaS).

5.4. Single Choice) Which of the following is not a characteristic of current cloud computing? ()

A. Privatization
B. Super-scale
C. High reliability
D. Virtualization.

5.5. (Single choice) Which of the following is not one of the "4 V" features of Big Data? ()

A. Value
B. Variety
C. Velocity
D. Various
E. Volume.

5.6. (Multiple choice) What are the changes in traditional data analyses compared with big data analyses? ()

A. Changes in data format
B. Changes in data relationships
C. Changes in processing methods
D. Changes in processing costs.

5.7. (Single choice) Which of the following descriptions is incorrect for data storage and management of Big Data? ()

A. The method of traditional centralized storage is relatively monotonous in the type of stored data and stable in structure, which is suitable for early traditional Internet data
B. Distributed storage distributes data on multiple storage nodes, and each node is interconnected and managed through network interconnection, which solves the limitation of local storage systems in terms of storage capacity and maintenance cost
C. According to the openness of the cloud, the cloud storage can be divided into three categories: public cloud storage, private cloud storage and hybrid cloud storage, and according to the format of storage objects, it can be divided into another three categories: block storage, object storage and file storage
D. NoSQL generally refers to a relational database, which mainly solves ultra-large-scale and high concurrent data in the era of 5G and big data, and it has attracted wide attention in the application of big data.

5.8. (Single choice) Which method is not included in data screening? ()

A. Missing data processing
B. Noisy data removal
C. Consistency checking
D. Duplicate data record processing.

5.9. (Single Choice) Which of the following descriptions is incorrect regarding the classification of artificial intelligence? ()

A. Artificial General Intelligence (AGI) refers to a machine with human-like level intelligence, and a few technology companies have launched products with this level
B. There has been no internationally recognized standard for the definition of artificial intelligence research. If divided according to the strength of intelligence, intelligent machines are now commonly classified into two categories: Artificial Narrow Intelligence and Artificial General Intelligence
C. Artificial intelligence technology can be divided into traditional machine learning, deep learning, computer vision, natural language processing, and recommendation systems
D. Traditional machine learning algorithms can be divided into supervised, unsupervised, reinforcement, and other learning.

5.10. (Single choice) How can the relationship between Artificial Intelligence, Machine Learning, and Deep Learning be described? ()

A. Artificial Intelligence > Deep Learning > Machine Learning
B. Machine Learning > Artificial intelligence > Deep Learning
C. Artificial Intelligence > Machine Learning > Deep Learning
D. Artificial Intelligence = Machine Learning > Deep Learning.

5.11. (Single choice) The basic process of machine learning includes data collection, data screening, ________, ________, model evaluation testing, and model deployment and integration ()

A. Data compression, model training
B. Feature extraction and selection, model training
C. Feature consolidation, model training
D. Data compression, feature consolidation.

5.12. (Single choice) 5G and AI are two disruptive technologies. Although each brings changes and new experiences to industries, integrating 5G and AI will bring real change. Which of the following descriptions is false? ()

A. 5G has the characteristics of high bandwidth, large connection, and low latency, which can provide a higher transmission rate and comprehensively improve the capacity and reliability of mobile networks, providing a strong communication infrastructure platform to support the development of AI technology
B. The application of AI to 5G networks and devices will lead to more efficient wireless communications, longer battery lifespan, and better user experience
C. In terms of usage scenarios, the high speed of 5G technology will enrich the usage scenarios of AI technology. Still, the instability of 5G technology is currently a major obstacle for global operators and related industries to put 5G technology into practice
D. The mutual integration and empowerment of AI and 5G technologies brings both the effect of AI and 5G and the multi-effect of AI and 5G.

III. Short Answer Questions

5.13. Which two technologies does the integration come from that gave birth to ICT and what industries does ICT encompass?
5.14. Why is 5G an important link between the physical world and the digital world? What does 5G mean for smart cities?
5.15. What will be the trend of ICT innovation?

Chapter 6
5G Private Local Area Network Scenarios and Solutions

5G private LAN is a communication network built within some industries, departments, or companies to meet their needs for organizational management, safe production, scheduling and command, etc. 5G private LAN is the only way for 5G to reach all walks of life. Although the development of 5G private LAN is getting better and better, it still faces many challenges if it is to truly enable the digital transformation of industries.

This chapter focuses on 5G private LAN solutions for four industries: iron and steel, ports, coal, and 3C manufacturing.

Learning Objectives

- To master the application and solutions of 5G in the iron and steel industry
- To master the application and solutions of 5G in the port industry
- To master the application and solutions of 5G in the coal industry
- To master the application and solutions of 5G in 3C manufacturing

6.1 5G+ Iron and Steel Industry Applications and Solutions

With the rapid development of the 5G network, 5G enabling thousands of industries is the new development trend of the network. The iron and steel industry, as an important sector for the economy and people's livelihood, is a key breakthrough for the progress of 5G-enabled industries. The following section introduces how 5G enables digital transformation in the iron and steel industry, from insights into the iron and steel industry, to an analysis of 5G application scenarios and network requirements, scenario-based solutions, and application cases.

L. Song et al., *5GtoB From Theory to Practice*,
https://doi.org/10.1007/978-981-97-8453-0_6

6.1.1 *Iron and Steel Industry Insights*

Introduction to the Iron and Steel Industry

The iron and steel industry mainly produces and sells iron and steel products. Iron and steel is an alloy of iron (Fe) with C (carbon), Si (silicon), Mn (manganese), P (phosphorus), S (sulfur), and a small number of other elements. In addition to Fe, the content of C plays an important role in the mechanical properties of iron and steel, making it an iron-carbon alloy. Iron and steel are the most used metal materials of highest dominance in national economies.

The upstream, midstream, and downstream ecology of the iron and steel industry is shown in Fig. 6.1. The upstream of the chain mainly involves the production and supply of various raw materials; the midstream of the chain involves iron and steel smelting and the production of various iron and steel products; the downstream of the chain involves various sectors in need of iron and steel. Next, connections between each part and its components will be described separately.

1. **Upstream of the Iron and steel Industry Chain**

The end-to-end iron and steel production process consists of two main types: long process and short process. The long process mainly uses iron ore as the source of iron elements, and after steps such as reduction melting, oxidation refining, and secondary refining, the molten steel is solidified into continuous casting billets (or ingots) and then rolled into steel. The short process mainly uses various kinds of scrap as the

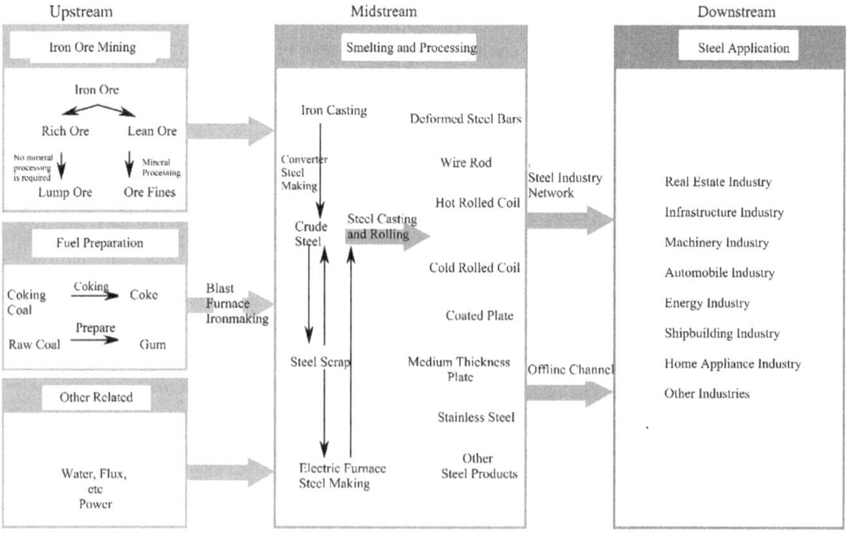

Fig. 6.1 Distribution of steel industry chain

main source of iron for steel making, and the scrap is melted into steel, then solidified and rolled into steel.

The upstream of the long process of iron and steel production involves the most important raw materials, such as coke, iron ore, limestone, etc. However, the supply of raw materials directly restricts iron and steel production. For example, uneven distribution of raw materials causes an increase in the cost of iron and steel production, thereby decreasing competitiveness. Among these raw materials, limestone has relatively abundant reserves, but its consumption is small. The following focuses on the supply of coal and iron ore.

First, coal supply. The distribution of coal in China is extremely unbalanced, abundant in the north and west but scarce in the south and east. Therefore, a large amount of coal is transported from the north to the south and from the west to the east of China. However, long distances are an obstacle to production and transportation. Coal is also one of the main raw materials for iron and steel production, so the supply of coal restricts the cost and the production efficiency of raw materials for iron and steel.

Second, about the pain points of iron ore supply. There are in China two problems for iron ore resource: First, there are many poor ores. Reserves of poor ores with less than 20% iron content account for about 80% of the total reserves. Second, there are complex ores with multi-element symbiosis and complex bodies. For some poor iron ore deposits, their upper part is hematite, and the lower part is magnetite.

China's iron and steel industry heavily relies on foreign iron ore for its upstream processes. The international situation, trade, and policies are prone to cause greater uncertainty in the cost of steel production. Statistics of China's iron ore and its concentrate imports in Fig. 6.2 show China's iron ore imports from 2014–2020, of which the total imports of iron ore and concentrate reached 1.17 billion tons in 2020, with a growth rate of 9.5%. There are mainly poor ores in China, so concentrated and rich ores supply relies on imports. If a large amount of poor iron ore is imported into the country, it will eat into the profits of China's steel mills.

2. **Midstream of the Iron and Steel Industry Chain**

The midstream part of the long process of the iron and steel industry chain mainly involves iron and steel smelting, production of iron and steel products, etc.

Among them, the long process is shown in Fig. 6.3. Its main steps include coal distribution, ore distribution, coking and sintering, iron making, steel making, continuous casting, and rolling. The iron ore and other materials are crushed through the crusher and mixed at a certain ratio before being sintered. The sintered ore with sufficient strength and proper particle size can be used as a clinker for iron making. Then, iron is made with a blast furnace. The raw material for iron making consists of iron ore and pellet ore, fuel (coke), solvent (limestone), and other parts. Iron production is realized by high-temperature reduction reaction to get iron from iron oxides.

The iron is pretreated and then passed through an electric furnace or converter for steel making. The pretreatment is mainly for desulfurization, dephosphorization, and desiliconization of the iron produced by the blast furnace. After refining, the steel is

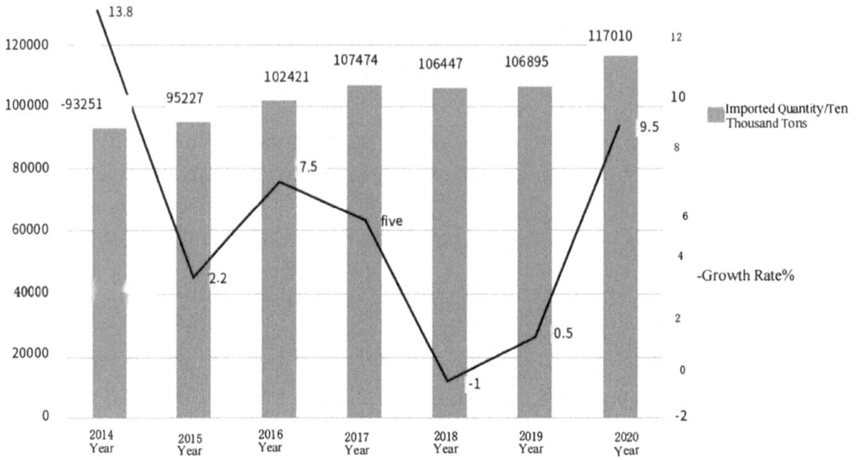

Fig. 6.2 Import statistics of iron ore and concentrate in China

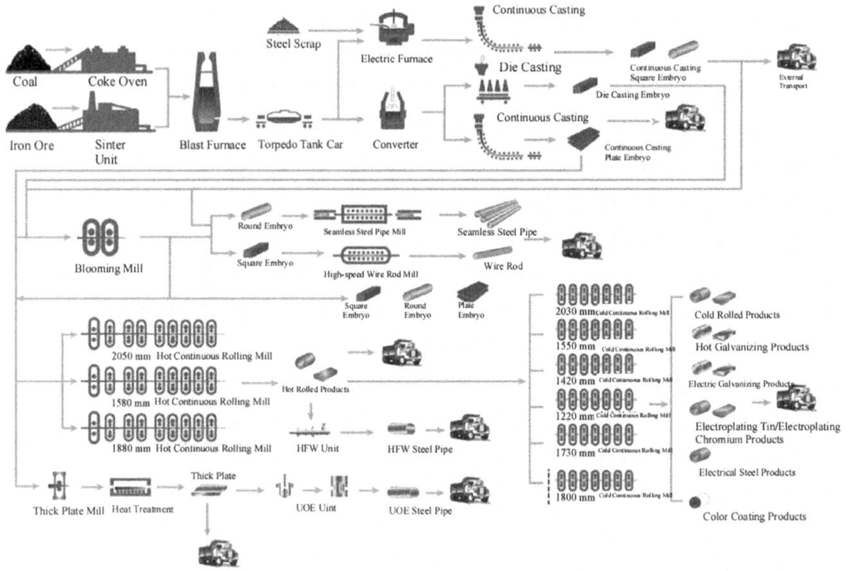

Fig. 6.3 Long process production process of iron and steel

passed through a continuous casting machine to complete solidification and cut into billets. Then, it is rolled according to different iron and steel products. Various iron and steel materials can be produced by rough rolling and finishing rolling.

Iron and steel products can be divided into different types of profiles, plates, pipes, and metal products according to the form of steel. Among them, profiles mainly refer to the metal formed through plastic processing with a certain cross-sectional shape and size of solid straight bars. Profiles have different specifications and diverse applications, such as heavy rail, light rail, large, medium and small sections, wire, and other different types and models. Plate mainly refers to metal plate and coil forged, rolled or cast, mainly including wide and thick plates, hot-rolled plates and, continuous-rolled plates, etc. The pipe is the material used to make pipe fittings, including seamless steel pipe and welded steel pipe, etc. Metal products include cold and hot-drawn steel wire, steel wire rope, and steel strand, etc.

Steel products can also be classified in other ways. For example, by thickness, plate, and coil can be divided into a thin plate, medium plate, thick plate, and extra thick plate; by production method, it can be divided into hot-rolled steel plate and cold-rolled steel plate. Other methods are based on application, surface characteristics, etc. Different kinds of steel can be used in different ways.

Although China's iron and steel production has exceeded 50% of the world, covering almost all varieties available in the world, and none of them is completely dependent on imports, there is still a shortage in high-quality iron and steel. For example, there is a big gap in variety and output in terms of special steel. The gap is much larger for extremely special steel, which has mainly relied on imports for a long time.

3. **Downstream Part of the Iron and Steel Industry Chain**

The sustainable development of the iron and steel industry, the foundation of China's industrial economy, is inseparable from the development of its downstream parts, such as real estate, machinery manufacturing, home appliances, automobile industry, shipping, infrastructure, and other industries. The iron and steel industry mainly provides raw materials for the downstream industries. The fast development of the downstream industries means a high demand for steel, which leads to the rapid growth of the iron and steel industry and vice versa.

Among them, the real estate industry has the largest demand for iron and steel, accounting for roughly 40% of the total iron and steel consumption. Steel prices largely depend on the rise and fall of the real estate industry. Infrastructure construction is also an important consumer of iron and steel. By the end of 2020, China's total road mileage reached 5,198,100 km, including 161,000 km of highway mileage; railroad mileage reached 146,000 km, including 38,000 km of high-speed rail. With the rapid development of transportation and other aspects, the supply of iron and steel has met higher requirements. The overall manufacturing industry (including automobiles, home appliances, ships, and other equipment manufacturing industries) has slowed down. Be the growth rate of end consumption or that of industrial investment. Both have gradually slowed down with the decline in economic growth. Nevertheless, the manufacturing industry remains a major consumer of iron and steel.

Three major industries, namely real estate, infrastructure, and industrial manufacturing basically consume about 90% of the total steel. However, as the economic growth rate continues to decline, the above three industries find it difficult to maintain a middle or high growth rate for the long run. Future iron and steel consumption is bound to further decline. Lower demand of downstream industries for steel directly leads to increased inventory and thus, lower prices, which in turn affects the profitability of steel companies. So, iron and steel companies need to adjust the overall structure of iron and steel products to maintain their competitiveness.

Trends in the Iron and Steel Industry

The development of China's iron and steel industry has gone through about five stages: Before 1977, it was a start-up stage mainly pursuing production increase and thus, a large amount of iron and steel was put into production. From 1978 to 2000, the development slowed down. The reform and opening up provided protection for the iron and steel industry in adopting technology and foreign investment. In 1996, iron and steel production exceeded 100 million tons. The years 2001 to 2010 saw a rapid expansion. After China's accession to WTO, the country witnessed rapid growth in the economy and explosive increase in the iron and steel industry; 2011–2015 saw a slowdown because of rising costs, economic crisis, policy regulation, and other external factors, resulting in a decline in iron and steel production growth; from 2016 till now, the country has been in supply-side reform phase. In the period of the 13th Five-Year Plan, the iron and steel industry has a direct impact from cutting overcapacity, structural adjustment, and green and smart manufacturing.

According to the data from the World Steel Association, global crude steel production reached 1.864 billion tons in 2020, down 0.9% year-on-year. China's crude steel was 1.053 billion tons, up 5.2% year-on-year. China's share of global crude steel rose from 53.3% in 2019 to 56.5% in 2020, making it the world's largest iron and steel producer and gradually moving towards a steel power. Since the start of the 13th Five-Year Plan, the equipment of China's major iron and steel enterprises has reached the internationally advanced level. The application of intelligent manufacturing in iron and steel production and manufacturing, enterprise management, logistics and distribution, and product sales has been strengthened. The numerical control rate of key processes exceeded 65%, and the equipment rate of enterprise resource planning (ERP) exceeded 70%. Informatization has been developed by leaps and bounds.

China is a large steel producer, but there is still a certain distance from being a strong country in iron and steel. The main gap is reflected in the following four aspects:

First, there is an uneven development. At present, mechanization, electrification, automation and informatization coexist, and the development level of different enterprises varies greatly.

Second, the industrial foundation is weak. Overall, intelligent manufacturing is at its initial stage. Therefore, intelligent manufacturing standards, software, and information security foundation are weak and lack industry standards. Common key technologies need to break through. Key software, underlying operating systems, development tools, and other technical areas are basically monopolized by foreign countries due to the lack of core patents.

Third, it is difficult to quantify the return on investment, and intelligence has not yet become the main production mode. Along with the increasing labor costs, demands of enterprise employees for the operating environment, labor comfort, and professional respect are constantly rising, and the demand and trend of remote and automated production are becoming increasingly urgent.

Fourth, there is a lack of core intellectual property rights and a low proportion of original innovation applications. Regarding research and development, a technological innovation system with the deep integration of industry, academia and research has not been formed yet. Tepid enthusiasm on original innovation is waiting for enhancement by policies.

With the continuous development of technology, China's iron and steel industry has improved its competitiveness by continuously adjusting its industrial structure. In order to better develop the iron and steel industry, China has introduced a series of policies to support it. The main policies include removing production capacity, deleveraging, reducing pollution, and optimizing structure. It provides a direction for the development of the iron and steel industry.

Because of uneven development, industrial restructuring and elimination of backward production capacity have become a new trend in the iron and steel industry. The State Council issued *Guidance on the Promotion of Mergers and Acquisitions in the Iron and Steel Industry to Dispose of Zombie Enterprises*, which is a top-level guide for the iron and steel industry in excessive capacity reduction and structural optimization. The document mentions that by 2025, the concentration of production capacity (Concentration Ratio, CR) of the top 10 enterprises will reach 60% to 70% of the whole industry. Steel enterprises will improve their efficiency of resource utilization, enhance their competitiveness, and put an end to steel overcapacity through mergers and acquisitions, capacity replacement, and other means.

In the process of iron and steel making, there is heavy energy consumption and environmental pollution. So national policies have also put forward new requirements for energy saving and emission reduction, which require the transformation of iron and steel production lines to ensure that they can meet the requirements of energy saving and emission reduction and eliminate backward production lines. Through carbon peak and carbon neutrality to force the iron and steel industry to reduce,adjust and upgrade, in the next few decades, "carbon reduction" will become the focus of industrial transformation and energy structure adjustment.

In a new era, the needs for iron and steel factories will become more flexible and customized. It is difficult for equipment on the original production line to meet flexible business demands. Besides, the original steel production line needs to be intelligently upgraded through 5G+ AI and other means to achieve refined management. Rapid production of multi-species, small-lot, high-quality products will be achieved in new production models. At the same time, predictive manufacturing will give rise to the demand for new technologies, including AI prediction of demand of downstream customers, equipment status, and upstream raw material market.

China is vigorously promoting ecological civilization, and the iron and steel industry has stepped into a stage with both quality improvement and quantity control.

Intelligent, green development will become the main direction of the future development of the iron and steel industry and also the key to the transformation and upgrading of the industry. In the future, the iron and steel industry will continue to develop high-quality, high-efficiency, intelligent, and green products instead of just focusing on output. This will also be the long-term development trend of the iron and steel industry.

Pain Points for the Iron and Steel Industry

Iron and steel production involves different scenarios and needs to produce better quality products in a more efficient, low-cost, safe and environmentally friendly manner with limited raw materials. Iron and steel production has the following characteristics:

(1) Large in number of workers being involved;
(2) Concentrated, large and technologically advanced equipment;
(3) Long in-production process involving a wide variety of production processes;
(4) High throughput of raw materials, auxiliary materials, and semi-finished goods;
(5) High energy consumption, high pollution, high risk, and harsh working environment.

Social progress has led to higher worker requirements for the production work environment. However, in the iron and steel industry, there are many inevitable harsh conditions, such as high temperature, dust, and noise. These seriously damage the health of employees and also cause increasing difficulties for steel companies in recruiting workers.

The industry is developing rapidly and is undergoing intelligent transformation. It is hoped that certain technologies can be used to meet the needs of enterprises and to solve or reduce problems encountered in the iron and steel industry.

6.1.2 Analysis of 5G Application Scenarios and Network Requirements

The iron and steel industry has the richest and most complex application scenarios in the field of industrial manufacturing, such as ultra-high temperature, ultra-low temperature, ultra-fast and ultra-slow equipment movement requirements, intelligent detection of a variety of equipment, high-precision positioning, and other industrial application requirements. 5G supports smart steel scenarios, which are broadly divided into three categories: remote control, video monitoring, and machine vision. Remote control mainly includes remote overhead crane control, unmanned overhead crane remote automatic control and (plus slag) robotic arm remote control; video monitoring mainly provides video backhaul for work monitoring, focusing on business status supervision; machine vision mainly includes steel surface detection and automatic steel transportation with more emphasis on AI identification through high-definition video.

Remote Overhead Crane

In the end-to-end process of iron and steel production, raw materials, intermediate products, and final goods are involved in lifting operations, mainly including the lifting of raw materials, intermediate products, finished products, and scrap generated during production. Therefore, overhead cranes are necessary during almost all steps of steel production, including raw materials entering the plant, iron-making, steel-making, hot rolling, cold rolling, finished goods storage, and materials handling in and out of scrap storage.

Figure 6.4 illustrates links involving overhead cranes in the end-to-end process.

The overhead crane is a core production equipment during the production process. Although industrial automation and information integration are improving continuously, overhead crane handling operation is still a major bottleneck that significantly disrupts the entire information chain. Nevertheless, overhead crane handling in traditional iron and steel industry has a large consumption of manpower and costs, and its high-risk operating environment leads to low job satisfaction, together with decreased efficiency and benefits. It is urgent to reduce the number of workers involved in overhead crane control or to realize unmanned control.

Current pain points of overhead crane operation include:

(1) High temperature, dust, and noise damaging employees' health.
(2) Special in work, with harsh working environments, continuous operation, and fatigue driving are prone to cause human operation errors, resulting in safety and production accidents.
(3) Narrow environment of the site boarding leaves potential risks. It takes a long time to get off the crane, which lowers work efficiency.
(4) Divided work positions lead to low management efficiency.

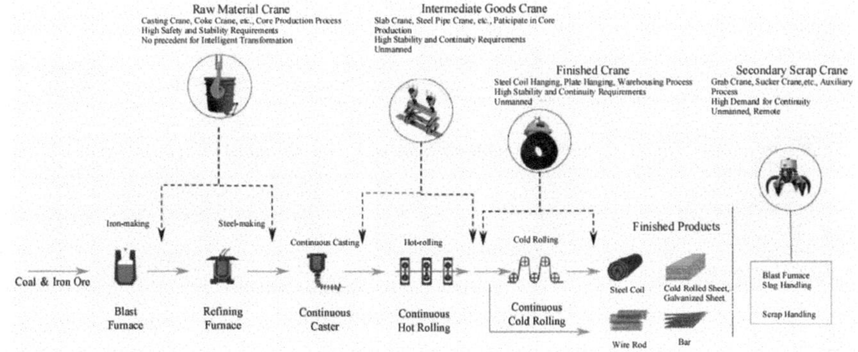

Fig. 6.4 Steel production involving overhead cranes

(5) One overhead crane needs more than one person to be in charge, which is a big waste of human resources.
(6) Traditional optical fiber connection is difficult and costly to deploy, while the Wi-Fi/microwave is poorly anti-interference and unstable, so the requirements for transformation cannot be met.

Therefore, it is a common demand of the industry to improve remote control of overhead cranes, thus improving the working environment of front-line operators and increasing the efficiency of operation.

The schematic diagram of the remote video control system on the overhead crane is shown in Fig. 6.5, which shows that remote overhead crane control can be realized with 5G network. The left side of the figure shows that after the overhead crane is connected to the equipment, the video is backhauled in real time, mainly through the HD camera. Then the data is forwarded through the 5G network. There are two types of traffic between the overhead crane control system and the overhead crane: video stream and programmable logic controller (PLC) control stream. Through the dual-end wireless way, the overhead crane control room and the overhead crane can get access to the equipment, and the driver can also control the overhead crane from the central control room. Figure 6.6 shows realistic pictures of an overhead crane transformed with 5G.

The overhead crane remote control system comprises three parts: control chair (including PLC), 5G network, and overhead crane (PLC, camera). First, the intelligent transformation of the overhead crane includes connecting PLC and 5G network links. Then, a 5G dedicated circuit is connected to the control end. A video camera is necessary for operating the relevant panoramic and partial high-definition video monitoring. The ultra-low latency provided by the 5G network is able to meet the real-time issuance of commands and remote control of the overhead crane.

During the overhead crane transformation, workers are required to work in shifts and operate at the site with noise, dust, and high temperatures. The working environment and working hours of the employees are demanding, and the working conditions

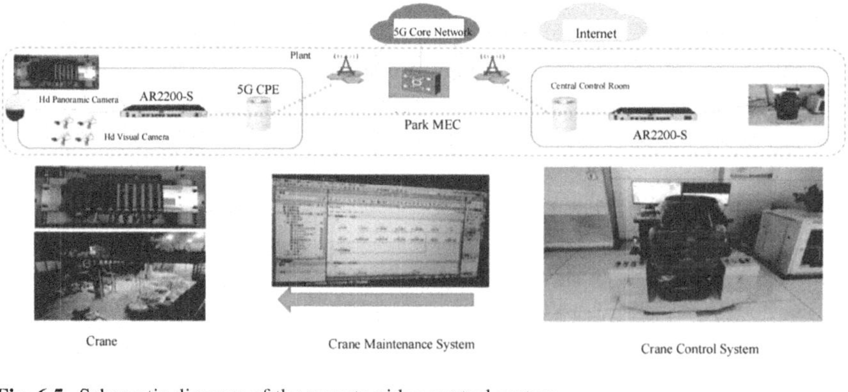

Fig. 6.5 Schematic diagram of the remote video control system

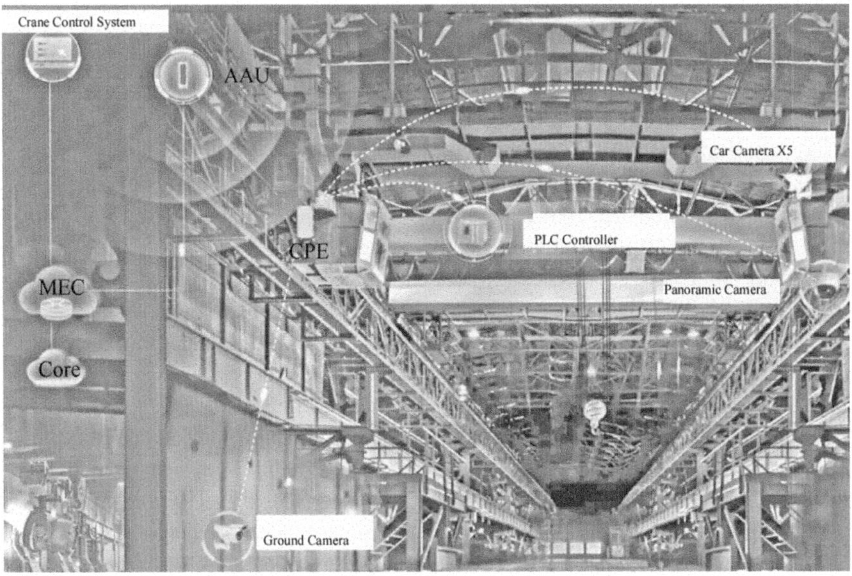

Fig. 6.6 Actual view of a crane

cannot be guaranteed, so the efficiency cannot be improved. After the 5G intelligent transformation, there are several advantages:

(1) Greatly improved production working environment; workers working in a more decent manner and having a sense of belonging and enhanced social status; reducing employee turnover rate.
(2) Improving overhead crane operation efficiency, reducing failure rate and energy consumption, reducing workers' labor intensity and improving safety.
(3) Wired connection switching to wireless access at any time; greatly improved deployment efficiency; monitoring blind areas being cleared efficiently and quickly.
(4) Alleviating abrasion caused by human irregularities on the equipment, improving the service life of the equipment, and reducing the intensity and cost of an overhaul.
(5) With big data analysis technology, an overhead crane can judge the most economical lifting method and the optimal path by itself.

Unmanned Overhead Cranes

Similar to remote overhead cranes, industry-leading iron and steel companies are also implementing a more automated, fully unmanned overhead crane solution on their production sites. Unmanned overhead cranes are more efficient and have more accurate operational data because they are docked to inventory management systems and IT systems, and they have a higher degree of automated control of site operations through a richer set of sensors. In large steel enterprises that emphasize efficiency, cranes have more urgent application needs. In the early stage, due to the lifting algorithm (unable to accurately and intelligently identify objects to be carried), some scenes mainly use remote overhead cranes. Remote control can solve the defects of the initial algorithm, for example, the irregular object loading and unloading efficiency is not high, the emergency scene response is not timely, etc. So, in the early stage of application, remote overhead cranes and unmanned overhead cranes can be used. If the lifting material's shape is regular and the working environment is safe and controllable, it is recommended to deploy the unmanned crane; if the lifting material's shape is irregular or there are dangerous movements, such as people moving in and out, it is recommended to deploy remote crane at the initial stage.

The architecture of a typical unmanned, fully automated bridge crane is shown in Fig. 6.7. The northbound bridge crane is commanded and dispatched by the inventory management system (Warehousing Management System, WMS), and the southbound control of PLC is achieved through the communication link, thus realizing the control of input and output devices through PLC. Therefore, in the whole system, the communication link of the north–south communication is done by the core module.

Based on the unmanned bridge crane plan, its control does not rely on video surveillance data. At the same time, the current control scheme dependent on wireless communication links is mainly the control communication between the ground control center PLC and bridge PLC, so the 5G network requirements by the unmanned bridge crane are focused on the control link.

The control process of an unmanned bridge crane includes three layers: bridge crane site control (site layer), control of crane bridge and ground control center (control layer), and control of ground control center and WMS (operation layer).

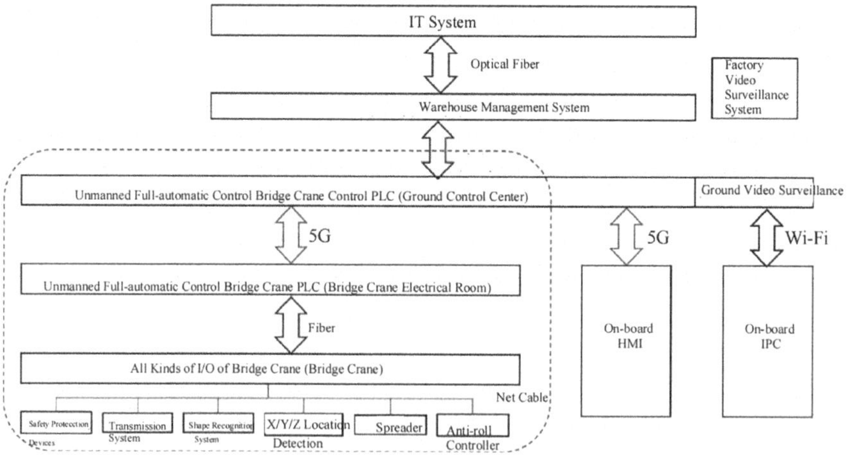

Fig. 6.7 Architecture diagram of unmanned fully automatic control bridge crane

(1) Site layer: Detecting site data by deploying cameras, laser, and radar scanners on site to achieve local closed-loop functions, such as environment modeling and collision avoidance.
(2) Control layer: The traveling car PLC periodically transmits input and output data to the control system, including information on the status and precise position of large cars, small cars, reels, spreaders, etc. Based on the lifting task from WMS, the control system plans and sends control commands to the traveling crane after calculation.
(3) Operation layer: WMS sends operation tasks to the ground control center based on the operation tasks of the business system.

The unmanned overhead crane system consists of three parts: collector (including 3D scanner, laser rangefinder, codec, and camera), 5G network, and PLC controller. The scanner collects horizontal and vertical information, and the rangefinder collects distance information to obtain information and images of surrounding materials, pit materials, vehicles, car hopper height, and loading and unloading positions and transmits the data to the edge cloud for data processing in real-time to establish a 3D data model of the site. At the same time, AI intelligent algorithm constructs an action command set and sends it to the overhead crane for execution, thus realizing unmanned and autonomous production of the overhead crane.

Slag Adding Robotic Arm

In the steel production process, the continuous casting process requires adding protective slag to the crystallizer to cover the steel surface and prevent the steel from being oxidized. However, the slagging workshop is high in temperature and dense in dust with slag splashing. Manual slagging on site in the past has safety hazards and high labor intensity. The thickness of the protective slag is controlled by the operator's experience, which will lead to uneven thickness of the protective slag on the surface of the molten steel. A comparison between before and after the installation of the slag-adding robotic arm is shown in Fig. 6.8. Before the deployment, the working environment is difficult, while after the deployment, workers can get away from the difficult working environment through the remote control of robotic arms.

Fig. 6.8 Comparison before and after installing a slag manipulator

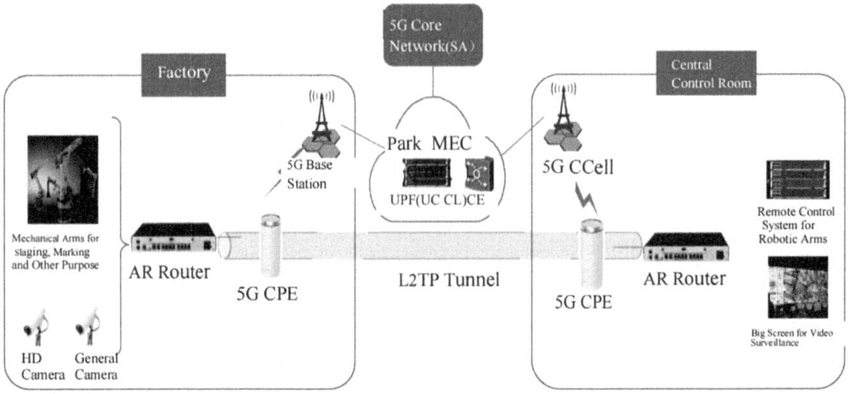

Fig. 6.9 Automatic slagging network

The automatic slagging network consists of three parts: PLC controller, cameras, and 5G network, as shown in Fig. 6.9. The central control room can analyze backhauled video from the camera and realize one-touch slag addition through the robotic arm's remote control system for remote automated processing. This reduces labor and manpower costs, ensures employee safety, and also minimizes the thickness error caused by manual slag addition.

Video Monitoring

There are safety risks in inspecting high-temperature hazardous areas, but through high-definition real-time video monitoring, the control center can grasp the site dynamics in a timely and accurate manner. It can also give real-time warning and early intervention, and can reduce the number of personnel in work inspection, improve efficiency, and avoid safety production accidents.

The production environment of steel requires a large number of backhaul videos on a small production area. 5G can be flexibly adapted to meet the needs of such video backhaul. The HD real-time video monitoring network is shown in Fig. 6.10. The camera can backhaul video to the video monitoring platform in real-time through 5G CPE. High-risk scenes are promptly identified through the platform's software to reduce casualties and property damage.

Steel Surface Detection

Quality problems, such as indentations, holes, and other surface defects during transportation and manufacturing, are inevitable, as are rust spots and other problems during long-term exposure to air. Steel surface detection uses high-definition video backhaul by cameras and machine vision algorithms to automatically identify a series

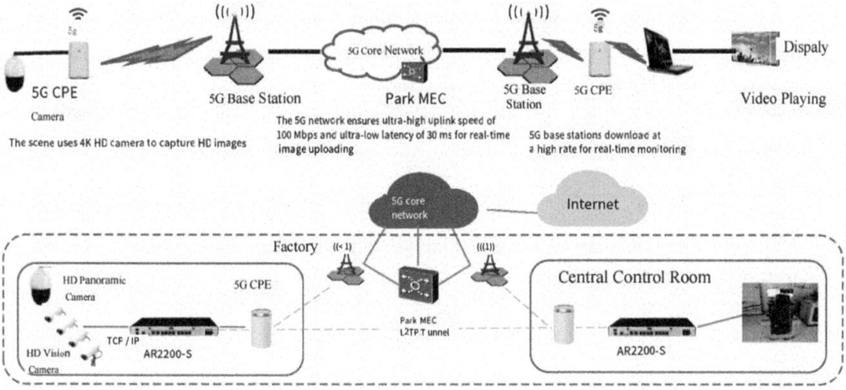

Fig. 6.10 Networking of HD real-time video surveillance

of problems existing on steel surfaces. This can greatly improve the efficiency of steel surface quality inspection, increase the quality of steel products, and reduce the work intensity of detectors.

Figure 6.11 shows the working principle of steel surface detection. The HD video is analyzed by the field-distributed inference engine, and the detection results are obtained. Test results can be taken from onsite application software. Platform training is for updating the test model, which supports customers to update the model autonomously. The more AI algorithms are used, the more accurate they become to meet the needs of long-term application evolution.

The following advantages will be obtained after the transformation of steel surface inspection.

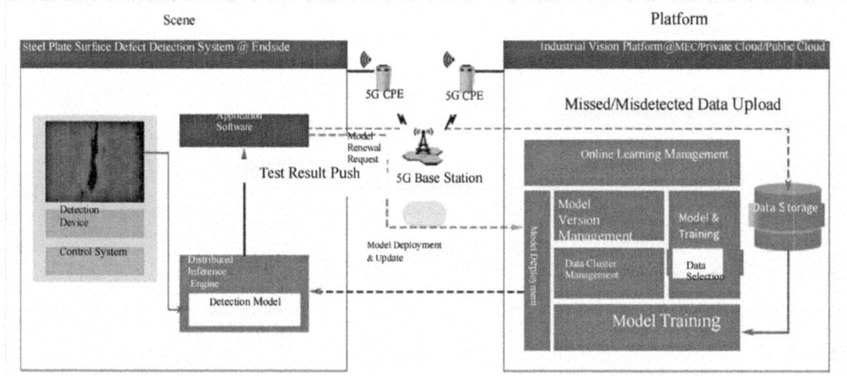

Fig. 6.11 Checking principles of a steel meter

(1) Enhancement of sampling rate: Manual detection has a low sampling rate; after machine vision detection, real-time detection can be realized, and the sampling rate can reach 100%.
(2) Non-contact processing: avoid secondary damage and improve reliability.
(3) Reducing cost: improve production line automation and increase production efficiency.
(4) Improving quality: stable product quality and enhanced brand competitiveness.
(5) High-precision detection: Detection accuracy can be increased to μ level, which cannot be detected by human eyes but can be done by machines.
(6) Classification of defects for grading of steel and fine pricing.

Automatic Steel Conversion

In the process of wide and thick plate rolling, the slab is rotated 90 degrees according to the rolling process and then fed into the mill for rolling after the billet enters the steel conversion area of the mill. Before the deployment of automatic steel conversion, as shown in Fig. 6.12, manual steel conversion is required on-site, which is a harsh working environment. However, after the improvement, it can be operated remotely, but the efficiency is still not high because there may be a decrease in efficiency in manual operation caused by inaccurate operation and decreased energy. Therefore, it is imperative to deploy automatic steel conversion.

An example of automatic steel conversion networking is shown in Fig. 6.13. Automatic steel conversion uses HD cameras for backhaul video. The platform-side AI software automatically identifies the direction and angle of the slab and docks PLC control system of the mill based on the visual recognition input results, thereby controlling the steering and rotational speed of rollers. This allows for a 90-degree automatic rotation of the slab.

The promotion of automatic steel conversion can effectively reduce the work intensity of steel conversion operators and lower the labor cost of enterprises; the roughing mill can roll one more plate per hour through automatic steel conversion, and the efficiency of the roughing mill increases, thus to increase production capacity.

Fig. 6.12 Before deploying automatic steel converter

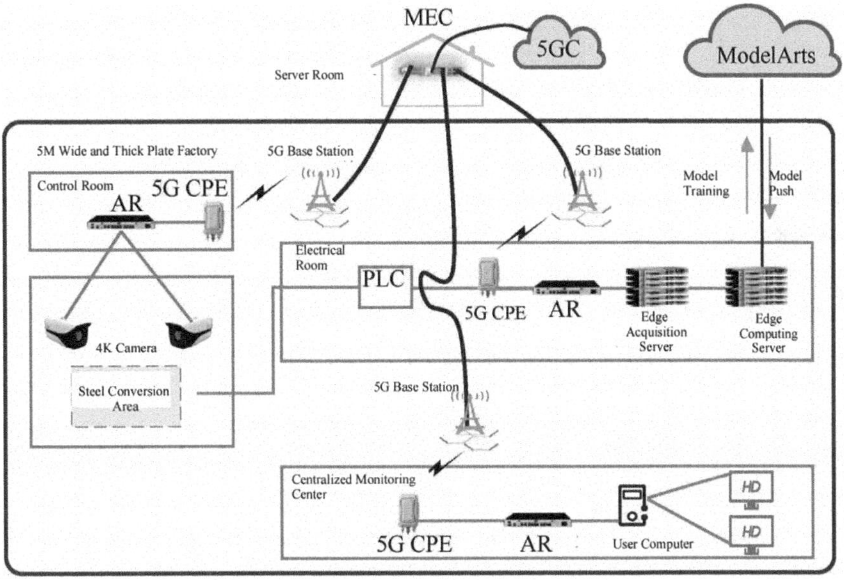

Fig. 6.13 Networking example of automatic steel conversion

6.1.3 Scenario-Based Solutions

The 5G steel private network is not positioned to replace the original wired network (like fiber optic network, etc.) of the steel plant but as an extension or supplement to the existing network. For a new steel plant site, the 5G private network can serve as a unified network for it. Fiber, Wi-Fi, or microwave networks can be used as a backup for certain important scenarios, thus meeting the high-reliability requirements from customers.

Smart steel 5G private network service builds a fully connected wireless network with 5G+ MEC technology. The MEC platform is deployed at the network's edge close to the business area, using the core network UPF sinking method. This provides a guaranteed low latency network with the wireless side resource block (RB) reservation and other features in the network capacity. On this basis, it can deploy intelligent control applications such as video monitoring automation analysis, vehicle automation scheduling, and security protection, which significantly improves user experience and data security.

Network Architecture

Based on the analysis above, a 5G private network for the iron and steel industry should contain wireless network, transport network, core network, etc. Overall private network architecture of a typical 5G iron and steel industry can be seen in Fig. 6.14.

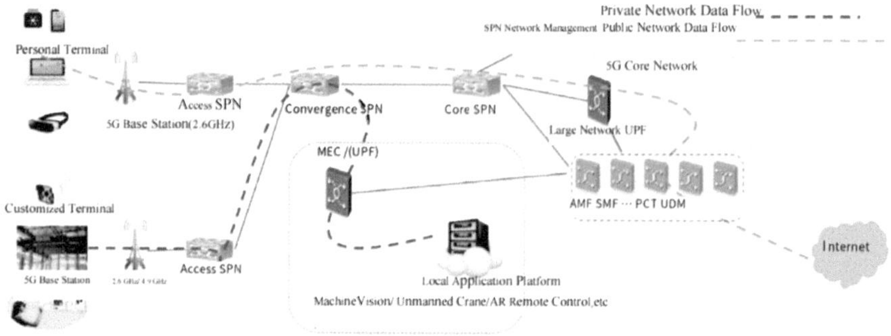

Fig. 6.14 Overall network architecture of a 5G iron and steel private network

The figure takes the 5G steel private network architecture of an operator as an example. First, the wireless side of the same base station implements access toB customized terminals and toC personal terminals. A personal terminal is mainly accessed through 2.6 GHz for download services. The custom terminal carries out upload service through 2.6 GHz/4.9 GHz different frequency networking. The intermediate transport side shares the converged sliced packet network (SPN) equipment; the personal service connects to the large network UPF for various Internet services through the core SPN; the customized terminal gets connected to MEC through the converged SPN. The internet is involved in a locally closed loop, and the service will not get out of the park.

Networking Scheme

The steel scenario-based private network involves three sub-networks: wireless network, transmission network, and core network.

1. **Wireless Networking Scheme.**

Almost all "5GtoB" projects are mainly for uplink services, whose bottleneck is in the uplink while the downlink capacity is generally sufficient. Therefore, the capacity coverage design should focus on the uplink. Considering a steel scenario being one with an indoor semi-enclosed area and a relatively spacious interior, it is recommended to deploy active antenna unit (AAU) modules in a row for network traffic below 200 Mbps to guarantee coverage and enhance capacity. Later, for some hot spot areas, Easy Macro modules will be deployed with multiple channels to achieve capacity supplementation.

When the uplink network traffic is below 200 Mbps and the station spacing is greater than 50 m, it is recommended to use AAU modules for inter-frequency networking to build the basic coverage capacity of the plant. When inter-frequency interference is large, such as when uplink noise power lifts to 20–30 dB, it is recommended to give priority to introducing inter-frequency networking. Generally, it is

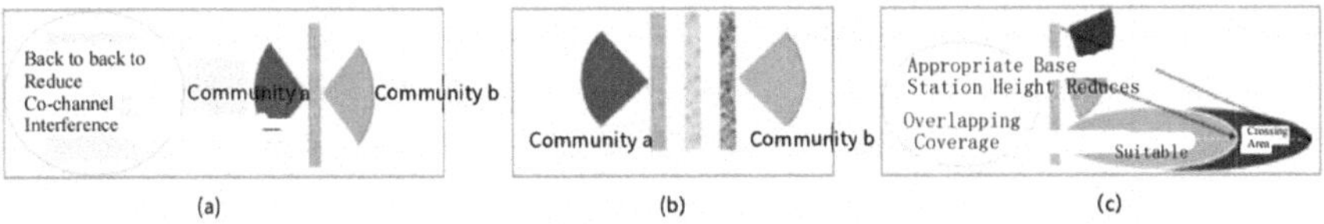

Fig. 6.15 Deployment rules for base stations

recommended to prioritize adding inter-frequency within inter-frequency networking to reduce the new cost of the cell and introduce different frequency points by using the broadband feature of existing base stations.

The network coverage design first designs the basic network according to the coverage objectives; then, the network overlay design is based on capacity requirements and applications such as low latency and high reliability.

Wireless site deployment needs to meet the following five principles:

(1) Balanced principle: The number and capacity of communication terminals covered by each cell/station are relatively consistent, and avoiding as much as possible overload or light load in some cell capacity;
(2) Avoiding sparring of inter-frequency stations: refer to the back-to-back method to reduce inter-frequency interference, as shown in Fig. 6.15a;
(3) Antenna installation location: such as avoiding metal object shading for AAU main flap coverage direction, setting the appropriate downward tilt angle, and letting the main flap align with CPE direction.
(4) Taking advantage of the physical environment to isolate coverage: For AAU point selection, try to use buildings in the plant to achieve physical isolation of inter-frequency signal to reduce interference, as shown in Fig. 6.15b.
(5) Station height selection: Priority is given to the location with higher height, but the station height should not be too high, and overlapping coverage needs to be considered, as shown in Fig. 6.15c. If the height of the overhead crane is about 3 m, in order to avoid the overhead crane blocking, it is suggested that the station height is 2–3 m higher than the height of the overhead crane to ensure that there is no obvious obstacle blocking the target coverage area.

2. **Transmission Networking Scheme**

The transmission network deployment strategy requires careful consideration of 5G base stations in the steel factory park, the user development needs of the iron and steel industry, and integrated planning of the end-to-end transmission private network.

The wireless side can connect to the access transmission device based on user requirements. The 5G base station in the park can access the transmission equipment through C-RAN mode. The forward network scheme adopts the form of optical fiber direct drive/passive wave division. For iron and steel enterprises involving remote control, safety production, emergency communication, and other special scenarios requiring network priority protection, slice selection and priority scheduling can be carried out by means of customized communication cards. Bind DNNs to different transmission channels to meet different service scenarios (multistage transmission devices must have the slicing function). For example, scenarios such as HD video backhaul in the iron and steel industry, unmanned and remote crane control, augmented reality (AR) remote assembly and machine vision quality inspection can match different transmission paths through different configurations, as well as matching different transmission slices. As shown in Fig. 6.16, different levels of SPNs are used to forward traffic based on various requirements of steel enterprises. Meanwhile, aggregation SPNs and core SPNs can be used to connect to the control surface of 5G core network.

3. **Core Networking Scheme**

The core network adopts 5G SA technology architecture and combines the edge computing and network slicing technologies unique to the 5G standard. This is done

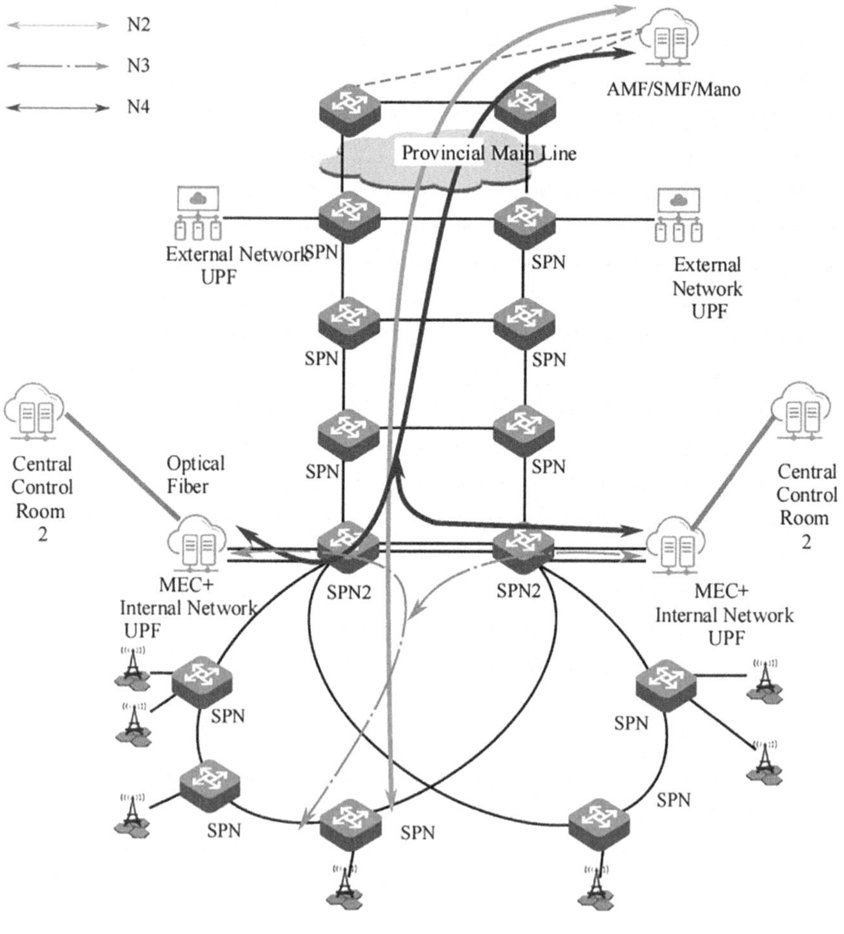

Fig. 6.16 Transmission network architecture

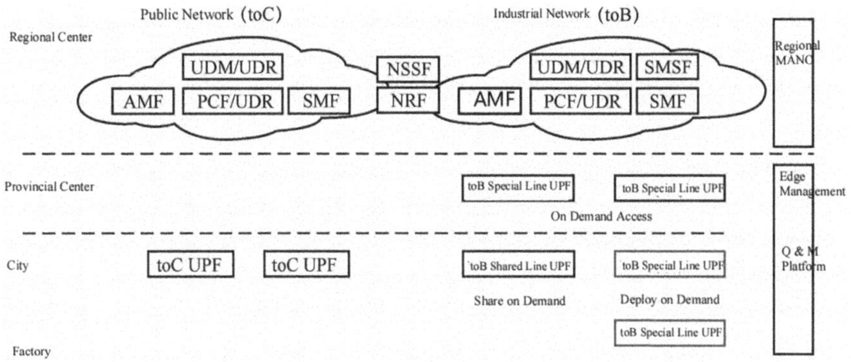

Fig. 6.17 Overall architecture of a 5G iron and steel private network

to meet the high reliability, low delay, large bandwidth, data security, and other business guarantee requirements of the iron and iron and steel industry.

Figure 6.17 shows the overall architecture of a steel 5G private network. For example, a carrier builds a network cloud data center in a regional center. At the same time, 5G SA industry network (toB) slices are planned for vertical industries in a unified manner, and the core network control plane net is deployed in virtualization mode. The iron and steel industry can selectively deploy or use dedicated or shared UPFs based on different business requirements. Control planes use the common part of toB, and user planes are isolated from each other according to the contract slice properties.

5G SA Industry Network (toB) slice can provide industry users with flexible and customized network deployment solutions based on the characteristics and business models of the iron and steel industry. Local aggregation of service traffic and data division is realized through user plane sinking on demand. User plane subsidence, as a prerequisite for the realization of 5G edge computing, provides more targeted network functions for the iron and steel industry. Edge computing provides task-specific cloud services at the edge of the mobile network for edge applications by offering computing, storage, and other infrastructure by being close to the data source or users and by IT architecture and cloud computing capabilities based computing capability.

As industrial automation and informatization are improving, large iron and steel enterprises put forward more stringent requirements for data security. At the same time, typical applications like remote crane control, video surveillance, optical character recognition, slag-adding robotic arm, UAV blast furnace temperature measurement, heap weight measurement, and AR remote assistance in the iron and steel industry require that edge computing can provide data protection, extremely low delay, save transmission and capacity opening.

(1) Data protection: Based on the local network, data collection and calculation are no longer transmitted to the cloud. Important and sensitive information can be

transmitted without an external network, and data can not leave the factory park, effectively avoiding the risk of data exposure in the transmission process.

(2) Extremely low delay: The server is deployed on the enterprise side and network edge, such as the access ring and aggregation ring near the client terminal device, so that the terminal can directly access the content source locally. The end-to-end service response latency is reduced from the data transmission path.
(3) Saving transmission: 5G network realizes local unloading of data traffic through the flexible deployment of user plane function UPF at the edge of the network. The system avoids detouring local service traffic, reduces transmission resource usage caused by massive data transmission, and effectively restrains network congestion.
(4) Capacity opening: Realize the business innovation of operators and vertical industries. The enterprise network operation and maintenance management information, location information, service enablement control, and rich industry customized services are provided to industry users by means of an open Application Programming Interface (API).

As the user plane of edge computing, UPF can realize the function of forwarding edge data. The edge computing platform provides a running environment for the edge application and realizes the management of the edge application. As shown in Fig. 6.18, in the 5G network architecture, edge UPFs can divert services from inside and outside the park and serve as private service anchor UPFs for enterprises to unload local traffic. At the same time, the local shunt in the plant shall cover all base stations in the factory park.

In a private 5G network for the iron and steel industry, operators can provide three important types of network capabilities such as local service assurance, data not leaving the park, and edge nodes. They all need to be satisfied by MEC (edge UPF).

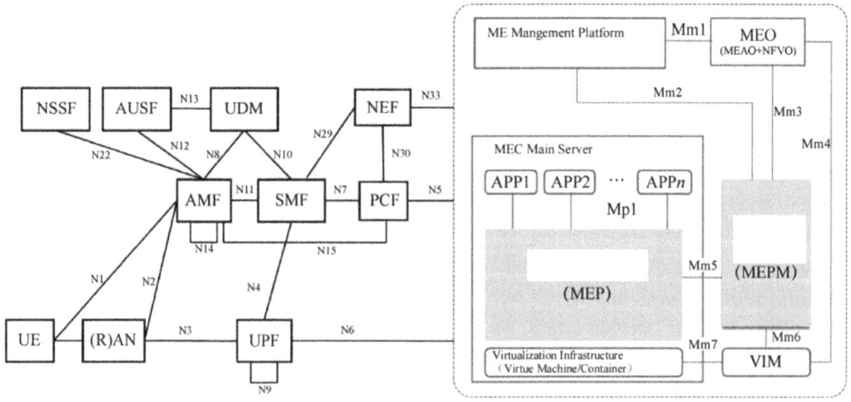

Fig. 6.18 Logical architecture of 5G edge computing system

6.1.4 Application Cases

Take the Hunan X Steel project as an example. Hunan X Steel joined with operators and Huawei to launch the Smart Steel 5G project in July 2019. The pilot phase was conducted at a 5 m wide and thick plate plant in Hunan X Steel. By providing Hunan X Steel with private + premium network services, typical application scenarios, including 5G unmanned overhead crane remote control, 5G AR remote assistance and 5G HD video monitoring in high-risk areas, were realized.

Project Background

Hunan X Steel has the complete steel production process and equipment technology of coking, sintering, iron making, steel making, refining, rolling, etc., which are advanced in the world today. Their products cover more than 400 varieties in three categories: wide and thick plates, wire rods, and bars. Thirty-three kinds of products have been awarded the title of national, ministerial, and provincial quality products, and 12 kinds of products have won the Golden Cup Award for the physical quality of national metallurgical products. Moreover, Hunan X Steel shipbuilding plates have been certified by nine national classification societies. After 60 years of relentless efforts, Hunan X Steel's total assets have reached 42 billion yuan, with a business income of over 60 billion yuan and 10,204 employees on duty. It is the largest single state-owned enterprise with the strongest comprehensive strength in Hunan Province, and its overall competitiveness is at the forefront of the industry.

Hunan X Steel will be supported by five industrial clusters, namely the main iron and steel industry, steel deep processing, circular economy, heavy equipment manufacturing, logistics trade, and services. The company aims to establish itself as a dual-type demonstration enterprise, being a 10-million-ton high-quality manufacturing enterprise, a responsive and comprehensive service enterprise, a highly efficient operation enterprise with fine management, a diversified development enterprise with complementary industries, and an intelligent energy-saving and ecological.

Pain Points Analysis

Hunan X Steel Pain Points analysis:

(1) 24-h shift operation;
(2) Manual operation in the overhead crane suspension control room results in high labor costs;
(3) High-risk operating environment makes work harder with many sources of danger, such as high temperature, dust, noise, and hazardous gases;
(4) High in the height of the overhead crane (20~50 m);
(5) Low in the level of intelligence in the workplace.

Fig. 6.19 Harsh working environment

"Increasing production and reducing consumption" is the primary demand for intelligent upgrading of enterprises. The factory labor cost is a big expense, and the working environment of employees is hazardous. It is urgent to realize automated unmanned operations. The traditional optical fiber connection mode is difficult to deploy and costs high, and the Wi-Fi mode is poor in anti-interference and instability, which cannot meet the requirements of transformation. As shown in Fig. 6.19, the environment of iron and steel smelting is extremely harsh, and the network communication environment is poor, making it difficult to deploy communications.

Solutions

As shown in Fig. 6.20, the network services in Hunan X Steel Park are analyzed, including non-production and production services out of the park. In the actual deployment process, the business of Hunan X Steel Park is currently focused on production applications in the park without considering non-production applications.

There are many dimensions to consider when determining the business guarantee of the park. Differentiated "planes" guarantees need to be realized in accordance

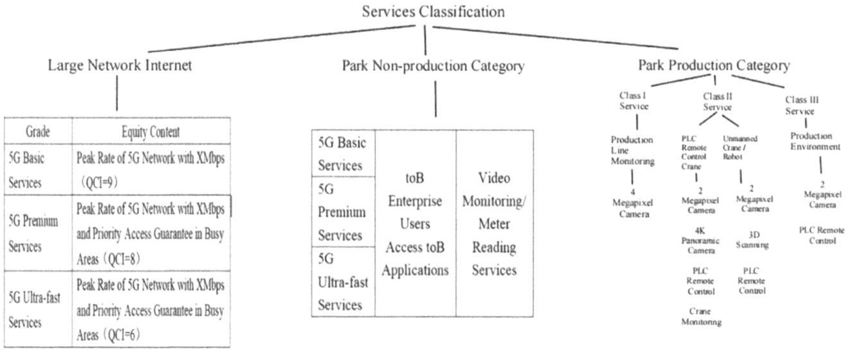

Fig. 6.20 Business analysis of Hunan X Steel

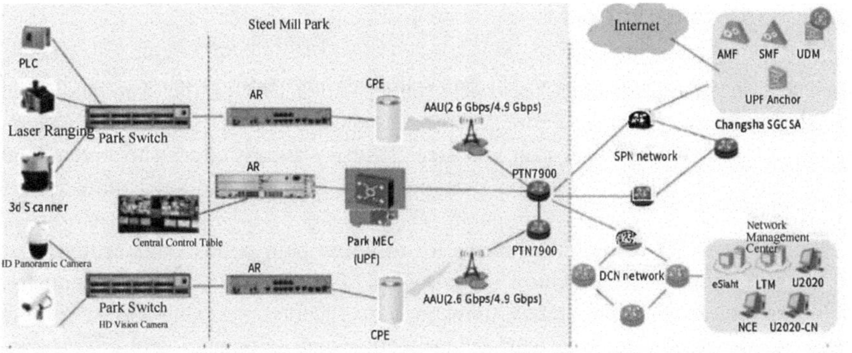

Fig. 6.21 Overall network architecture

with key areas, and differentiated "points" guarantees need to be realized in accordance with different businesses in key areas (such as PLC monitoring, camera video backhaul, etc.). This poses great challenges to network planning, service delivery and security.

As shown in Fig. 6.21, Hunan X Steel Project builds a fully connected wireless network through 5G+ MEC technology to realize a comprehensive perception of elements in key monitored areas. By doing so, it can meet main application requirements such as remote control of equipment, AR remote operation and maintenance assistance, machine vision, equipment data acquisition, high-definition video surveillance, 5G data acquisition, and no leave of data out of the park. The MEC platform is deployed at the edge server room near the monitoring area of Hunan X Steel. In addition to providing guaranteed low latency, it can also deploy remote intelligent driving control applications such as video surveillance automation analysis, vehicle automation scheduling, and security protection, significantly improving user experience and data security.

Business Value

Through 5G network application, Hunan X Steel realizes remote control of overhead cranes, working environment improvement and an efficiency increase by 25%. The efficiency of remote semi-automatic overhead cranes is expected to increase by 60% (one person controls three cranes). Unmanned overhead cranes are digitally inputted and executed, enabling unmanned operation and a 25% increase in machine efficiency. In this way, labor costs are reduced, the working environment is improved, and production efficiency and production safety are greatly enhanced. An "unmanned factory" can be expected in the future.

6.1.5 Summary (of 6.1)

This section first introduces the background knowledge of the iron and steel industry, including the steel production process and the iron and steel industry chain ecology. Then, trends in the iron and steel industry are introduced to understand its development. Finally, common pain points in the iron and steel industry are introduced.

Huawei provides different solutions for different pain points. This section first explains in detail the application scenarios of 5G in the iron and steel industry, including remote overhead cranes, unmanned overhead cranes, slag-adding robotic arms, video monitoring, steel surface inspection and automatic steel conversion. Then it introduces the overall solution of a 5G private network for the iron and steel industry, including the networking strategy of wireless, transport network/transmission network, core network, etc. Finally, practical application cases and business values are introduced with the Hunan X Steel project, etc.

By reading this section, readers should have a certain understanding of 5G solutions in the iron and steel industry and know their progress and application scenarios.

6.2 5G+ Port Industry Applications and Solutions

As the transportation hub, ports play a pivotal role in promoting international trade and economic development. In the context of digital transformation, ports are also currently undergoing fully automated and digital transformation and upgrading. With the global deployment of 5G, 5G will drive ports to continue to upgrade automation and intelligence and create a "green, environmentally friendly, and efficient" smart port.

The following introduces how 5G can enable the port industry with insights from the port industry, 5G application scenarios and network requirements analysis, scenario-based solutions, and application cases.

6.2.1 Port Industry Insight

Introduction to the Port Industry

Shipping is the most important mode of transportation in international logistics, accounting for 70% of the global trade volume.

Ports are important distribution centers for shipping trade, with port enterprises providing essential facilities and services for ships, passengers, and cargo. These services include ship entry and exit, docking and maneuvering, cargo handling, warehousing, and barging within the port area.

1. **China Playing a Crucial Role in Promoting the Growth of Global Shipping**

With the Belt and Road Initiative, China has emerged as the first in trade and a key driver of global shipping growth. The following data highlights China's significance in this field:

(1) Approximately 90% of China's total imports and exports are transported by sea.
(2) China boasts the largest maritime cargo volumes globally.
(3) China is home to 15 of the top 20 ports in global throughput and 7 of the top 10 container ports worldwide.

2. **Port Layout**

The following describes the layout of China's ports.

(1) Number of ports: In 2019, there were 145 freight ports in China, of which 55 were coastal ports and 90 inland river ports, distributed in 21 provinces and cities.
(2) Number of berths: There are 23,919 berths in China, of which 2,444 are above 10,000 tons.
(3) Capacity: There are 35 ports with annual throughput of over 100 million tons, including 25 coastal ports and ten inland river ports.
(4) Five major coastal port clusters: In 2007, the Ministry of Communications issued the *National Coastal Port Layout Plan* and *National Inland Waterway and Port Layout Plan*, forming five major port clusters in the Bohai Rim, Yangtze River Delta, Southeast Coast, Pearl River Delta, and Southwest Coast.

3. **Port Operation Process**

Container is called "the box that changes the world", and it realizes the standardization of logistics systems in the world, such as ships, ports, roads, transfer stations, bridges, and tunnels.

The flow of containers in a port involves various systems, such as shore bridges, field bridges, container trucks, gates and so on. The transportation system of a port can be divided into two categories: horizontal transportation and vertical transportation. The main tools of horizontal transportation are container trucks, AGVs, and straddle carriers, and the main tools of vertical transportation are bridge cranes, rail cranes, and tire cranes.

The container terminal operation area is divided into three major areas: loading and unloading, yard operations, and gate operations. A port unloading operation process is shown in Fig. 6.22.

Unloading in the terminal operation process consists of four steps.

(1) A ship arrives at the shore and unloads the cargo through the shore bridge (vertical transportation).

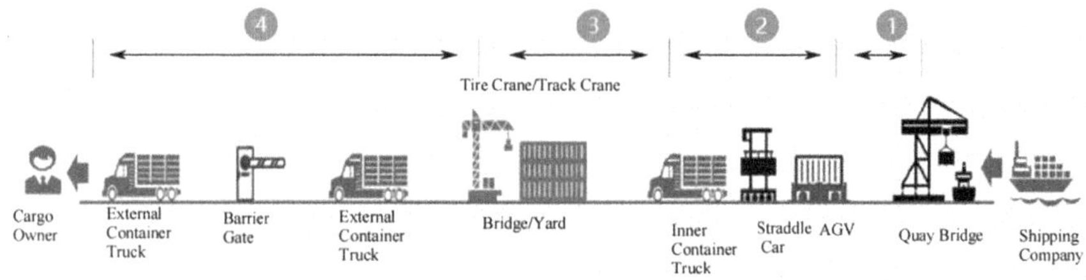

Fig. 6.22 Port unloading process

(2) Internal container trucks, AGVs, and straddle trucks transport the cargo to the yard (horizontal transportation).
(3) A yard crane (rail crane, tire crane) loads the cargo to an external container truck (vertical transportation).
(4) An external container truck pulls the cargo from customs (horizontal transportation).

4. **Support System**

The supporting system consists of three parts: vertical transportation, horizontal transportation, and terminal operation.

(1) Vertical Transportation System

Vertical transportation system equipment includes a quay crane and a yard crane.

Yard crane is classified into rail cranes and tire cranes, with their respective modes of operation being the main difference. Tire crane can realize transfer operation and its single-machine utilization rate is higher than that of rail crane. The rail crane has a fixed track with limited movement.

Quay crane is the core of container terminal operation and determines the productivity of the container terminal.

(2) Horizontal Transportation System

Horizontal transportation system equipment mainly includes manual driving, internal container trucks and AGV.

(3) Terminal Operating System

Terminal Operating System (TOS) works as a unified scheduler, realizing coordination and management between dispatching and monitoring centers with three on-site operation areas, namely, loading and unloading, container yard, and gate. Video surveillance systems and trunking calling systems are important cooperative tools.

Port Industry Trends

In November 2019, nine departments, including the Ministry of Transport and the National Development and Reform Commission, jointly issued *Guiding Opinions*

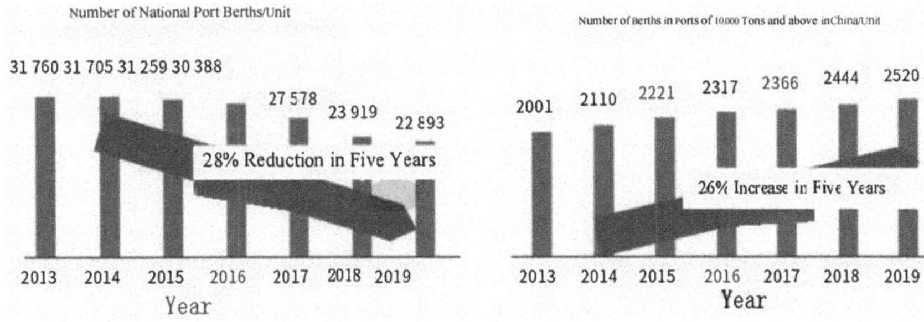

Fig. 6.23 Number of port berths and number of those of 10,000 tons

on the Construction of World-Class Ports, aiming at "By 2025, significant progress will be made in the construction of world-class ports, major breakthroughs will be achieved for major ports in green, intelligent and safe development, and significant improvement will be realized in the level of specialization and scale of regionally important ports and general ports." Therefore, the integration of resources, large-scale specialization, and efficiency improvement of ports have become the development trend of the port industry.

1. **Integrated in regional resources to avoid homogeneous competition**

The Ministry of Transport, in response to the current situation of overcapacity and homogeneous competition, will optimize the overall layout of ports and increase the macroeconomic control of ports. This will be followed by prudent and orderly development of new port areas and strict approval of shoreline use.

2. **Being large-scale and professional**

The terminal structure is upgraded and optimized, and the modern port system is formed. The number of port berths and 10,000-ton port berths is shown in Fig. 6.23. In 2019, there were 22,893 port berths nationwide in China, a 28% decline in five years since 2013; in 2019, there were 2,520 port berths of 10,000 tons and above nationwide in China, a 26% increase in five years since 2013.

3. **Improving operational efficiency and decreasing cargo dwell time in port**

Container loading and unloading operations reflect port efficiency, and excellent ports can reach 30 TEU/hour. This directly impacts the stopping time of goods.

Pain Points for the Port Industry

At present, global ports are facing such difficulties as harsh working environments, high labor intensity, many manual maloperations, difficulty in improving operational efficiency, and complicated network operation and maintenance. Global ports have become in demand to reduce costs and increase efficiency through automation transformation.

1. **Harsh operating environment**

The harsh operating environment of ports is usually manifested in three aspects of: job hazards, hard work and occupational diseases.

(1) Job hazards: The driver's room is 30 m from the ground, requiring workers to climb up a vertical ladder.
(2) Hard work: Workers are exposed to noise and vibration at the site, and they cannot take breaks on the ground for several hours. Eating or drinking is also prohibited.
(3) Occupational diseases: Due to long-term work with head down, workers are prone to developing conditions like cervical spondylosis. Generally, at the age of 45, they need to switch to another position.

2. **Manual maloperations and frequent accidents**

In commercial insurance claims at container terminals, claims of accidents caused by collision during operations of rail cranes, bridge cranes, stacking, and container truck crane arms, accounted for about 50%.

3. **Difficulty in improving the efficiency of manual loading and unloading of containers**

Manual loading and unloading efficiency is at 25 to 27 boxes per hour, while automated terminals can increase efficiency by over 30%.

4. **The port wireless communication system is a complex network that supports multiple standards and requires careful operation and maintenance**

Figure 6.24 shows that this system includes various communication technologies such as 5.8 GHz Wi-Fi, 2.4 GHz Wi-Fi, 1.4 GHz private network, and 400 MHz trunking, each playing a specific role in their respective fields. The 5.8 GHz Wi-Fi is utilized for backhauling video monitoring data, while the 2.4 GHz Wi-Fi is responsible for backhauling operational status information, fault information, wind speed, and other sensor data. The 1.4 GHz private network serves as a means of sending control commands from the driver to the trailer, and the 400 MHz trunking is employed for voice dispatching throughout the port's different areas.

6.2.2 Analysis of 5G Application Scenarios and Network Demand

Through research, discussion, and analysis of network requirements, three major smart port application scenarios have been selected that have potential demand for

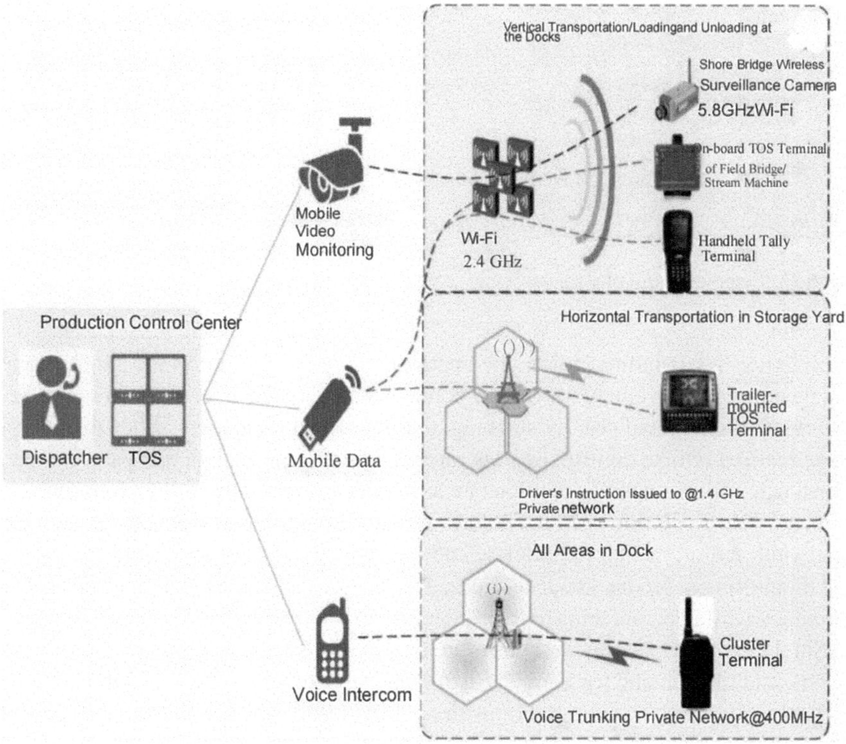

Fig. 6.24 Port wireless communication system

wireless communication, can be empowered by 5G in the future, and have typical representative significance.

(1) Remote control of gantry cranes.
(2) Optical Character Recognition (OCR) intelligent cargo handling.
(3) AGV and Intelligent Guided Vehicle (IGV) assisted driving.

Remote Control of Gantry Crane

In the container terminal, two kinds of gantry cranes, rail cranes and tire cranes, are widely used in port machinery and equipment. At present, tire cranes are mostly used in stock terminals and rail cranes are mostly used in new terminals. Tire cranes account for a relatively high proportion of stock terminals.

After the modification of the remote control, cameras, and PLCs are installed on the gantry cranes. The gantry crane remote control is shown in Fig. 6.25. The driver's operation changes from an on-site one to a remote one in the operation room. After the operator obtains the dispatching task issued by TOS, according to the HD video and audio data backhauled from the gantry crane at the container yard in real-time,

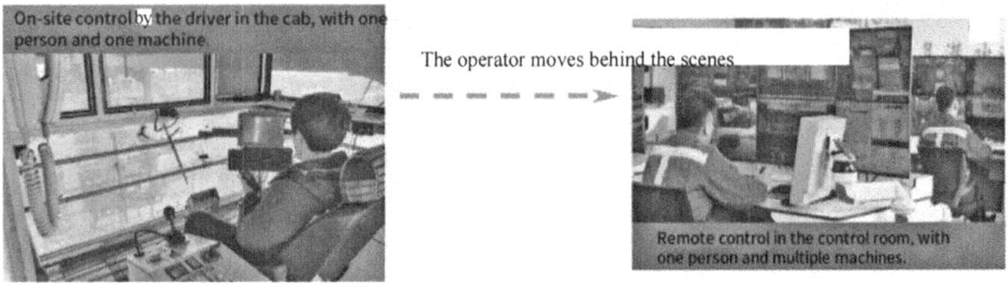

Fig. 6.25 Remote control of a gantry crane

through joystick communication, the operator can remotely control the gantry crane and its gripper's movement for grasping, releasing, and other operations in real-time, achieving efficient and orderly stacking and transfer of containers. After the gantry crane realizes remote control, one operator can control more than one gantry crane, which can improve operation efficiency and reduce safety risks at the same time.

The wireless solution needs to ensure sufficient continuous coverage and meet the bandwidth requirements, time delay, and packet reliability of PLC control service and monitoring video backhaul services. This ensures enhanced container operation efficiency while guaranteeing safe production.

Single gantry crane remote control generally needs to return 5 to 18 monitoring videos, and the demand for bandwidth at 1080P resolution is 15 to 54 Mbps.

The analysis of the time delay in the gantry crane remote control is shown in Fig. 6.26. PLC send cycle is 16 ms, but if PLC sends and receives network time delay within 20 ms, it can guarantee good operation of the service. If in more than 3 cycles, i.e. 48 ms, no data packet is received at the receiving end. Then, the receiving end will judge that the link is interrupted, which will cause the crane to stop immediately.

The 5G remote control scheme of the gantry crane is shown in Fig. 6.27, and the scheme is described as follows:

(1) Currently, the yard crane and other harbor equipment use L2/L3 protocol mixed transmission, which requires a new industrial gateway or AR router at the front end of CPE to establish a GRE tunnel.
(2) To reduce the control signal time delay, the monitoring video and control signals are transmitted separately through different tunnels (CPE1, CPE2) in the networking scheme.
(3) The yard crane remote control delay requirement is high (within 20 ms). Currently, the 5G yard crane remote control scheme reduces the end-to-end delay of control signals in the park by deploying MEC.

OCR Intelligent Cargo Handling

OCR plays two roles: one is text recognition and the other is shape recognition. The principle of text recognition is that electronic devices use optical methods to convert the text in an object into an image file and then convert the text in the image into text

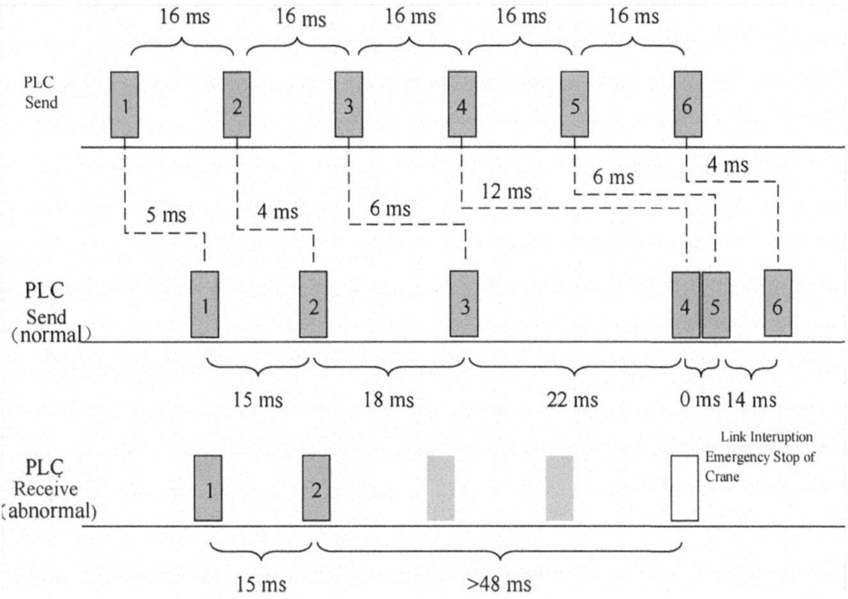

Fig. 6.26 Delay analysis of eemote gantry control

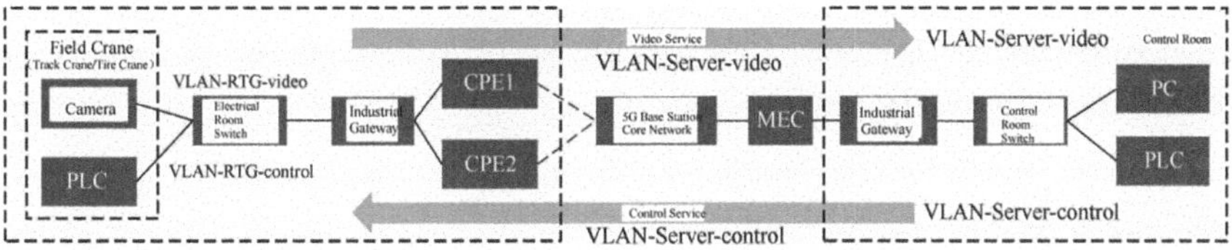

Fig. 6.27 5G remote control solution for gantry cranes

format through software. The principle of shape recognition is that the electronic device recognizes the shape of an object by detecting its light and darkness and then translates the shape into computer text.

OCR technology can be widely used in loading and unloading, stowage, cargo handling, disability inspection, container lifting and customs clearance. It is used to identify container box number, loading and unloading and lifting status, with or without lead seal, the degree of box residual damage, etc.; the collection, sending and statistics of these data directly affect the flow efficiency of the whole big data of logistics. Shipping enterprises and ports are very focused on the collection and analysis of container data.

At present, OCR cargo handling has many pain points. For example, each quay crane is equipped with ten cameras, and these videos are backhauled with Wi-Fi.

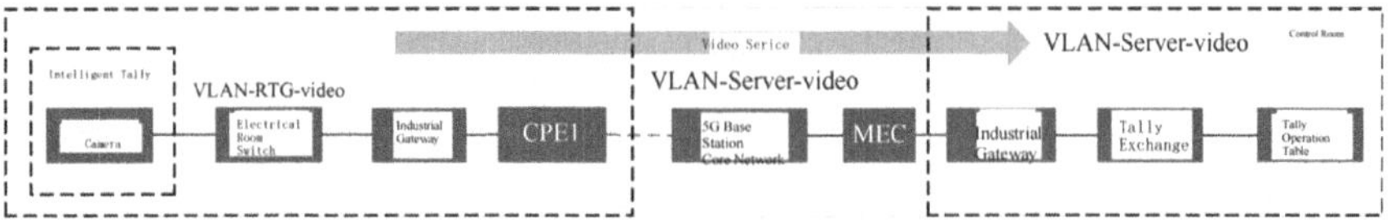

Fig. 6.28 Networking scheme of 5G intelligent cargo tally

However, the stability is poor, and the background cannot access the video in real-time; if the background can not access the data, staff members need to climb up to the quay crane dozens of meters high to copy video data.

The 5G+ AI intelligent cargo handling application can accurately control the shooting of container box number pictures. Then, high-definition images and video data are quickly and real-time backhauled to the AI system in the cloud through a 5G network. The 5G+ AI intelligent cargo handling system, with an intelligent AI system, automatically identifies and writes off box numbers, box losses, trailer numbers, and other massive repetitive labor work, realizing automatic collection of cargo handling operation information and improving the accuracy and efficiency of the operation. 5G+ AI intelligent cargo handling requires a 5G network uplink bandwidth rate of 40–60 Mbps and a time delay of less than 100 ms.

5G Cargo Handling Networking Solution is shown in Fig. 6.28.

The smart cargo handling scenario controls the camera for capturing by synchronizing PLC data, and the video is transmitted in the same way as PLC in the networking scheme. Therefore, only 1 CPE needs to be deployed to establish a GRE tunnel for transmission.

AGV and IGV Assisting Driving

AGV vehicles use a wireless network for centralized control and scheduling, enabling fully automated operation in the channel with a magnetic nail positioning system installed throughout the port. AGV self-driving control is automatically controlled by the network scheduling center, aiming to prevent safety accidents resulting from emergency stops and collisions between vehicles. This places strict requirements on the wireless control network for high reliability and low latency but also guarantees accurate and reliable AGV control and dispatching instructions.

AGV automatic guided vehicle relies on magnetic pegs deployed on the ground, an inertial guidance system on the vehicle body, and a wheel encoder to achieve autonomous travel at a speed of approximately eight m/s. TOS perceives the current position and status of the box through the device management system and issues control tasks to AGV based on the box's final destination. AGV automatically travels to the designated location according to the set route, where it can load and unload goods automatically or manually.

Currently, existing AGVs use 5.8 GHz Wi-Fi wireless technology for vehicle control, which has problems such as low communication reliability, susceptibility to

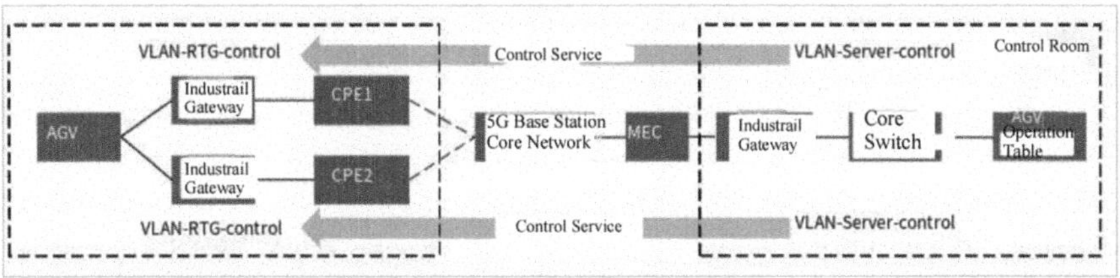

Fig. 6.29 Networking plan of 5G AGV intelligent scheduling

wireless interference, and insufficient network depth coverage. 5G can help realize AGV trunking management, automated command issuance, and status uploading of AGVs, as well as guarantee service stability and delay reliability. The AGV standalone demand for 5G network upstream and downstream bandwidth rate not less than 500 kbps, and maximum delay not more than 200 ms.

5G AGV intelligent dispatch networking solution is shown in Fig. 6.29. AGVs adopt a redundant architecture driven by dual motors. Therefore, although there are two CPEs in the actual network, both transmit control signals, and they have high delay requirements (within 50 ms).

IGV will be an important development direction for port horizontal transportation in the future. Port IGV unmanned container truck is transformed from commercial container truck chassis, integrated lidar, millimeter wave radar, camera and so on. It also uses high-precision maps and positioning, and the control center monitors IGV's position, attitude, power, and load and delivers vehicle planning information to realize autonomous driving. Unlike AGV, IGV does not require embedded magnetic nails. It can be applied to both new ports and storage ports at low cost.

IGV applications need to support vehicle management platform scheduling and remote takeover under abnormal conditions. IGV driving requires high communication delay, and video backhaul is required when multiple IGVs are taking over remotely. The delay and bandwidth of traditional wireless networks cannot meet the requirements of autonomous driving, so a 5G network is needed. Currently, IGV requires a response time of less than 50 ms for 5G network downlink control commands.

6.2.3 Scenario-Based Solution

Network Architecture

The port environment is special. For example, in the remote control scenario of the gantry crane, CPE is installed on the top of the gantry crane, so the coverage is good, but there will be more overlapping coverage of the cell and frequent terminal switching. As for road covering, it is easy to form weak road covering due to the shielding effect of the container box. Therefore, wireless signal coverage planning and optimization face great challenges in the port scenario.

The first wave of 5G smart port application is mainly based on the remote control of the gantry crane, which belongs to the scenario of HD video backhaul + PLC remote control. Port 5G network planning and design is strongly related to service application scenarios and distribution. The traditional planning thinking that takes base stations as anchor points needs to be transformed into wireless network planning and design that takes service scenarios and terminal deployment as reference points. Three major challenges need to be addressed.

(1) Simultaneous use of multiple gantry cranes requires a high uplink capacity for a 5G network, so multi-band networking needs to be considered.
(2) Overhead signal line-of-sight transmission, cell interference and switching are difficult to control.
(3) The signal coverage on the road will be easily blocked by the container, thus leading to weak coverage.

The current ports have all built data centers, which can meet the current and future requirements in the next few years, with less demand for public cloud. For the demand with AI, it can be deployed on its own private cloud. Port 5G private network overall architecture is shown in Fig. 6.30.

The port 5G service scenario is mainly realized by focusing on the construction of a 5G wireless network, terminals, transmission, and core network (local MEC deployment). The network and system deployed in the port area mainly have the following two parts:

(1) Network equipment, including 5G base stations, transmission and MEC, etc.
(2) Terminals and routing equipment, including 5G CPE and Access Router (AR) for establishing tunnels.

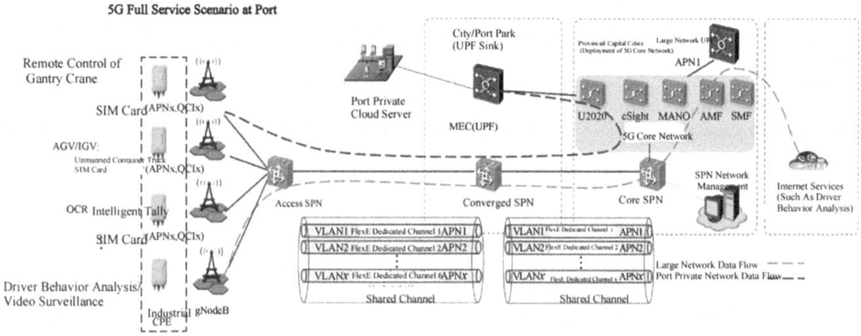

Fig. 6.30 Overall architecture of the 5G port private network

Networking Scheme

The port scenario-based private network involves three sub-networks: wireless network, transport network and core network.

1. **Wireless Networking Scheme**
   (1) Site Deployment Principle

The base station deployment principle is shown in Fig. 6.31. Mutual interference should be avoided in inter-frequency station planning. If there are lighthouse sites in the container yard, back-to-back can be referred to reduce the same frequency of interference. When choosing antennas, AAU takes into account the scenarios of yard and shore cranes, and tries to use wide horizontal and vertical wave flap antennas. Overlapping coverage needs to be considered when selecting station height, which should not be too high or too low and can be deployed on the upper level of a nearby lighthouse to guarantee height and line-of-sight coverage.

2. Terminal Deployment Principles
   (1) The terminal should not be too close to the base station.
   When selecting a CPE point, it is essential to consider coverage and uplink power saturation, and it cannot be too close to the base station. It is verified by experimental tests that if the level is high (e.g., RSRP is greater than -68 dBm), the saturated uplink power will decrease the average uplink rate. Therefore, it is necessary to consider adjusting the terminal location, ensuring, for example, that the terminal is at least 30 m away from the base station.
   (2) The terminal is assigned to a specific cell in order to maintain stability and avoid frequent switching.
   The container yard uplink needs fixed points, so the location of toB terminal must also be relatively stable. Through frequency locking, locking PCI, and other ways to maintain access to a fixed cell, the impact of cell capacity can be reduced due to terminal signal fluctuations. This ensures the stability of cell capacity and meets the basic service rate requirements

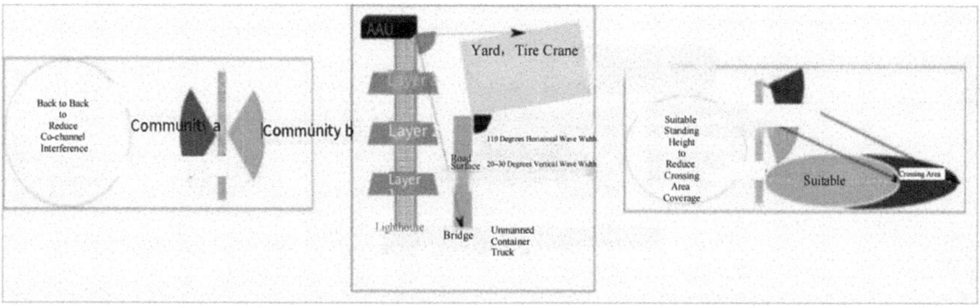

Fig. 6.31 Deployment principles for base stations

of each terminal. For terminals with a large mobile range, cell merging network could be considered to be used to avoid excessive delay caused by cell switching.

(3) CPE points should be as few as possible.

For the gantry crane and self-driving container truck, double CPE should be used to meet the demand. Deployment could be performed as required.

For video surveillance services, the number of cameras converging on a single CPE should be limited to 10 to 20 so as to reduce uplink interference level from neighboring cells.

3. A design scheme with high reliability and low delay should be adopted

The design solution of high reliability and low latency depends on the network deployment model. End-to-end latency evaluation mainly covers two scenarios: UE and cloud service communication scenarios, as shown in Fig. 6.32. This is a typical remote data acquisition and remote control application. The second is the UE and UE communication scenario, as shown in Fig. 6.33. This is a typical application of near-end data acquisition, mobile control, and near-end logic control.

Experimental results show that the mean delay and delay reliability of remote, medium, and near deployment locations are very different. Especially at remote deployment, the deterioration is more obvious. Therefore, for the low-latency solution, medium- and near-point deployment is required.

4. Sub-scene Wireless Solution Description

(1) Container yard gantry crane network solution

AAU should be deployed on the reinforced light pole or platform in the yard, and the locations of AAU and CPE should be as close to each

Fig. 6.32 Communication scenario between UE and cloud service

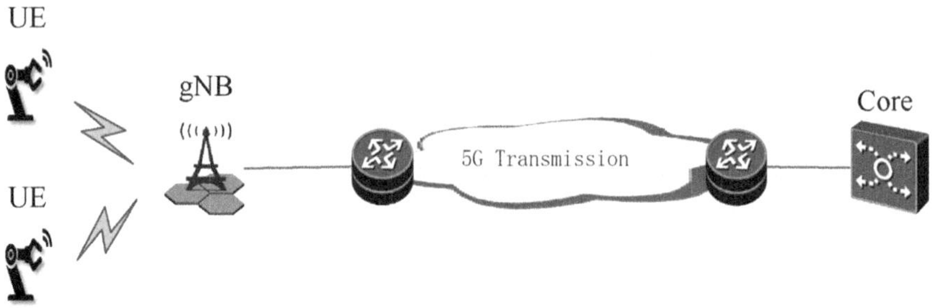

Fig. 6.33 UE and UE communication scenario

other as possible so that the line-of-sight coverage can be fully guaranteed. Therefore, for a single yard, from the coverage point of view, it is sufficient to use one AAU to cover the gantry crane.

According to the single-cell AAU capacity caliber, if the demand capacity is lower than the single-cell capacity caliber, it is enough to use a single AAU coverage when multiple gantry cranes are concurrent; if there are more concurrent gantry cranes in the yard and the single cell capacity is lower than the demand capacity, then the opening of a large uplink feature can be considered at this time.

If the gantry crane's service cell uplink interference is large due to the surrounding neighborhood coverage, further inter-frequency deployment can be considered to ensure the low latency and high-reliability requirements of remote control.

(2) Yard AGV network solution

The height of AGVs in the yard is about 3 m, and the container has a metal structure, which has a large wear loss and the number of layers changes in real-time, which directly affects the coverage. In order to ensure the coverage of AGVs in the yard, two AAUs can be considered for deployment in the 400 m × 400 m yard, and the station needs to be built by relying on the high light pole position.

When the number of AGVs deployed in the yard is small, two AAUs can be deployed to cover the whole yard; when the number of AGVs is large, the capacity of two AAUs cannot meet the requirements, then the large uplink feature can be considered to activate.

(3) OCR intelligent cargo handling network solution

Intelligent cargo handling has no demand for time delay, and AAU cells can be deployed according to the actual installation environment to cover the whole berth area.

If the same cell is used to cover both unmanned container trucks and quay cranes, it is necessary to consider whether the capacity can meet the claim. When the capacity is insufficient to meet both services, the large uplink feature must be activated.

5. **Networking Scheme of Transport Network**

The bearer network deployment strategy requires careful consideration of 5G port base station planning and user development needs of that area. Coordinated planning is also necessary for the end-to-end transmission network construction. To ensure reliable operation of the 5G network, the 5G base stations in a park are connected to the transmission equipment with C-RAN/D-RAN, and the front transmission scheme adopts the form of fiber direct connection/passive wavelength division.

The wireless side Baseband Unit (BBU) is docked with the undertaking access equipment, which can be adapted to the user's demand. In particular, port enterprises involved in remote control, safety production, emergency communication, and other special scenarios need a network priority guarantee. This can be achieved by slicing, selecting, and prioritizing customized communication cards. Different transmission pipes can be used to bind DNNS to meet the applications of different business scenarios.

The transport network in operation can adopt different transmission technologies. If SPN technology is used, the equipment is a Packet Transport Network (PTN) product; if IP Radio Access Network (IPRAN) technology is used, the equipment is an Access Transport Network (ATN) product. The port industry needs to choose the corresponding transmission equipment according to the operator for industrial park transmission networking.

According to the size of the park and the distribution of base stations, a choice could be made between SPN/IPRAN one-layer networking and two-layer networking. Because BBU is CRAN networking, one-layer networking is recommended; if the BBU is a DRAN scenario and spans multiple buildings in different locations or if the park is relatively large, two-layer networking is recommended. See Table 6.1 for specific recommendations.

3. **Networking Scheme of Core Network**

Operators generally build network cloud data centers in regional centers, plan 5G SA industry network slices unifiedly for vertical industries, and deploy core network control plane elements with virtualization.

5G SA industry network slices can provide flexible and customizable solutions for deploying user plane network elements tailored to the needs of industry users, taking into account port industry characteristics and business models. Local convergence of service traffic and data streaming is achieved through the on-demand sinking of a user plane. This sinking, as a prerequisite for 5G edge computing, ensures rich and efficient network performance in the port industry. Edge computing provides cloud services for edge applications running at the edge of the mobile network and running specific tasks by providing infrastructure, such as computing and storage close to the data source or users, based on IT architecture and cloud computing capabilities.

As the user plane of edge computing, UPF realizes the data forwarding function. The edge computing platform system provides the running environment for edge applications and realizes the management of edge applications. In 5G network architecture, the edge UPF can realize the service diversion inside and outside the port and can be used as the anchor UPF of the private network to complete the local

Table 6.1 Transport networking design

BBU position	Number of BBU	Networking level	
Concentric machine room	X < 8	Layer one	
	8 < X < 12	Layer one	
	X > 12	Layer two	Access layer
			Convergence layer
Cross machine room	Unlimited	Layer two	Access layer
			Convergence layer

traffic unloading. The local traffic diversion within the port needs to cover all the base stations in the port.

In the context of continuous improvement of industry automation and informatization, large ports have put forward more stringent requirements for data security. Meanwhile, typical applications in the port industry, such as gantry crane remote control, AGV/IGV, and intelligent cargo handling, put forward demands of data protection, very low latency, transmission saving, and open capability for edge computing.

Equipment Selection

5GtoB stations need to be reasonably selected in the deployment scenario by combining with the deployment limitations of the coverage scenario. The main equipment capabilities of 5GtoB base stations are shown in Table 6.2.

The port is a wide area outdoor scene, so the wireless environment is relatively open and the beam is easy to expand. Therefore, AAU or pole stations can be considered as the main deployment options, and the equipment selection principles are as follows:

(1) The container wear loss in container yards is relatively large where there are pylons, lighthouses or reinforced lighthouse-type yards. The distance required to cover is far, and the distribution of services is both high and low, so it is recommended that AAU be used.

Table 6.2 5G base station equipment capabilities

Porduct form	Technical specification	Multiple antennas and pairing capabilities	Features and applicable scenarios
AAU	Number of antennas: 64T64R Power: 320W Volume and weight: large volume and high load-bearing requirements	Support multi-user, multiple-input, multiple-output, MU-MIMO pairing Support uplink 4-stream pairing, and have large cell peak capacity The strongest anti-interference ability	Large installation space and high load-bearing requirements Coverage distance is far horizontally and wide vertically, which is suitable for relatively empty scenes outdoor and indoor
Pole station	Number of antennas: 8T8RPower: power is less than AAU Volume and weight: the volume is smaller than AAU, with certain bearing requirements	MU-MIMO poor pairing ability	The installation space is small, which has certain bearing requirements Coverage distance is far horizontally

(2) In key areas that need to fill in the blindness, or only a partial area (100–200 m) needs covering, pole stations can be deployed on light poles with limited load-bearing capacity.

6.2.4 Application Cases

Zhejiang province houses China's largest and extra-largest deep-water berths at one of its ports. There are 19 port areas with more than 620 production berths in the port.

In 2018, the cargo throughput exceeded 1 billion tons, solidifying its position as a super port of over 1 billion tons. Furthermore, it ranked first among global ports in ten consecutive rankings. The annual container throughput exceeded 26 million TEUs.

Project Background

The following describes the current network status of the port:

(1) Fiber optic is dominant: the effective coverage of Wi-Fi is small (about 30 m). With many typhoons along the coast, AP is unstable, so it is difficult to deploy outdoors on a large scale. As a result, the network transmission in the port area is dominated by fiber optics.
(2) Layer 2 networking: The industrial control PLC system on the gantry crane in the port area is a Layer 2 protocol, and PLC signals and camera monitoring traffic are interoperable with the control room through wired access to the enterprise L2 network.
(3) Network isolation: The production network, security network, and office network are isolated from each other with an independent chimney system.

Pain Points Analysis

The following are the current pain points analyzed:

(1) Each crane needs to deploy PLC control and monitoring video, and cable deployment costs about 2 million RMB. Moreover, the traveling gantry crane requires replacement every two years due to its need to move flexibly and quickly wear out.
(2) Each crane requires operators to work in shifts 24 h a day, with high manpower costs for multiple operators.
(3) Working conditions for crane operators are difficult.
(4) Wi-Fi APs have weak lightning resistance, high failure rate and poor network performance.
(5) Bridge cranes block each other, making it easy to form a network blind spot.

(6) Video backhaul currently uses a wired connection, which has high cable costs and poor scalability.

Solution

The gantry crane remote control scenario was selected for transformation and verification of remote control performance. Gantry crane remote control mainly consists of two business streams: control stream and video stream. Control streams have the requirements of low latency and high reliability, while video streams needs to guarantee their certain capacity and maintain the smoothness of video. A total of 18 1080P HD cameras are deployed on each gantry crane, and the video is backhauled to the central control room to assist the staff in realizing remote control such as precise movement and container grabbing through the PLC remote control lever.

In a port area, which is the only all-electric gantry crane terminal, there are 28 bridge cranes and 91 tire cranes. It is planned to make a wireless remote control transformation of 6 tire cranes in Pile 6 and Pile 7.

1. **Overall Solution**

For this port, the overall solution is as follows:

(1) Key parameters of gantry crane remote control: PLC transmission period is 16 ms, Watchdog delay is 224 ms, and video uplink bandwidth is 30 Mbps/station.
(2) Wireless networking scheme: The port has a high demand for uplink capacity, so AAU macro stations are used for continuous coverage. There are 13 cells which can meet the uplink capacity. To avoid strong uplink interference in neighboring areas caused by toB large uplink service, inter-frequency networking is used to mitigate interference.
(3) Low latency and high reliability: The gantry crane control flow has high requirements of low latency and high reliability. In order to improve the reliability performance, the router is used for the AR dual-transmission selective reception mechanism.
(4) 5QI planning: Dual CPE is used to carry two types of services, with control stream 5QI being 6 and video stream 5QI being 8.
(5) CPE terminal selection: CPE selection takes into account the high requirements of port business delay stability, preferably matching the industrial grade CPE with a + 24 V unit POE power supply module.
(6) CPE lock cell: For the same container yard or business that occurs in the same 5G coverage cell, it is recommended to lock the cell and frequency in CPE to avoid affection on the service due to frequent switching of CPE between 5G cells.

Table 6.3 List of equipment for the solution of gantry crane remote control

Domain	Name	Quantity
Terminal	5G CPE	2
Wireless network	AAU	8
Wireless network	BBU	4
Bearer network	SPN	3
Core network	MEC	6
Core network	5GC	1
Data communication	Switch	2
Data communication	Switch	1
Data communication	Fire wall	2
Data communication	VXLAN	2
Third-party package	PLC control system	1

2. **Measured Index**

The measured round-trip time (RTT) is about 10.6 ms, with an average delay jitter of 1.7 ms, which meets the delay requirement of being less than 18 ms for PLC remote control.

The average uplink rate of 161 Mbps can meet the uplink transmission requirements of a single gantry crane camera, ensuring that the monitoring video is clear and lag-free and can also accurately assist the remote operation of the gantry crane.

3. **List of End-to-End Equipment for the Solution**

Gantry crane remote control transformation involves a wireless network, transport network, core network, and third-party supporting software. A list of equipment for the solution of the gantry crane remote control is shown in Table 6.3.

Business Value

Gantry crane remote control business can save on the cost of cables, equipment, manpower, and other aspects, as well as improve work efficiency.

1. **Cable Saving**

Each gantry crane wired fiber optic cable + cable needs to be replaced every two years, and the total replacement cost is about 2 million RMB yuan, which is about 1 million RMB yuan/unit/year.

2. **Equipment Saving**

After wireless replaces wired, multiple roving cranes can work in parallel, improving working efficiency by 50%. The purchase quantity of gantry cranes is reduced. According to the price of 5 million RMB yuan per unit, in a 10-year depreciation calculation, a saving of 250,000 RMB yuan/unit/year can be achieved.

3. **Manpower Saving**

Traditional control requires three persons to rotate for each unit. After remote control is adopted, one person can control 3 to 6 gantry cranes at the same time, which significantly reduces the labor cost. Based on the calculation of simultaneous control of 4 units and a labor cost of 100,000 RMB yuan/person/year, a saving of 275,000 RMB yuan/unit/year can be achieved.

6.2.5 Summary (of 6.2)

Relying on 5G + cloud + AI technology, by 2025, China will make significant progress in the construction of world-class ports, achieve major breakthroughs in the green, intelligent, and safe development of major ports, and realize significant improvement in specialization and scale of regionally important ports and general ports. By 2035, the overall level of national ports will jump up, with generally planned world-class levels for main ports. Several hub ports will be built into world-class ports, leading the green and intelligent development of global ports.

This section first introduces trends and business pain points of the port industry and then explains in detail the application scenarios of 5G in ports, the overall solution of 5G port private network, and industry practice cases.

By learning this section, readers are expected to have a certain understanding of 5G solutions in the port industry and know the progress and application scenarios of the industry.

6.3 5G+ Coal Industry Applications and Solutions

China has now developed a 5G-based system for mining and obtained initial application in underground coal mines, showing good results and broad application prospects. The integration of 5G technology in coal mining will definitely change the traditional production organization and workforce structure of the industry and break the constraints of safety and environment on coal development and utilization.

The following describes how 5G enables the coal industry from coal industry insight, 5G application scenarios and network demand analysis, scenario-based solutions and application cases.

6.3.1 *Coal Industry Insights*

Introduction to the Coal Industry

The coal industry refers to an industry that focuses on the mining and transportation of coal resources, and China is facing a pattern of "Rich in coal, poor in oil, and little in gas". Because of the relative abundance of coal reserves, it is one of the main sources of national energy and one of the important pillars of China's economy.

1. Coal Consumption

China is a large producer and consumer of coal, and coal consumption is spread all over the power generation, iron and steel, building materials, and other industries, as shown in Fig. 6.34, which is difficult to replace in the short term.

2. China's Coal Distribution

In 2019, there were seven coal companies with raw coal production of more than 100 million tons in China, namely National Energy Group, China Coal Energy Group, Datong Coal Mine Group, Shaanxi Coal Group, Yankuang Group, Shandong Energy Group, and Shanxi Coking Coal Group. Together, these companies accounted for approximately 1.46 billion tons or 38% of the country's total coal production. This is illustrated in Fig. 6.35.

In addition to coal enterprises with over 100 million tons, China's major coal enterprises include Jinneng Group, Jizhong Energy Group, Yangquan Coal, Shanxi Lu'an Mining Group, Henan Energy and Chemical Group, China Huaneng Group, Huainan Mining, Shanxi Jincheng Anthracite Coal Group, Inner Mongolia Yitai Group, Huadian Coal Group, Inner Mongolia Huolinhe Opencast Mining, Inner Mongolia Huineng Coal and Power Group and Heilongjiang Long Coal Holding Group, etc.

The distribution of coal resources in China is extremely unbalanced, with richer coal resources in the north and west and fewer coal resources in the east and south. Thus, the basic pattern of "transportation of coal from north to south" and "transfer coal from west to east" has been formed.

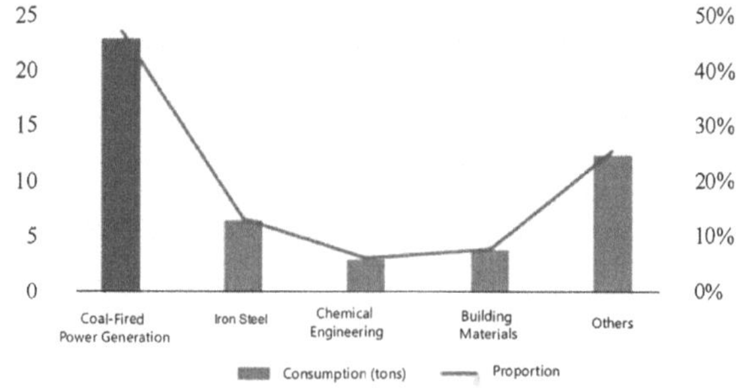

Fig. 6.34 Situation of coal consumption industries

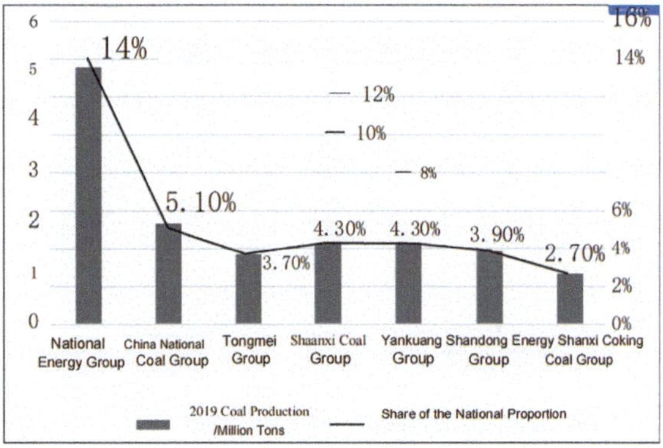

Fig. 6.35 Seven major coal enterprises in China

3. Coal Mine Operation Process

Coal mines are generally divided into two categories: open-pit and shaft coal. Open-pit coal mining involves removing the overburden and surrounding rock that cover the upper part of the ore body and its surroundings, transporting the waste rock to the dumping site, and extracting the ore directly from the open-pit mine. Open-pit mining is preferred over underground mining when the ore body is shallow or exposed at the surface. The work of removing the overlying rock and soil is called stripping. The ratio of overburden removed to the amount of ore extracted is called the stripping ratio. Open-pit mines with a high stripping ratio have high operating costs and should consider switching to underground mining methods. Open-pit coal mines are further divided into mountain surface coal mine (Fig. 6.36) and sunken open-pit coal mine (Fig. 6.37).

A shaft mine is a type of underground coal mine where the mining face is located underground and can have multiple mining faces. It is connected to the surface by vertical or inclined shafts, which serve as a transportation route for personnel and ore. A diagram of a shaft mine is shown in Fig. 6.38, and a real-life image of a shaft mine is shown in Fig. 6.39.

The basic business process of coal mining is shown in Fig. 6.40, which mainly includes geological exploration, open-pit development or underground development, open-pit or underground mining, coarse processing, transportation, processing according to different functions and comprehensive utilization, etc.

(1) Geological exploration: It takes 2 to 5 years to explore an ore deposit.
(2) Exploration: It takes 3 to 7 years to develop a large-scale mine and 2 to 3 years to develop a medium-sized mine.
(3) Mining: The production cycle takes 30 to 50 years.

The process of open-pit mining mainly includes piercing, blasting, shovel loading, transportation, crushing, and waste rock disposal, as shown in Fig. 6.41.

Fig. 6.36 Real-life image of a mountain surface coal mine

Fig. 6.37 Real-life image of a sunken open-pit coal mine

(1) Piercing

Piercing is the first process of open-pit mining to prepare for blasting by providing a hole for explosives. The equipment used for perforation mainly includes drilling machines, jumbos, rotary drilling rigs, and down-the-hole drills.

(2) Blasting

Blasting refers to filling the hole with explosives and detonating them, aiming to break the ore rock to a certain level to facilitate subsequent shovel loading. The hardness

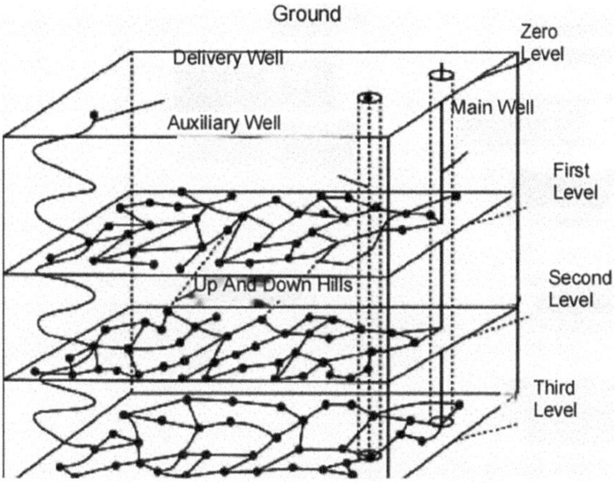

Fig. 6.38 Schematic diagram of a shaft coal mine

Fig. 6.39 Real-life image of a shaft coal mine

and weathering of the ore determine whether blasting is required. Typically, ore with higher hardness requires blasting, while ore with lower hardness can be excavated directly with an excavator and does not require blasting.

(3) Shovel Loading

Shovel loading refers to the use of loading machinery to excavate ore from the blasted piles and load it into the compartments of transport machinery, with the primary equipment being excavators.

(4) Transportation

Transportation means transporting the extracted ore to the ore processing plant and the rocks to the dumping site, primarily using mining trucks.

(5) Crushing

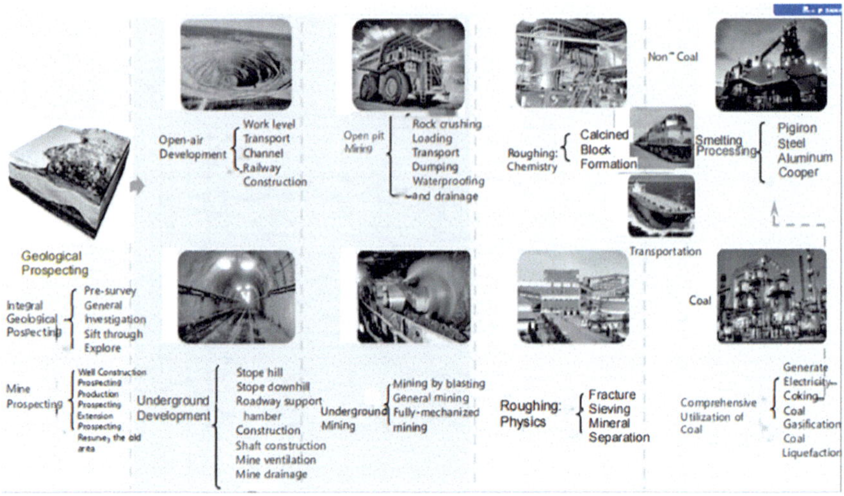

Fig. 6.40 Fundamental operational process of coal mining

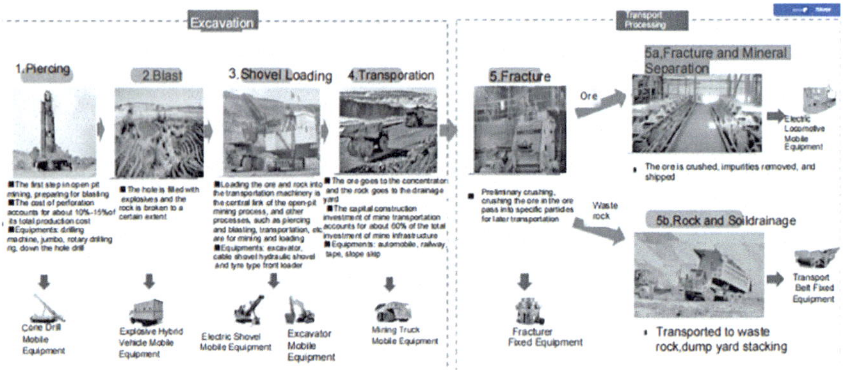

Fig. 6.41 Mining process of open-pit mine

The ore needs to be initially crushed to a specific particle size, and impurities need to be removed for later transportation. The typical crushing equipment is a crusher.

The operational interface of a shaft coal mine is shown in Fig. 6.42.

(1) Main tunnel: The tunnel is typically ten to hundreds of kilometers in length and can usually accommodate two vehicles.
(2) Pipelines: Pipelines are laid along the main tunnel, including electrical cables, fiber optics, gas extraction pipes, etc.
(3) Fully mechanized mining face: Coal mining working face, usually 2 to 4 m in height (depending on the thickness of the coal seam), 300 m in width, and more than 2 km in length.

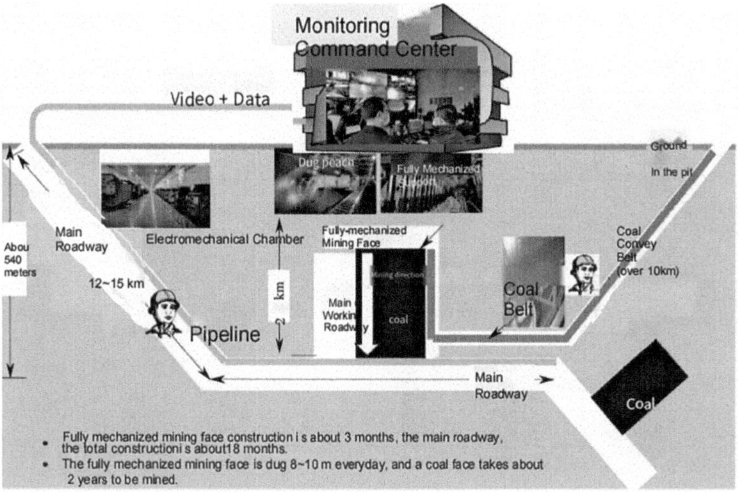

Fig. 6.42 Operational interface of shaft coal mining

(4) Main working tunnel: Underground production system monitoring center and maintenance tunnel.
(5) Coal conveyor belt: It transports coal to the surface coal washing center in real-time.
(6) Electromechanical chamber: To store critical electrical equipment, etc.
(7) Ground monitoring and command center: Remote monitoring and control.

The business process of shaft coal mine mainly includes ore falling/mining, transportation, and ore processing, as shown in Fig. 6.43.

(1) Ore falling/mining: Typically, comprehensive mining equipment, continuous mining equipment, and blasting methods are used to cut the ore into small pieces and allow it to fall automatically onto the transportation platform below.
(2) Transportation: After the ore falls, it is usually automatically transported out of the shaft by a belt conveyor system.
(3) Ore Processing: The ore is transported by the belt conveyor system to the ore processing center, where it undergoes washing and sorting processes to remove impurities and is then transported out based on its quality.

4. Introduction of Coal Mine Equipment

The common equipment in the shaft coal mine includes boring machines, mine supports, coal mining machines, belt conveyors and, shuttle cars, etc.

A boring machine is a machine used to excavate tunnels on flat ground, as shown in Fig. 6.44, which mainly consists of walking mechanisms, working mechanisms, loading mechanisms, and transfer mechanisms. As the walking mechanism advances, the cutting head in the working mechanism continuously breaks the rock and transports the crushed rock away.

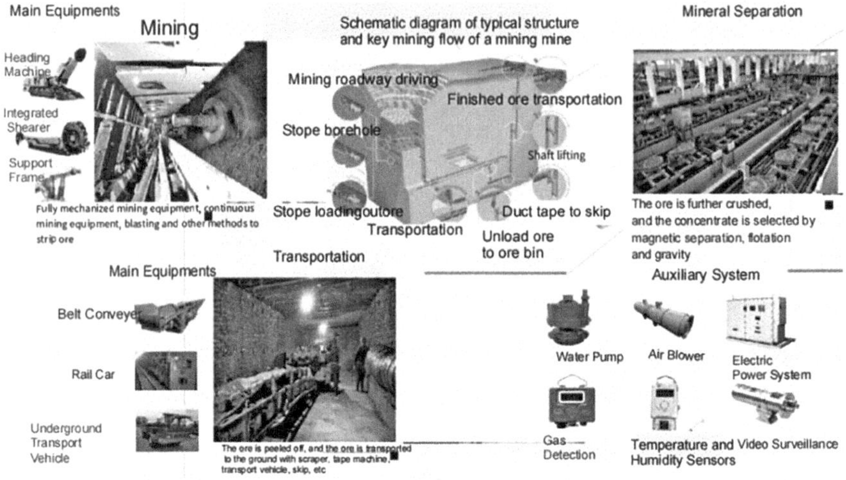

Fig. 6.43 Mining process of shaft coal mining

Fig. 6.44 Boring machine

Mine support refers to the equipment installed in the mine tunnel to support the surrounding rock, as Fig. 6.45 shows. Its role is to control the movement of the surrounding rock and prevent the rock from falling to protect the tunnel's safety.

A coal mining machine is a large and complex system integrating mechanical, electrical, and hydraulic components, generally consisting of cutting, loading, traction, electric motor, operating control system, and auxiliary devices, as shown in Fig. 6.46. On the coal mining face, the coal mining machine dislodges the coal from the coal seam (coal cutting) and loads it onto the working face conveyor (coal loading). The coal mining machine travels at a specified traction speed (traction) to ensure that the cutting and loading processes can proceed continuously.

Fig. 6.45 Mine support

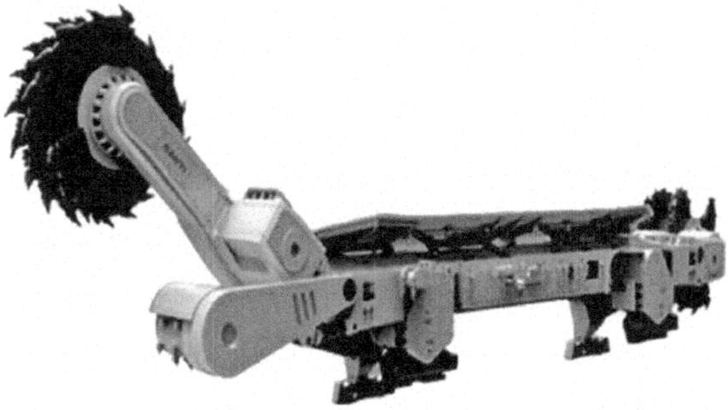

Fig. 6.46 Coal mining machine

A belt conveyor is a kind of underground conveyor machinery used in coal mines, employing the endless motion of a belt to transport materials, as shown in Fig. 6.47. The materials are continuously delivered to the conveyor belt and move along with the conveyor belt, thus realizing the transportation of the material.

Fig. 6.47 Belt conveyor

Fig. 6.48 Rubber-tired vehicle

The rubber-tired vehicle is a special transportation equipment for underground equipment and material transportation, as shown in Fig. 6.48. It is easy to operate and is an important transportation tool for underground vehicles.

The common equipment in open-pit coal mines mainly includes excavators and mining trucks.

Common excavator structures include a power unit, working device, swing mechanism, control system, transmission system, traction system, and auxiliary facilities, as shown in Fig. 6.49. They are mainly responsible for various excavation operations in mines.

A mining truck is a heavy-duty dump truck used in open-pit mines for rock and ore stripping and transportation tasks, as shown in Fig. 6.50.

Fig. 6.49 Excavator

Fig. 6.50 Mining truck

Coal Industry Trends

Coal has always been the primary source of energy in China, and it is mainly used for electricity generation, heating, and industrial purposes. However, as China undergoes economic transformation and places great emphasis on environmental protection, the extensive production method in the coal industry is no longer sustainable. These methods not only cause a significant waste of human and material resources but also lead to serious damage to the ecological environment. Therefore, promoting the transition from old to new forms of energy and upgrading the energy consumption structure have become an inevitable trend in the coal industry. Specific measures include merging small coal enterprises to achieve centralized production capacity, improving energy efficiency and emission reduction to promote cleaner coal resources, accelerating the integration of technology with the coal industry to facilitate the intelligence of coal mines.

In 2020, eight government departments, including the Ministry of Industry and Information Technology, the Ministry of Science and Technology, and the National Development and Reform Commission, issued *Guiding Opinions on Accelerating the Development of Intelligent Coal Mines.* These opinions call for expediting the structural reform of the supply side of the coal industry, promoting the integration of intelligent technologies with the coal industry, and enhancing the level of intelligence in coal mines.

The three-phased goals for the development of intelligent coal mines are shown in Fig. 6.51. By 2021, multiple types and different models of intelligent coal mines will be built. By 2025, large-scale and high-risk coal mines will basically achieve automation. By 2035, various types of coal mines will largely realize intelligence, constructing a coal mine intelligent system with multiple industry chains and integrated systems, ultimately achieving unmanned coal mine production.

Pain Points in the Coal Industry

The traditional coal industry faces the problem of harsh working conditions and a high risk of personnel safety. Recruitment difficulties are common in the industry,

Fig. 6.51 Three phased goals for the development of intelligent coal mines

leading to high transportation costs for coal and low efficiency due to manual labor. Coal mining areas often lack effective supervision, resulting in high costs and low efficiency in human patrols. Coal mining areas still rely on traditional, costly "chimney-style" networks for information systems, severely impeding data communication.

(1) Harsh working conditions and prominent safety issues

Coal mining operations occur in adverse environments, with the presence of hazards, such as gas, coal dust, and water-related incidents. Currently, roof accidents and gas accidents are the main sources of accidents in China's coal mines, and coal mining faces the primary location for roof and gas incidents, constantly posing a threat to the safety of workers. The economic losses caused by safety accidents include direct economic compensation and indirect economic losses caused by production stoppages and rectification.

In recent years, mine safety has significantly improved, with a 98.9% decrease in the ten-year fatality rate per million tons of coal produced. While this places China at a moderate level of coal production safety worldwide, there is still room for improvement.

Therefore, directly reducing the number of personnel at the coal mining face is an important means to reduce injuries and ensure safety, as well as an urgent requirement for safe coal mining operations. This situation can be fundamentally changed only by freeing workers from hazardous, harsh, and noisy working environments, and allowing machinery to replace human labor in coal mining.

(2) Labor shortages and high ore transportation costs, along with low efficiency of manual operation

Miner salaries are relatively low compared to the labor intensity of the work, with 36.5% of employees barely satisfied and 32.8% dissatisfied with their salary. The working environment is harsh. In Inner Mongolia, for example, temperatures can reach around 40 °C in the summer and drop to approximately -40 °C in the winter. Some large mining areas have poor commercial and recreational conditions. The coal industry has a high incidence of occupational diseases, as inhaling mineral dust can lead to pneumoconiosis. At the same time, the jolting of vehicles and irregular dietary and sleeping habits make workers susceptible to gastric illnesses and herniated discs.

Several factors, including the relatively low salary concerning miner labor intensity, harsh working environment, and high risk of occupational diseases,, contribute to labor shortages in mining areas. It is challenging to attract young workers, resulting in an average employee age of over 45. High turnover rates mean almost all workers need to be replaced every 1.5 years. There is also a shortage of skilled labor, and mining jobs have low social acceptance, with 44.6% of college graduates dissatisfied with their coal mining work background, and they believe that their work status affects their social status and quality of life.

Ore transportation costs are high, accounting for approximately 50% of the ore's overall cost. Mining transportation labor accounts for more than 50% of the total labor, and stricter safety regulations in mining areas have led to increased wages and training expenses for workers. Take ore transportation costs as an example. In terms of labor costs, a driver's annual salary is 150,000 RMB yuan, with four drivers per vehicle, resulting in labor costs of 600,000 RMB yuan per vehicle per year. Regarding material loss costs, manual operation leads to significant equipment wear and a high accident rate.

Low manual operation efficiency can be exemplified by comparing manual and autonomous driving. A manually driven mining truck typically works for 5,500–6,000 h per year, while an automatically driven mining truck works for 7,000 h per year.

(3) Comprehensive regulations and rules exist in the coal industry, but there is a lack of effective oversight tools, resulting in high costs and low efficiency of manual inspections

The coal industry has a complete set of relevant regulations and rules. However, relying on manual inspections is costly and inefficient due to the dispersed and numerous mining sites and a lack of effective oversight tools.

(4) Currently, in ICT systems, "chimney-style" networks, high deployment and maintenance costs severely hinder data flow. This is mainly manifested by:

- Point-to-point integration of mine applications with high coupling and challenges in managing and maintaining system interfaces, resulting in poor data integration timeliness.
- Difficulty in connecting numerous devices makes it challenging to merge IT and OT and analyze and model massive equipment data.
- Adapting to cloud strategy potentially introduces cross-cloud interfaces, increasing message integration complexity, with reliability and security becoming challenges.

- System cross-network integration between groups, suppliers, and refineries introduces challenges related to integration security and protocol conversion.

6.3.2 *Analysis of 5G Application Scenarios and Network Requirements*

5G will boost the intelligent transformation and upgrading of the coal industry. 5G-enabled smart coal mines mainly include six categories of applications, namely high-definition video surveillance, unified network access, remote equipment control, intelligent inspection, ubiquitous communication support and precise positioning.

High-Definition Video Surveillance

Most of the current video surveillance systems were deployed by mining enterprises in the early stage and had basic monitoring functions, but there are several common problems: Fixed-line transmission—the cameras are inflexibly deployed in limited locations; Low camera resolution—this results in unclear video footage or images, making it difficult to make accurate judgments; Monitoring for mobile scenarios requires manual inspections, incurring labor costs.

Real-time monitoring of personnel, environmental conditions, and production safety in mining areas requires the deployment of high-definition cameras in key positions to realize visible and manageable mining operations. Typically, 30–50 cameras are deployed every 2 km in coal mining tunnels, 50–100 cameras every 20 km on coal conveyor belts, and 20 to 30 cameras in fully mechanized mining faces. In order to ensure clear visibility in the mining area, 1080P HD cameras are used, requiring 2–8 Mbps of bandwidth per camera.

Unified Network Access

There are six underground systems, as shown in Fig. 6.52, including the coal mining, transportation, tunneling, drainage, ventilation, and electromechanical systems. Equipment manufacturers of each system integrate their own communication systems for signal acquisition and transmission. These systems are independently networked, lacking a unified integration platform, resulting in complex maintenance and challenging evolution.

Facing a plethora of sensors, communication between mining maintenance personnel, video surveillance, and other services within the mining area, 5G, as a unified communication network, can eliminate information silos. It leverages 5 G's ultra-large connectivity capabilities and vast bandwidth to facilitate the massive upload of data from people, machinery, and the environment.

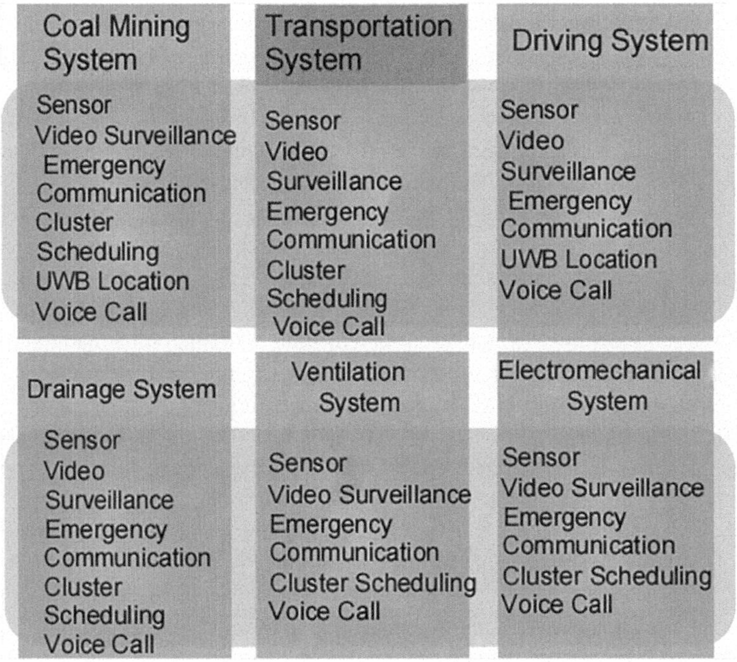

Fig. 6.52 Six underground systems

Remote Equipment Control

In shaft coal mining, the equipment that needs remote control mainly includes the boring machine and the coal mining machine. In open-pit mines, equipment that requires remote control mainly includes electric shovels and mining trucks.

The remote-control system for boring machines can be divided into video streaming, status streaming and control streaming. Video streaming: 6 HD cameras are installed locally on the boring machine. The monitoring host of the remote control platform pulls relevant images in real time based on the operating scenario and displays the images from the six cameras. Status streaming: The Human Machine Interface (HMI) of the industrial computer on the remote control platform displays sensor data received from the boring machine's PLC and radar in real time. The main information includes the machine's deviation angle from the tunnel axis, pitch angle, and tilt angle. Control streaming: Combining the HMI screens and video images on the remote control platform, the operator executes control action, generates a PLC control command, and sends to the local PLC of the boring machine to realize real-time control.

In order to provide reliable manual intervention and emergency stop function, the total video + control delay is required to be no more than 500 ms, with an average delay of 450 ms for the video subsystem and a maximum RTT for the control link of no more than 50 ms. It is recommended that the control link be prioritized to avoid delays caused by network congestion during interventions against abnormities. Table 6.4 shows the network requirements for the remote control of a boring machine.

The remote control system schematic for coal mining machines, shown in Fig. 6.53, is also divided into video streaming, status streaming, and control streaming. Video Streaming: The coal mining machine locally installs two high-definition cameras. The monitoring host on the remote control platform pulls relevant images in real-time based on the operating scenario and displays the images from the two cameras in real time. Status Streaming: The industrial computer human–machine interface (HMI) on the remote control platform displays sensor data received from the coal mining machine PLC and radar in real-time. The main information includes the machine's deviation angle from the tunnel axis, pitch, and tilt angles. Control Streaming: Combining the HMI screens and video images on the remote control platform, operators execute control actions, generate PLC control commands, and send them to the coal mining machine's local PLC for real-time control.

Like remote control of excavators, remote control of coal mining machines requires a maximum round-trip time (RTT) for the control link of no more than 50 ms. The network requirements for remote control of coal mining machines are detailed in Table 6.5.

The coal mining area is the core operation area of an open-pit coal mine, which mainly includes bucket-wheel excavator mining, simultaneous mining of multiple electric shovels and mining truck transportation operations. 5G network capacity is

Table 6.4 Network requirements for remote control of a boring machine

	Type	Specification	Key quality indicators (KQI)	Basic network requirements
Uplink	Camera	6 channels	Standard 1080P (configurable 2 to 8 Mbps, at least 2 Mbps), video (300 ms jitter cache), no stalling	2 Mbps × 2.5 × 6 × 0.8 = 24 Mbps
	Laser radar	2	Sampling in local industrial computer, remote control visualization	1 Mbps
	Data acquisition	1 channel	Data acquisition of 50 ms cycle	500 kbps
	TCP ACK	1 channel	For downstream control information, ensure ACK transmission with priority and without network congestion	30 kbps
	Total uplink bandwidth			>26 Mbps
Downlink	Motor control	1 channel	50 ms cycle,1.4 KB	300 kbps RTT < 50 ms@99.99%

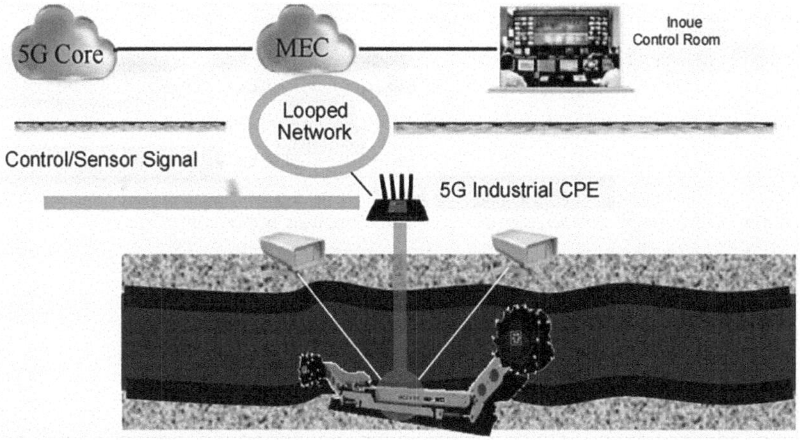

Fig. 6.53 Remote control system of coal mining machines

Table 6.5 Network requirements for remote control of coal mining machines

	Type	Number	Key qualities indicators (KQI)	Basic network requirements
Uplink	Camera	2 channels	Standard 1080P (configurable 2 to 8 Mbps, at least 2 Mbps), video (300 ms jitter cache) no staffing	2 Mbps × 2.5 × 2 × 0.8 = 8 Mbps
	Data acquisition	1 channel	Data acquisition of 50 ms cycle	500 kbps
	TCP ACK	1 channel	Ensure ACK transmission with priority and without network congestion	20 kbps
	Total uplink bandwidth			>9 Mbps
Down link	Motor control	1 channel	50 ms cycle, 1.4 KB	300 kbps, RTT < 50 ms @99.99%

highly required in the open-pit coal mining scenarios. Take a specific coal mine as an example, it operates primarily in two alternating coal mining areas, each of which is 900 m long, comprising a total of 3 layers. The upper coal layer is approximately 40 m deep, and the two lower layers are 15–20 m deep each. The total depth of the mining area is about 80 m.

There are 3 electric shovels working in the coal mining area before the coal seam is exposed, and 5 –6 mining trunks support each electric shovel to transport overburden. After the coal seam is exposed, four mining equipment operate simultaneously, requiring a total of 30 mining trucks to transport coal. If up to 5 mining trucks are in remote control mode at the same time, the total capacity of the mining area is required to be 400 Mbps. Table 6.6 shows the network requirements for remote control of coal mining areas.

Table 6.6 Network requirements for remote control of coal mining areas

Scenario	Coverage area	Service type	Service appeal	Number of jobs that can be remotely controlled simultaneously	Uplink capacity requirement	RTT
Coal mining district	900 m long 3 layers 80 m deep	Unmanned driving of mine trucks	8 kbps	4 electric shovels + 5 mining trucks	400 Mbps	/
		Remote control of mine trucks	40 Mbps			20 ms in average
		Remote control of electric shovels	50 Mbps			20 ms in average

The vertical height of each mining plateau is about 15 m, the horizontal width is 60–100 m, and the minimum distance between mining trucks and the inner side of each plateau is about 10 m. On each plateau, up to 4 electric shovels work simultaneously, with each electric shovel supported by 5–6 mining trucks to transport coal. Normally three trucks wait at the shovel, and no more than 5 trucks should be in line. If up to 5 mining trucks are in remote control mode simultaneously, the capacity requirement for the mining plateau area is about 400 Mbps. Table 6.7 shows the network requirements for remote control of the mining plateau area.

The backfill area is similar to the mining plateau area and has a topography characterized by descending large plateaus. The backfill area comprises approximately 10 plateaus, each with a height of 15–30 m and a width of 60–90 m. There is no electric

Table 6.7 Network requirements for remote control of mining flat areas

Scenario	Coverage area	Service type	Service appeal	Number of jobs that can be remotely controlled simultaneously	Uplink capacity requirement	RTT
Mining flat area	1800 m long 10 layers 15 m high for each floor, 60–100 m wide	Unmanned driving of mine trucks	8 kbps	4 electric shovels + 5 mining trucks	200 Mbps	/
		Remote control of mine trucks	40 Mbps			20 ms in average
		Remote control of electric shovels	50 Mbps			20 ms in average

shovel operation in the backfill area, and each layer has mining trucks to transport the overburden back for filling. Because the mining trucks in the backfill area need to dump overburden, remote control operations should be ensured. After dumping, the mining trucks switch to autonomous driving mode and leave the backfill area at about 30 km/h.

As coal mining operations progress, 5G base stations in the backfill area need to be adjusted, and the general relocation period is about 20 days. If up to 5 mining trucks are in remote control mode simultaneously, the total capacity of the backfill area must be 200 Mbps. Table 6.8 shows the network requirements for remote control of the backfill area.

The crushing area is the operational area where the coal stones transported by mining trucks from the mining area are crushed and conveyed back for processing through conveyor belts. The crushing area has a flat terrain, occupies a small area, and has no electric shovel operation. The main operations here include mining truck autonomous driving and remote control of the dumping of coal, with an average of about 4 mining trucks present at each crushing station. The capacity requirement for this area is about 160 Mbps. Table 6.9 shows the network requirements for remote control of the crushing area.

Table 6.8 Network requirements for remote control of the backfill area

Scenario	Coverage area	Service type	Service appeal	Number of jobs that can be remotely controlled simultane ously	Uplink capacity requirement	RTT
Backfill area	10 layers, 15–30 m high for each floor, 60–90 m wide	Unmanned driving of mine trucks	8 kbps	5 mining trucks	160 Mbps	/
		Remote control of mine trucks	40 Mbps			20 ms in average

Table 6.9 Network requirements for remote control of the crushing area

Scenario	Coverage area	Service type	Service appeal	Number of jobs that can be remotely controlled simultaneously	Uplink capacity requirement	RTT
Fracture area	–	Unmanned driving of mine trucks	8 kbps	5 mining trucks	160 Mbps	/
		Remote control of mine trucks	40 Mbps			20 ms in average

Intelligent Inspection

Coal transportation tunnels are mainly used to convey coal from underground to the surface via coal conveyor belts. The tunnels are usually long, typically exceeding 10 km, and the width and height are about 5–6 m respectively. As an important part of coal mine transportation, belt conveyor systems are prone to abnormalities such as belt idling, misalignment, coal accumulation, foreign objects, slippage, and belt breakage. These abnormalities can significantly impact equipment lifespan and reduce production efficiency and safety. AI-based intelligent video applications make it possible to automatically monitor and identify conditions like belt idling, coal accumulation on the belt, and belt misalignment. This enables alerts to be sent to management personnel or signals to be provided to the belt control system to initiate shutdown procedures, thus reducing the probability of accidents.

The monitoring in coal transportation tunnels mainly includes mobile inspection robots and fixed cameras. A fixed camera is deployed every 50 m, and a combination of mobile inspection robots and fixed cameras is typically used. Table 6.10 shows the network requirements for monitoring coal transportation tunnels.

Currently, mainstream cameras are still regular cameras, and the video is transmitted over the network to be processed for intelligent video using GPUs or AI processing cards deployed in the cloud. 5G cameras can be equipped with explosion-proof enclosures or utilize cameras with + 5G explosion-proof terminals for 5G backhaul. When multiple cameras are present in a local area, the output from these regular cameras is aggregated at intelligent edge nodes for processing and then backhauled via 5G explosion-proof terminals.

In recent years, software-defined front-end intelligent cameras have been gradually deployed, equipped with built-in neural network chips with a certain computational capability, or aggregating data at intelligent edge nodes with built-in processing

Table 6.10 Network requirements for monitoring in coal transportation tunnels

Type	Service type	Transmission direction	Code	Average speed rate in seconds
Fixed camera	Basic video streaming	Uplink	1080P/H265	4 Mbps (2–8 Mbps application specifications variable, need to be determined with the application integrator and the owner)
Inspection robot	Basic video streaming	Uplink	1080P/H265	4 Mbps (2–8 Mbps application specifications variable, need to be determined with the application integrator and the owner)

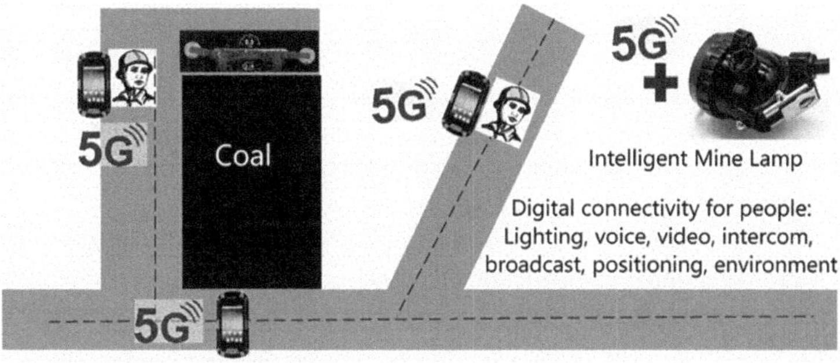

Fig. 6.54 5G communication support

power. The system operates in a collaborative mode between the edge and the cloud, enabling local target detection and cloud-based retrieval. The neural network is used for object recognition, and the corresponding model is deployed at the edge and updated through the network.

Ubiquitous Communication Assurance

Currently, in the shaft coal mines, the communication between the surface and the underground mainly depends on fixed telephones, which do not support video calls. The maintenance of multiple systems such as landline phones, broadcasting and intercom systems is complex, and the underground workers cannot access real-time communication in an emergency.

5G ensures ubiquitous communication, as shown in Fig. 6.54. The first is to provide communication coverage for fully mechanized mining face and maintenance tunnels, enabling collaborative work between the surface and underground in the fully mechanized mining face and remote maintenance guidance. The second is to enable real-time communication throughout all tunnels and real-time scheduling of underground vehicles.

Precise Positioning

In shaft coal mines, a large number of personnel, vehicles, and equipment are distributed across various locations. It is one of the key technologies of coal mine automation to achieve precise positioning of mine personnel and vehicles. To prevent overstaffing in underground coal mines and reduce the occurrence of major accidents in coal mines, *Coal Mine Safety Regulations* stipulate that coal mines must be equipped with a mine personnel location monitoring system. The underground personnel positioning system plays an important role in preventing overstaffing in critical areas such as underground mining faces, minimizing the occurrence of major accidents, locating trapped personnel, preventing individuals from entering blind tunnels, and monitoring underground workers' attendance. The precise positioning technology for mine personnel and vehicles will also be used to prevent unauthorized use of conveyors, avoid vehicle-related accidents and collisions, and enable unmanned driving for underground rubber-tired vehicles and electric locomotives in coal mines.

The early underground personnel positioning system of coal mines primarily relied on Radio-Frequency Identification (RFID) technology, which could only identify the location of problems within specific identification zones. As underground personnel positioning technology developed, systems based on technologies such as ZigBee and Ultra-Wideband (UWB) were successfully developed, as shown in Fig. 6.55. These systems reduced positioning errors from several meters to the current accuracy of 0.2 m. But coverage was limited to about 200 m, and these positioning systems were isolated, creating information silos.

With 5G technology, positioning accuracy can reach meter-level in 3GPP Rel-16 and sub-meter-level in Rel-17. 5G integrates communication, monitoring, and positioning capabilities as a unified communication system. Leveraging the coverage advantages of 5G networks, precise positioning of mine personnel and vehicles in the mining area can be achieved more effectively.

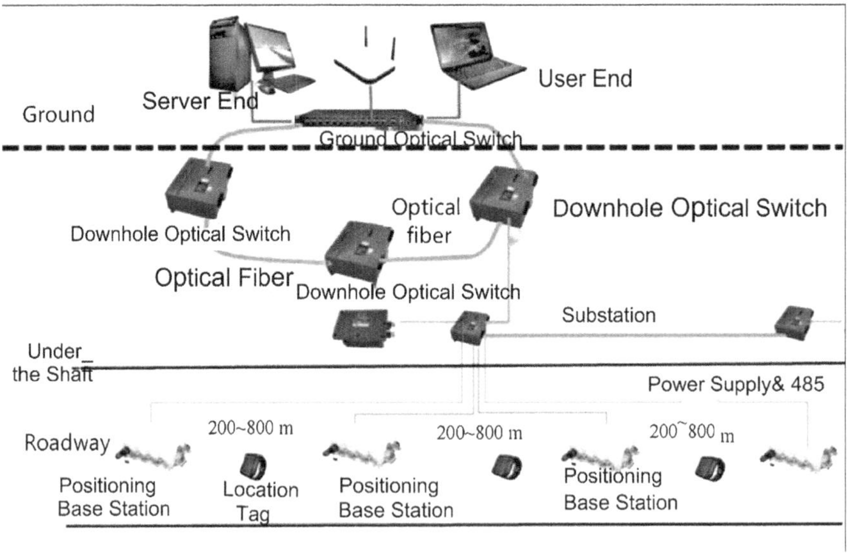

Fig. 6.55 UWB positioning system

6.3.3 Scenario-Based Solution

The scenario-based solution of 5G in coal mines involves the overall network architecture, core network networking scheme, bearer networking scheme, wireless networking scheme, and equipment selection. The following three aspects of network architecture, networking scheme, and equipment selection are introduced.

Network Architecture

The 5G communication network in coal mines is an industry-specific network system built based on the 5G core network of the network operators, as shown in Fig. 6.56. 5G networks mainly include 5G core networks centrally deployed by the network operators, usually in large regions or provincial capital cities; UPF and transmission equipment deployed on the surface of the coal mine park; 5G base stations and transmission equipment deployed underground are either fixed in equipment rooms or as standalone 5G industry terminals. Local UPF deployment enables local services of coal mines to be diverted to their own servers and control platforms, ensuring that data does not leave the park. The transmission equipment, both above ground and underground, is deployed in a ring configuration to enhance the reliability of the coal mine link transmission. The 5G base station adopts the form of a low-power small station that meets the safety standards of coal mines, including BBU, Radio HUB (RHUB), and pico Remote Radio Unit (pRRU).

Networking Scheme

1. Core Networking Scheme

To meet the requirement of keeping customer data within the park, UPF is deployed in the coal mine park, as shown in Fig. 6.57. A distinct Data Network Name (DNN) is planned for the park data network. After user attachment and activation in SMF, SMF selects the park UPF as the primary anchor point based on the independent DNN to route park-specific services locally. Redundancy and backup are implemented by deploying two UPFs to share the load, ensuring disaster recovery capabilities.

2. Carrier Network Deployment Scheme

The recommended network architecture for underground transport equipment for coal mines involves creating a ring network with ground-based transport equipment. This design ensures that a fiber optic failure in any section will not disrupt normal data transmission, thereby enhancing data transport reliability. Ground-based transport equipment forms a convergence ring with the carrier's backhaul network equipment. The Mobile Edge Computing (MEC) network within the coal mine facility can be dual-connected to two ground-based transport equipment units, as shown in Fig. 6.58,

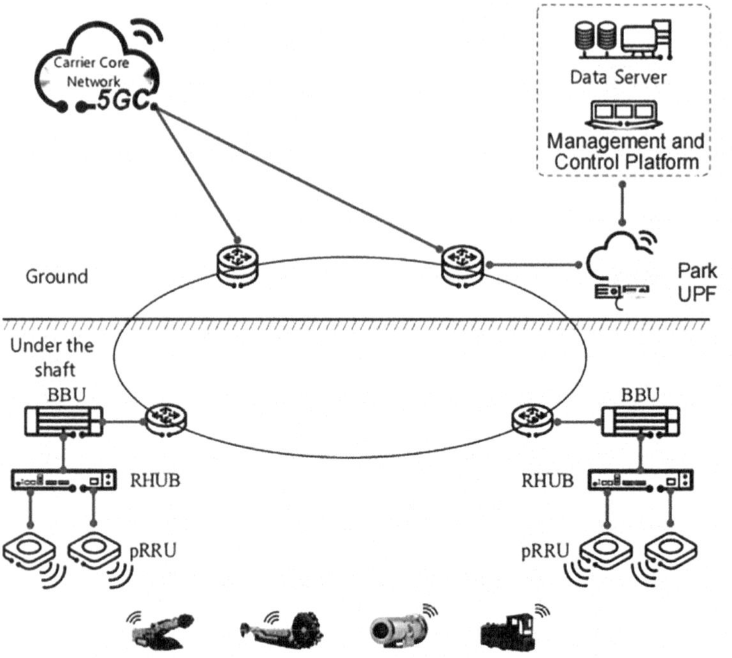

Fig. 6.56 5G private network architecture in coal mines

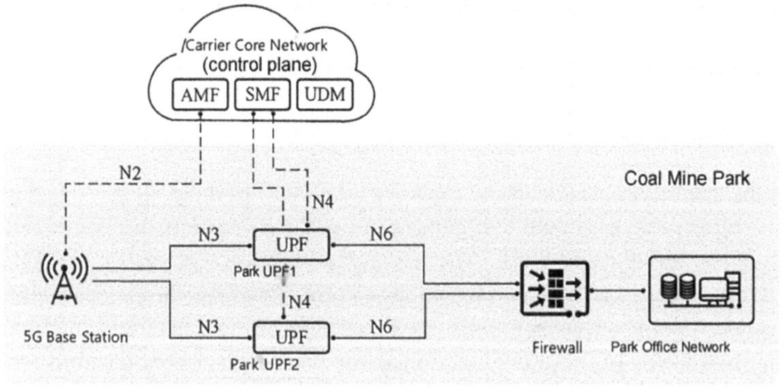

Fig. 6.57 Networking scheme of a 5G core network

or it can be connected to just one ground-based transport unit, depending on the number of ground-based transport units deployed.

For medium to large coal mines, multiple underground transport rings can be established based on the physical location of the underground areas. As shown in Fig. 6.59, these underground rings are interconnected with the ground-based transport equipment.

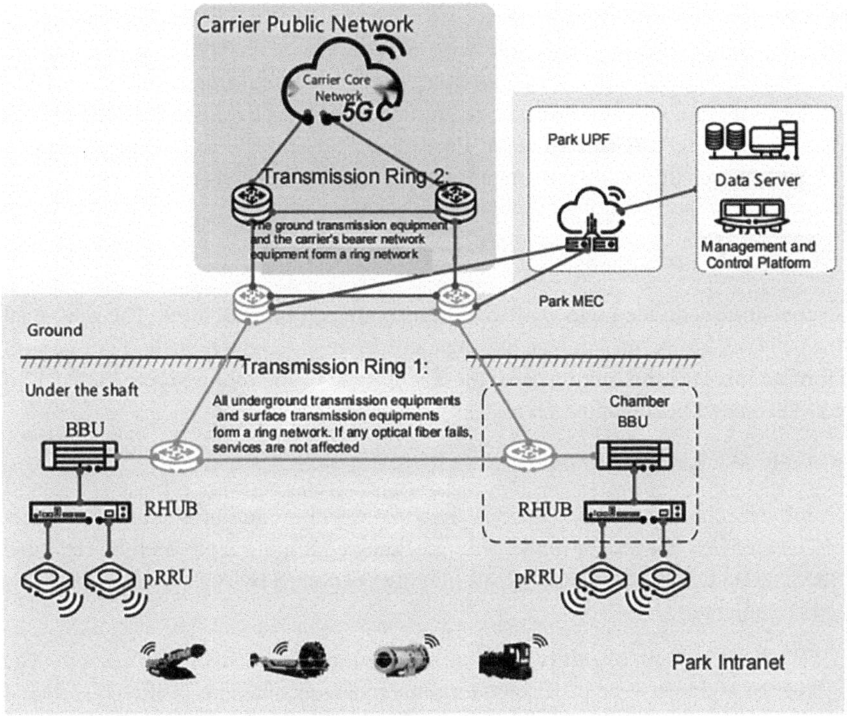

Fig. 6.58 Carrier network deployment scheme

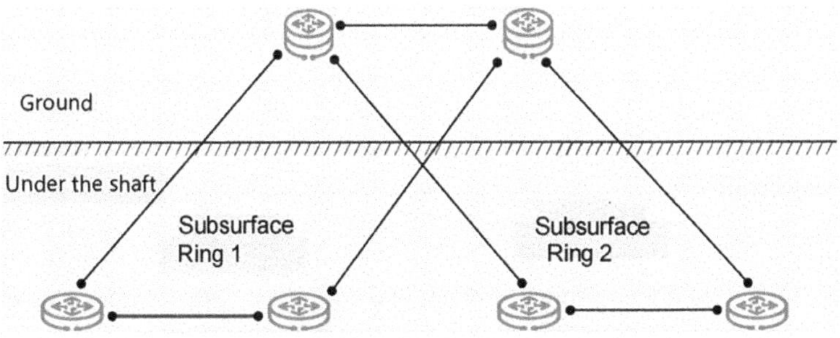

Fig. 6.59 Deployment of underground rings

The interface between the Baseband Unit (BBU) and underground transport equipment is via a 10 Gigabit Ethernet (GE) optical port. The transmission ring can operate as 10 GE ring network or 50 GE ring network, depending on the number of wireless base stations and data capacity requirements. To ensure low latency in data transport, MEC should have a connection to the ground-based transport equipment with an interface speed of at least 10 GE, which should match the speed configuration of the base station's transport interface. In terms of base station clock synchronization, it is recommended to use the Precision Time Protocol (PTP) based on 1588v2 synchronization technology, which provides synchronization among multiple BBUs. To prevent disruptions in network synchronization due to the loss of external clock source, a clock server with a Global Positioning System (GPS) backup should be deployed within the facility to maintain synchronization, ensuring uninterrupted service even in the event of a public network failure.

3. Wireless Networking Scheme

The coal mine is divided into shaft coal mine and open-pit coal mine. The shaft coal mine involves a fully mechanized mining face, driving face and tunnels. The open-pit coal mine involves the mining area, the flat surface of the mining area, the backfill area plateau and the crushing area.

(1) Fully Mechanized Mining Face

The fully mechanized mining face is the core working area of a shaft coal mine. Only when 5G is connected to the fully mechanized mining face can it be regarded as the real 5G underground mining. Achieving 5G connectivity in this area presents several challenges:

(1) The tunnel length of a fully mechanized mining face is usually 160–320 m. The line-of-sight between pRRU and terminal receiving antennas is often obstructed by large equipment like the shearer and hydraulic support, affecting the single-site coverage distance.
(2) The height of tunnels in the coal face is 2–3 m. but top plate support of hydraulic support is generally about 3 m from the ground, and wireless signal propagation is constrained within the tunnel space.
(3) The hydraulic support typically has a 30–60 cm diameter, and the spacing between them can vary. This can lead to the terminal antennas at the top of hydraulic supports being obstructed by nearby supports, affecting wireless signal quality.
(4) The coal mining machine moves back and forth in the tunnel, and the cable/liquid supply cable in the cable trough is relatively thick. The optical cable is easy to be damaged by moving back and forth, the optical cable is relatively expensive, and the replacement will affect at least two shifts.
(5) Stable low-latency communication is required for remote control of the coal mining machine, and Wi-Fi switching solutions are not suitable for this scenario.

To address these challenges, it is recommended to use the indoor distributed base station LampSite, including BBU, RHUB, and pRRU. 4T4R pRRU is used to enhance coverage, and a flat plate antenna is attached externally to every 2T2R channel. Select and deploy an appropriate number of pRRU according to the actual length of the fully mechanized mining face. For example, deploy one pRRU at each end of the fully mechanized mining face less than 100 m for covering the whole area. For fully mechanized mining faces with a length of 100–200 m, deploy one pRRU in

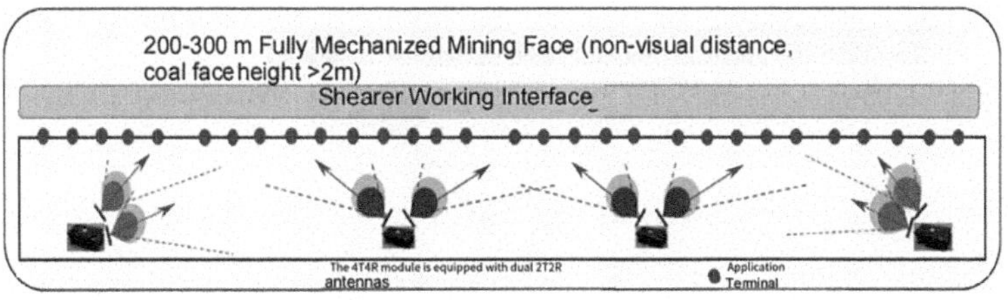

Fig. 6.60 Fully mechanized mining face overcover deployment

the middle. For fully mechanized mining faces with a length of 200–300 m, deploy two pRRUs in the middle and one pRRU at each end, as shown in Fig. 6.60.

(2) Face of Heading

Figure 6.61 shows the deployment of tunneling face coverage. The coverage requirements for the heading face should be as extensive as possible because the heading face is constantly advancing. LampSite 4T4R is recommended, with one flat panel antenna connected outside of every 2T2R channel, providing co-directional coverage on both sides.

(3) Tunnels

The base station mainly provides wireless network coverage within the tunnels, as shown in Fig. 6.62. Explosion-proof base stations for mining are installed in crucial locations such as main shaft, secondary shaft, personnel and vehicle areas, transport tunnels, conveyor tunnels, substations, and mining/excavation workforces. The 5G base station is mainly composed of BBU, RHUB and pRRU. BBU and RHUB are deployed inside the chambers adjacent to the tunnels, while pRRU is wall-mounted on tunnel walls and powered through cables. RHUB is connected to multiple pRRUs in a star topology with optical cables. Typically, one BBU is connected to three RHUBs, with each RHUB accommodating six pRRUs. Based on a spacing of 400 m between stations, one BBU can cover approximately 7.2 km of straight tunnels.

(4) Coal Mining Area

The coal mining area is covered by a macro station, and the equipment is BBU, RRU or AAU. There are usually three coverage schemes in the coal mining area, as shown in Fig. 6.63. Depending on actual requirements, one of these three options can be selected for site deployment.

(1) A small area is placed on each side of the flat plate above the coal mining area to cover the coal mining area.
(2) Position two sector stations in the coal seam of the mining area and cover the mining area back to back.
(3) If there is a higher demand for capacity, one base station can be deployed above the mining area and one in the mining area, and the two stations can cover the mining area with different frequency bands.

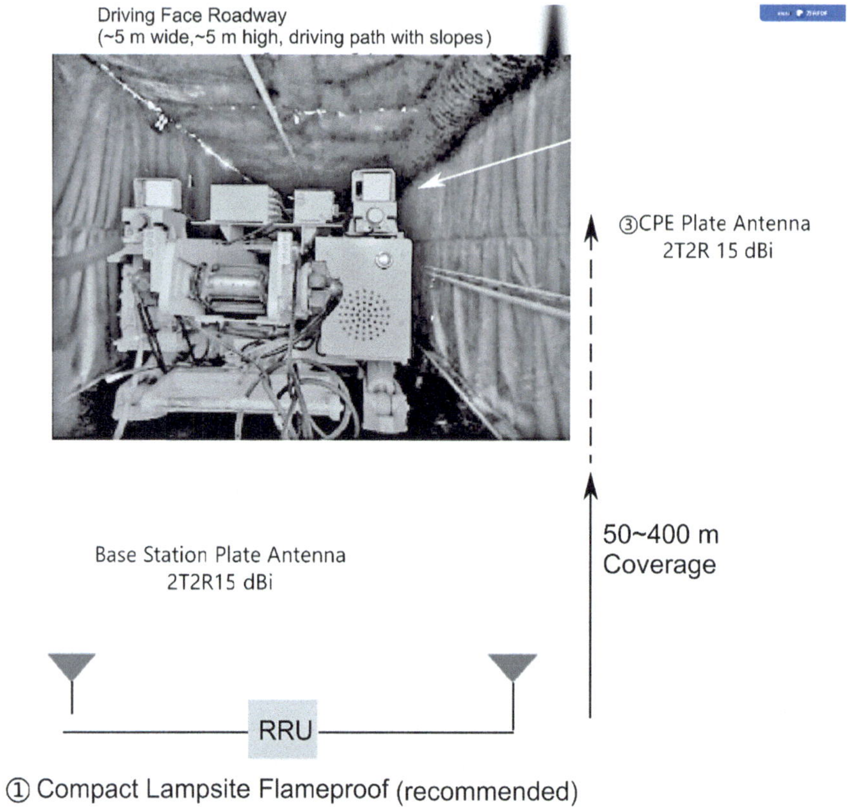

Fig. 6.61 Heading face coverage deployment

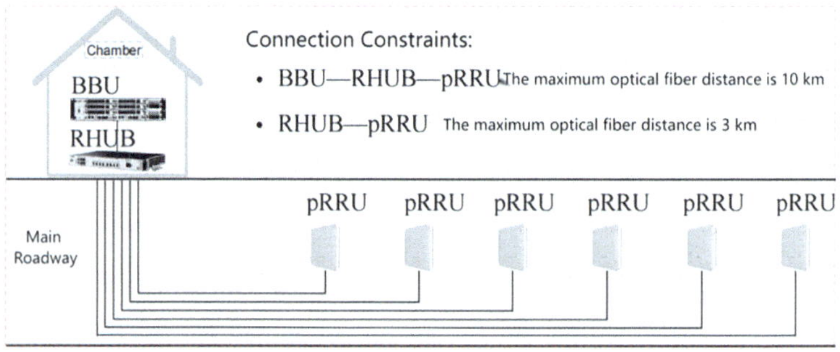

Fig. 6.62 Tunnel coverage deployment

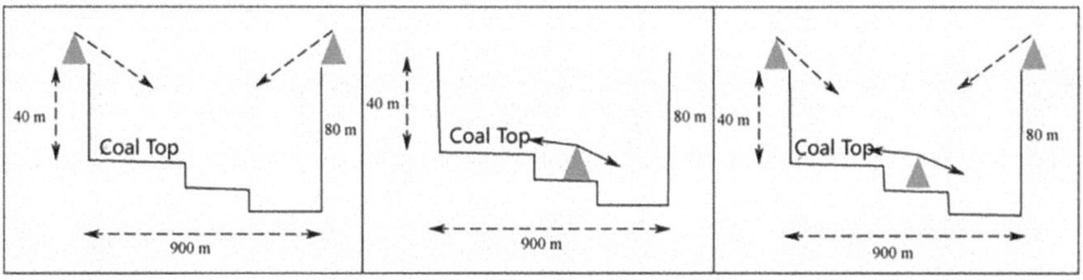

Fig. 6.63 Three coverage schemes in coal mining area

(5) Flat Surface in Mining Area and Flat Bed in Backfill

The same macro station deployment applies to the flat surface of the mining area and the flat surface of the backfill area, and equipment such as BBU, RRU or AAU is used. Due to the height differences, the signal coverage across layers may be at risk of being blocked depending on the placement of the site. Therefore, for the flat layer scene with such a gradient difference, the site deployment should be based on the following principles:

(1) The position of each station should have a certain vertical distance from the outer edge of the flat plate otherwise the signal of the lower flat layer may be blocked;
(2) Theoretically, each three-cell station can provide coverage across one flat layer;
(3) Each station should not provide coverage across two layers otherwise the signal of the third layer will be blocked.

(6) Crushing Area

The flat terrain in the broken area has a relatively small coverage area, with a maximum of 4 mining trucks staying at the same time. Their capacity demands are low, and it can be satisfied with a single AAU station coverage.

Equipment Selection

Coal mines have customized equipment requirements, including 5G base station equipment and mining terminal devices, as shown in Fig. 6.64.

Indoor distributed base station LampSite is adopted for fully mechanized mining face, heading face, and tunnels in shaft coal mines, and the equipment include BBU, RHUB, and pRRU. Coal mines place high demands on the safety of equipment. In 1990, China implemented a "Product Safety Mark Management System for Coal Mine Use", stipulating that "there must be a safety mark for products related to the safety of underground coal mine", namely coal safety certification. Coal safety certification is the certification for explosion-proof products used in underground coal mines (coal safety certification is also required for above-ground equipment directly connected to and communicating with underground equipment). Obtaining

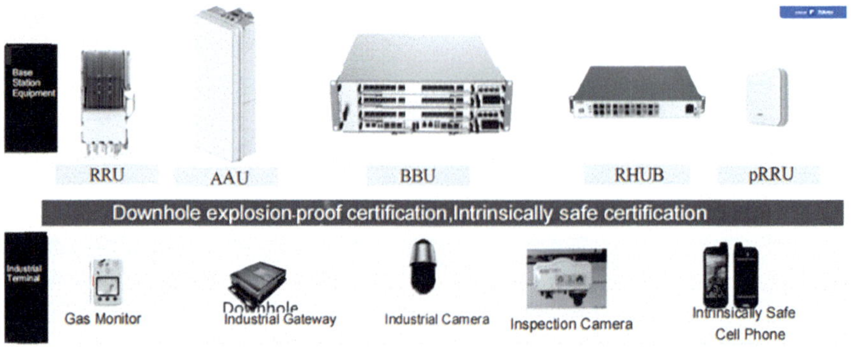

Fig. 6.64 5G devices in the coal mine scenario

coal safety certification is an industry entry requirement. In addition, some low-gas mines can use non-explosion-proof products, such as general mining equipment.

There are only two types of explosion protection for communication products in coal safety certification, namely "explosion-proof" and "intrinsically safe". The explosion-proof device, when in normal operation, encloses components capable of producing ignition sources within an explosion-proof enclosure. The explosion-proof enclosure can withstand the internal explosion pressure without damage, and can prevent the explosion from propagating outside the enclosure. Generally, as long as the power of the whole equipment is below 250 W and the transmission power of the wireless transmission equipment is less than 6 W, it can meet the explosion-proof protection requirements. Intrinsically safe equipment limits the electrical circuit energy through relevant technologies so that the spark or hot surface generated by the electrical equipment is insufficient to ignite substances in the explosive environment. This type of equipment achieves explosion protection through energy limitation. Generally, the voltage should not exceed 24 V DC, the power should not exceed 24 W, and the transmission power should not exceed 6 W (usually defaulted to 2 W).

The coal mining areas in open-pit coal mines, including the coal extraction area, flat surface in the mining area, backfilling area, and crushing area, use outdoor macro base stations. The equipment includes BBU, RRU, or AAU.

6.3.4 Application Cases

A certain coal and power group is a Fortune Global 500 company with an annual coal production of 30 million tons. It currently operates 27 mines and 8 coal preparation plants. The coal and power Group owns a large coal mine with an annual production capacity of more than 10 million tons, which serves as a national pilot project for intelligent standard mines.

Project Background

The current network status and project background of the coal mine are as follows.

(1) Video surveillance: The existing standard-definition cameras provide insufficient clarity. They use fixed-line transmission, and the deployment of cameras is inflexible and limited to certain areas, which cannot meet the basic requirements for unmanned mining operations.
(2) Network isolation: There are six major systems underground, and equipment suppliers of each system have independently integrated communication systems for signal acquisition and transmission. Each system operates independently without a unified integration platform.
(3) Remote control: Manual on-site control of coal mining machines is used at the coal mining face, either through handheld remote controls or on-site control stations. This poses a high risk for underground personnel.
(4) Device inspection: Manual inspection is required to check the operational data of the device periodically.
(5) Underground communication: Communication between the underground and the surface depends on the fixed telephone. In case of emergency accidents involving underground workers, real-time communication cannot be established.
(6) Precision positioning: There are a large number of personnel, vehicles, and equipment distributed underground, but the signal coverage is insufficient, making precise positioning difficult.

Pain Points Analysis

The current major business needs in shaft coal mines are as follows: one is to realize the real-time monitoring of the underground personnel, equipment, and environment; the other is to enable the remote control of the underground boring machine, coal mining machine, and other operations.

There is a substantial demand for underground networking involving a large number of personnel, devices, and environmental aspects, as shown in Fig. 6.65. A large number of sensors require the network to have high connectivity capability, and a large number of cameras require the network to have substantial uplink bandwidth capability, as shown in Table 6.11.

The crucial problems of real-time monitoring are as follows.

(1) A large volume of video data in mining areas needs to be backhauled from various locations within the mining area, leading to distributed and independent networks. At present, a wired connection is used, which results in high cable costs and limited scalability.
(2) The Wi-Fi network experiences poor quality with weak lightning protection capabilities for access points (APs), resulting in high failure rates and low reliability. Additionally, underground areas are prone to signal blockage, creating network blind spots.

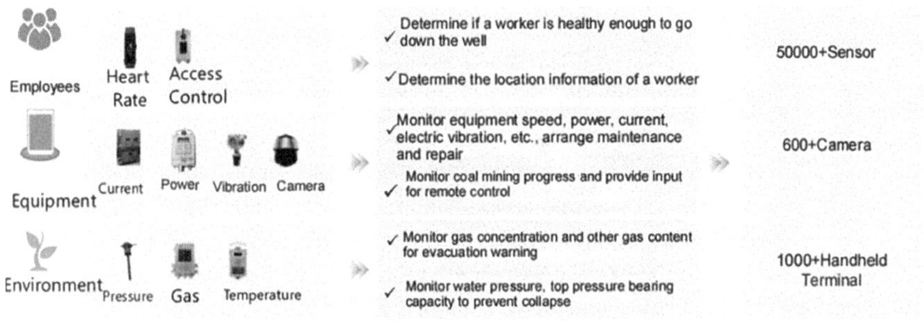

Fig. 6.65 Underground networking terminal

Table 6.11 Networking requirements for underground terminals

Environment perception, human body perception, equipment perception, monitoring equipment	<1000	<1 kbps	50,000+	
Coal lace, driving face, transport transfer point, transport field video	<10	20 Mbps	1 fixed camera + 4 mobile Cameras @200 m Laneway 80 cameras fully-mechanized mining	0.4 Gbps 1.6 GbDs

The crucial pain points for remote control are as follows:

(1) Coal mining machines and boring machines need to operate underground under harsh conditions with a high level of workplace risks. Multiple operators increase labor costs, so there is a desire for remote control.
(2) The Wi-Fi network experiences poor quality with weak lightning protection capabilities for access points (APs), resulting in high failure rates and low reliability. Additionally, underground areas are prone to signal blockage, creating network blind spots.
(3) Boring machines and coal mining machines are mobile equipment. If a wired connection is used, frequent cable renovations are required, which results in high costs and limited scalability.

Solution

Adopt "Four Ones Design Principles" to build an intelligent mining network framework.

(1) One Game Plan: Unified construction, operation and maintenance planning.
(2) One Network: Only one network for the same functions.
(3) One Map: Unified updates of underground digital blueprints.

(4) One Sheet: Seamless integration of all information for consistent data.

1. Comprehensive Solution

The coal mine utilizes an integrated 5G private network consisting of above-ground AAU&MEC and underground LampSite, as shown in Fig. 6.66. Subsequently, it supports network slicing to achieve physical and logical isolation between enterprise and public network data.

Wireless network: The underground coverage uses a distributed antenna system, while macro base stations provide coverage for office areas above ground.

Transport network: Within the industrial park, IP RAN is used to connect the distributed antenna systems and MEC, with MEC interfacing with the 5G core network via the transport network.

Core network: A dedicated enterprise DNN is established, with the control plane based on the operator's 5GC, and the forwarding plane utilizing a newly built edge MEC in the industrial park to achieve local traffic offloading, ensuring that data does not leave the park.

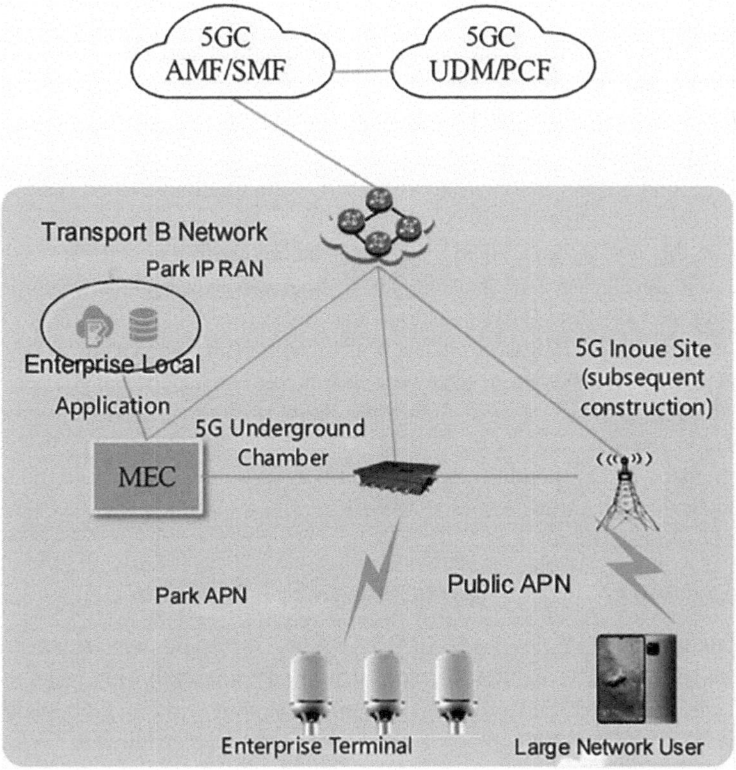

Fig. 6.66 End-to-end solution

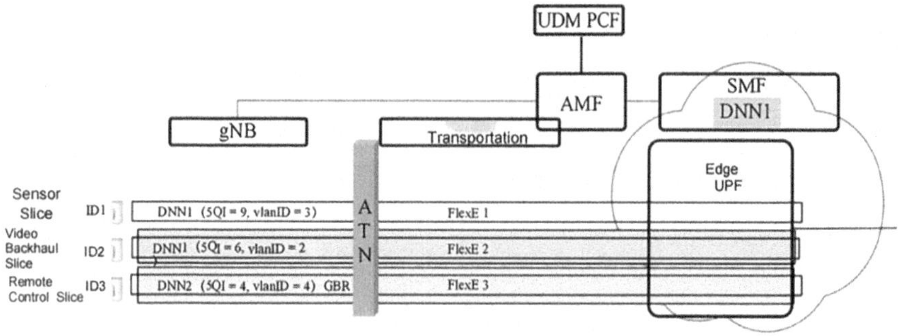

Fig. 6.67 Slice E2E

E2E slicing: The slicing management system deploys low-latency and high-bandwidth slice private networks, offering end-to-end service level agreements (SLAs) with different priority levels to ensure network isolation and guarantee, as shown in Fig. 6.67.

Slice 1: Sensor data collection, underground environment, and machine operation monitoring.

Slice 2: HD video backhaul, covering the working face, tunneling face, transport transfer point, and distributed video.

Slice 3: control information, remote centralized control of coal mining machines at the coal mining face.

2. Measured Indicators

At present, the 5QI + FlexE slicing solution is a combination of software and hardware slicing, and the measured results are in line with expectations. High-priority and Guaranteed Bit Rate (GBR) assurance are used for high-value remote control scenarios, while Non-GBR assurance is prioritized for other scenarios.

Non-GBR (5QI = 6/7/8) services will allocate air interface resources according to the scheduling factor ratio of 5QI when there is competition for air interface resources.

When there is competition for air interface resources, GBR and non-GBR services prioritize allocating resources to GBR services, with the remaining resources allocated to Non-GBR services.

3. Underground Equipment Meets Intrinsically Safe Requirements

For the underground deployment of 5G LampSite, the equipment includes BBU, RHUB and pRRU. BBU and RHUB, which have undergone explosion-proof modifications, are installed in the substation or chamber, as shown in Fig. 6.68. pRRU and antennas, also modified for explosion-proof compliance, are installed in the tunnels, as shown in Fig. 6.69.

Fig. 6.68 Installation of BBU and RHUB after explosion-proof modification

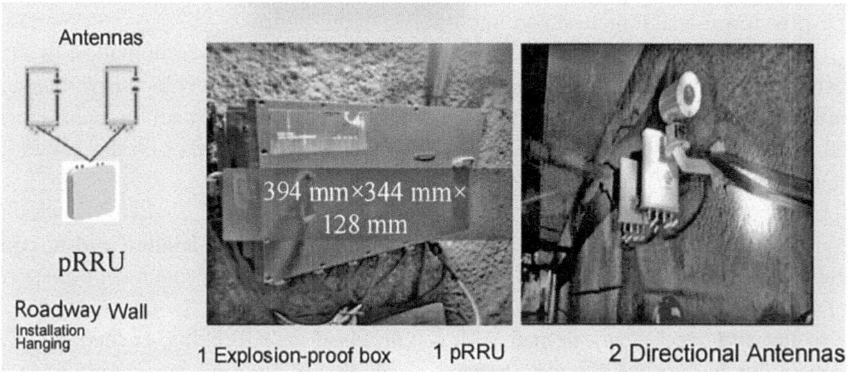

Fig. 6.69 Installation of pRRU and antennas after explosion-proof modification

Service Value

Smart coal mines monitor the operation status of underground equipment in real time through intelligent sensing. Through remote centralized control, it reduces the risks of personnel going underground. Intelligent decision-making is employed to enhance work efficiency.

(1) Intelligent sensing: Using technologies such as machine vision, speech recognition, and Internet of Things (IoT) to detect the operational status and health of conveyors, including issues such as coal pile-up on the conveyor, belt misalignment, foreign objects on the belt, abnormal motor operation, damaged and non-rotating idler rollers, fallen supports, belt frame collapse, belt tear, and roller slipping.
(2) Remote centralized control and integrated scheduling: Intelligent sensing enables on-site automatic control of the working face, coal mining area, tunnel

conveyor systems, and the entire coal flow on the conveyors. It also enables centralized control by surface control personnel and on-site processing and communication by inspection personnel.

(3) Intelligent decision-making: Utilizing machine vision for entire conveyor systems to recognize coal quantities, allowing for intelligent variable frequency speed adjustment of the conveyor belts.

6.3.5 Summary (of 6.3)

This section begins by introducing the current situation of the coal industry, including the pattern of the coal market in China, the process of coal mining operation, the primary equipment used in coal mines. Then, it highlights some crucial pain points currently faced by the coal industry, such as harsh working conditions, labor shortages, and low efficiency in manual operations. As a result, the coal industry is bound to evolve towards automation and intelligence in the future.

Based on the analysis of coal mine operation scenarios, the combination of the coal industry and 5G network mainly includes high-definition video surveillance, unified network access, remote equipment control, intelligent inspection, communication support, and precise positioning. The section also elaborates on the network requirements for different application scenarios.

Establishing networks in the coal industry presents unique challenges, which require scenario-based solutions. In this section, we provide detailed end-to-end solutions for coal mining scenarios, including core networking schemes, transport networking schemes, wireless networking schemes, and so on.

In the end, readers are helped with a real-world case to better understand 5G applications in the coal industry. After studying this section, readers should have a reasonable understanding of the solutions that 5G offers to the coal industry, as well as insights into industry development and application scenarios.

6.4 5G+3C Manufacturing Industry Applications and Solutions

With the rise of the new generation of information technology, 3C manufacturing enterprises have been venturing into smart manufacturing to strengthen their economic and technological capabilities and solidify their industry-leading positions. China's 3C manufacturing sector is in a period of rapid advancement, transitioning from a "manufacturing powerhouse" to a "manufacturing leader". 5G + cloud + AI aligns perfectly with the demand of 3C manufacturing industry for new ICT technologies.

In May 2018, the Ministry of Industry and Information Technology issued *Action Plan for Industrial Internet Development (2018–2020)*, which specified the upgrade and construction of industrial Internet based on 5G. In November 2019, the Ministry of Industry and Information Technology issued *512 Project Promotion Plan of 5G + Industrial Internet*, which specified at least 20 typical 5G industrial application scenarios.

6.4.1 3C Manufacturing Insights

Introduction to 3C Manufacturing Industry

The 3C manufacturing industry mainly provides end-to-end solutions for various 3C products, including design, development, manufacturing, testing, raw material procurement, logistics and after-sales service.

1. What are 3C Products

3C products refer to Computer, Communication, Consumer Electronic (3C) products, and innovative products that combine computer and communication technology with traditional household appliances, such as refrigerators, washing machines and, microwave ovens and other white goods.

2. Scale of 3C Manufacturing in China

China is a manufacturing powerhouse and also a major player in the production of 3C products. There are approximately 190,000 companies engaged in 3C manufacturing in China, accounting for more than 70% of the global production capacity. Among these, China is the source of more than 90% of the world's laptops, 96% of game consoles, and 70% of mobile phones. The top ten global consumer electronics manufacturers have invested and built factories in China, establishing the country's central position in the global consumer electronics manufacturing sector.

Revenue and investments in 3C manufacturing have grown steadily. In 2020, the primary revenue of China's 3C manufacturing reached 11 trillion RMB yuan, with fixed asset investments totaling 1.96 trillion RMB yuan. From 2014 to 2020, the primary revenue and fixed asset investment of the 3C manufacturing industry are shown in Fig. 6.70.

3. Production Mode of 3C Manufacturing

There are three production modes of 3C manufacturing, which are as follows:

(1) Original Brand Manufacturer (OBM), with its brand, design, and production;
(2) Original Design Manufacturer (ODM), OEM production, with design and production;
(3) Original Equipment Manufacturer (OEM): no brand, no design, only OEM production.

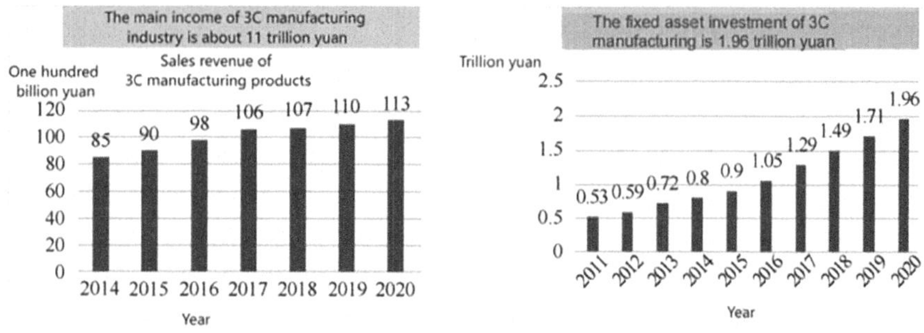

Fig. 6.70 Primary revenue and fixed asset investment of 3C manufacturing industry source: office for national statistics

4. 3C Manufacturing Scenario Classification

3C manufacturing industry is classified by scenario as follows.

(1) Production Line Scenario

The production line is the core production segment of a factory. It has high requirements for safety and data must not leave the park. The communication network for the production line should have the capability of flexible networking, high-precision positioning, high data transmission rates, and lower latency.

(2) Logistics Scenario

Factory logistics covers the storage and movement of raw materials, semi-finished products, work-in-progress items, and finished products in warehouses and production sites. The efficiency of logistics will impact the production process and operational costs. The logistics communication network should have the ability to locate AGV carts, personnel, and goods. At the same time, high-security requirements must be met, and data must not leave the park.

(3) Park Scenario

Park services include toB services, such as drone inspection, mobile robot patrol, video surveillance, and toC Internet services. The communication network in the park has high requirements on uplink bandwidth and business isolation. Additionally, toB data must not leave the park.

Trend of 3C Manufacturing Industry

The 3C manufacturing industry is facing three major trends: firstly, the trend of integrated cross-border activities is becoming increasingly prominent; secondly, vertical integration and horizontal specialization are deepening; thirdly, the market competitiveness of Chinese electronics manufacturing enterprises is gradually improving.

(1) Cross-Integration: Cross-border innovation characterized by cross-fusion has gradually become mainstream. For 3C products, innovation goes beyond single-point technologies such as AI, AR, big data, and breakthroughs in single

products. It involves the integration of various technologies such as sensing, communication, computing, and display.

(2) Business Model: There are two trends in the global electronics industry: one is the trend of vertical integration, while the other is the deepening trend of horizontal specialization. In this process, OBM considers design, marketing, and brand management as the core of its competitiveness and outsources the manufacturing process to ODM. OBM concentrates resources on core R&D and marketing, as well as sharing functions and risks with ODM. To better accommodate global consumer needs, ODM strives for lower costs and greater flexibility in order to fulfill production orders.

(3) Competitiveness of China's 3C Enterprises: China's 3C manufacturing enterprises consist of both OBM and ODM, releasing their own brands while also undertaking processing business for international brands. After years of development, China's 3C manufacturing enterprises have become increasingly competitive. Their strengths lie in their rapid market responsiveness, quality control advantage, cost leadership, customer resource advantage, and the ability to manage operations at a larger scale.

Pain Points of 3C Manufacturing Industry

The 3C manufacturing industry faces many challenges, including numerous safety hazards, insufficient flexible production, low logistics efficiency and difficulty recruiting workers. Addressing these challenges will help enterprises overcome the difficulties and seize the market opportunities.

(1) Safety production: In factory production, safety comes first. The 3C industry is a discrete sector with complex finished goods logistics and inter-factory logistics, which leads to various safety hazards due to the large number of vehicles. Factories also have high-risk, high-temperature, and dust-prone areas. Large machinery and stamping equipment can easily lead to production accidents.

(2) Product quality: Quality is the bottom line for business survival. Many aspects of quality inspection in the 3C industry still rely on manual processes, and there is room for improvement in reliability and consistency.

(3) Flexible production: Personalized production is becoming a trend. Traditional production lines lack flexibility in deployment and adjustments. At present, wired connections dominate communication, with some devices using Wi-Fi, which has obvious problems in reliability and stability, making it unsuitable for the collaboration of multiple robots and smart production equipment.

(4) Logistics turnover efficiency: Most material distribution and finished goods warehousing rely on manual dispatching, resulting in lower efficiency.

(5) Labor shortages and high labor costs: 3C manufacturing is labor-intensive. With China's demographic dividend gradually disappearing, labor costs are rising, leading to industrial outflow pressures. To address this, there is a need to enhance automation and information technology levels and leverage robots and AI to reduce the dependency on human labor.

6.4.2 Analysis of 5G Application Scenarios and Network Requirements

5G technology meets the application requirements of equipment interconnectivity and remote interaction for 3C manufacturing enterprises in their transformation towards smart manufacturing. 5G plays a key supporting role in six typical industrial application fields, including real-time control in industry, AI video surveillance, machine vision, AGV scheduling, data collection, and AR assistance.

Real-Time Control in Industry

Real-time control is the most fundamental application in manufacturing. Its core is a closed-loop control system through a communication network. The following introduces the role of remote real-time control in production.

1. Service Description

Figure 6.71 shows the remote control networking of industrial devices. The remote real-time control of industrial equipment requires a stable and reliable 5G communication network. The controlled object sends PLC status information and on-site video to the controller through the 5G network. The controller analyzes the received video and PLC status information, makes decisions, and issues commands through a control lever. The controlled object performs the corresponding action based on the received instruction.

2. Technical Requirement Analysis

The real-time remote control has the following requirements for the network:

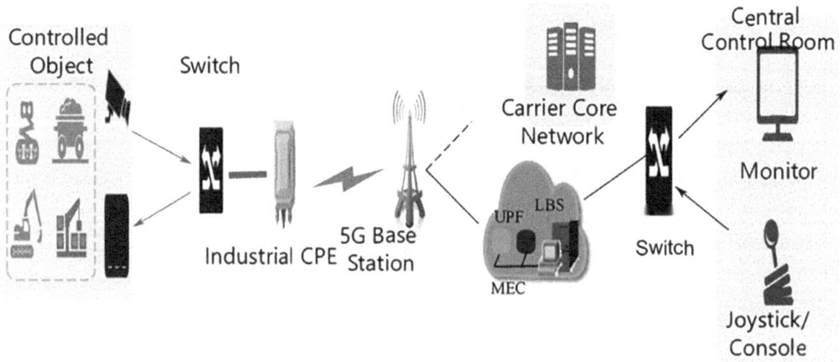

Fig. 6.71 Example of remote control networking for industrial equipment

(1) Stable and reliable data transmission. In remote control, the dependable transmission of control commands is very important. If control commands are transmitted with errors, it can lead to disruptions in production at best and serious safety incidents at worst. Current Wi-Fi communication technology is not sufficiently reliable and stable to meet the requirements of real-time industrial control.
(2) High uplink bandwidth. The operator requires high-definition video feeds from remote locations to ensure precision in control. Typically, a single 1080P video feed with a frame rate of 30 (Frames Per Second, FPS) requires about 4 Mbps of network bandwidth. In scenarios involving remote operation and control of over 30 devices or monitoring points, such a system would necessitate an uplink bandwidth of 120 Mbps. In practical applications, existing wireless communication networks struggle to reliably and stably meet such substantial uplink bandwidth requirements.

3. Value

The value of real-time remote control is as follows.

Real-time remote control can improve working conditions and reduce production accidents for workers. For enterprises, real-time remote control can improve work efficiency and reduce production costs. For flexible production, it can reduce the quantity of cables used and save time in production line adjustment.

AI Video Monitoring

A 3C manufacturing factory encompasses production and processing sites and various facilities such as offices, production areas, warehouses, restaurants, parking lots, and many other places. By integrating video surveillance systems with AI, real-time video content analysis can be performed to detect threats and abnormal activities. For example, within the industrial park, installing 5G+ AI cameras allows for intelligent monitoring of overspeeding vehicles and unauthorized operations. The control center can use high-definition video surveillance to promptly and accurately monitor on-site dynamics, enabling early intervention and the prevention of safety incidents.

1. Service Description

AI video surveillance network is shown in Fig. 6.72. Fixed cameras face challenges in covering dynamic scenarios such as robot inspection, drone surveillance, etc. The 5G + Cloud + AI video surveillance solution can fully cover monitoring scenarios and realize functions such as vehicle recognition, early warning of hazardous activities, equipment inspection, and fault diagnosis.

The park deploys 5G cameras in areas where wired coverage is difficult or in mobile scenarios. These cameras connect to the park's MEC (Mobile Edge Computing) video processing platform via the 5G network, enabling rapid cloud

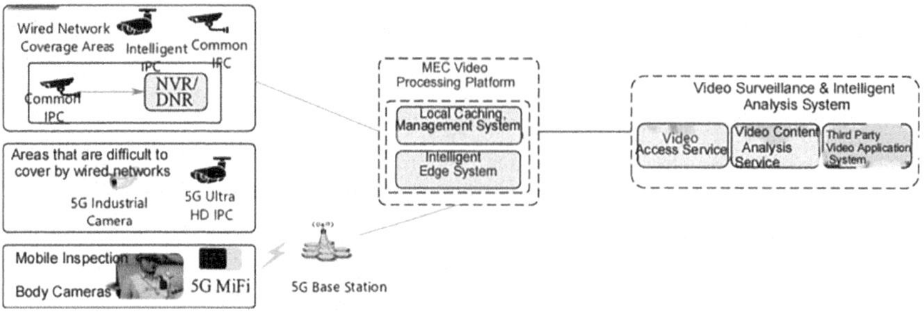

Fig. 6.72 AI video surveillance networking

access. The park's MEC video processing platform is near the user end, with lower latency for intelligent analysis feedback. The public GPU cloud host cluster can elastically support, allocate resources on demand, improve GPU utilization, reduce operational costs, and provide disaster recovery services. The video surveillance and intelligent analysis system deployed on the public cloud is compatible with the park's existing MEC management system to enable a hybrid cloud approach.

2. Analysis of Technical Requirements

The following describes the network requirements of AI video surveillance.

(1) Real-time computing requirements

In a factory, the information related to personnel and production areas is mutually interactive and continuously updated. These real-time video updates must be promptly transmitted to the cloud for computation and processing.

(2) Video transmission requires a large bandwidth and storage capacity

After undergoing various encoding processes, high-definition videos typically have bit rates ranging from 6 to 20 Mbps. With a duration of 1 h, a video file size is approximately 2.7 to 9 GB. When calculated for a storage duration of 2 months, the data storage capacity required for a single 6 Mbps video feed is around 3.89 TB (2.7 GB/h × 24 h × 30 days × 2 × 1 feed), which means that approximately 4 TB of net capacity is needed to store one video feed.

3. Value

The value of AI video surveillance is as follows.

Reduce major accidents: Enterprises strengthen the control of equipment and personnel and reduce production accidents caused by unauthorized operations.

Enhance emergency response capability: With full monitoring coverage, real-time awareness of the distribution of personnel and the scope of events is achieved. In an emergency, this valuable information enhances the scientific basis of decision-making and minimizes losses as much as possible.

Flexible Accessibility: Without the constraints of cables, employees can deploy video surveillance on demand.

Machine Vision

Machine vision is a branch of artificial intelligence that is rapidly evolving. Simply put, machine vision involves using machines to perform measurements and make judgments instead of human vision. Machine vision system captures the visual information of the target through machine vision products, converting it into image signals that are then transmitted to dedicated image processing systems. These systems extract shape information of the captured target based on pixel distribution, brightness, color, and other information, transforming it into digital signals. The image processing system performs various operations on these signals to extract the features of the target and, based on the results of this analysis, controls the actions of the equipment in the field.

Machine vision is a complex system that integrates various technologies, such as image, mechanics, control, lighting, optical imaging, sensors, and computing. The most basic feature of a machine vision system is to improve the flexibility and automation of production. When human-operated tasks are unsuitable due to hazardous work environments or when human vision cannot meet the requirements, machine vision is often used to replace human vision. At the same time, in high-volume repetitive industrial production processes, machine vision inspection methods can greatly improve the efficiency and automation level of production.

The application scenarios of machine vision in the industry include AI quality inspection, QR code recognition, object recognition, etc. Due to space constraints, AI quality inspection is mainly introduced as follows.

1. Business Description

The AI quality inspection (AI QA for short) process is shown in Fig. 6.73. With 5G enabling AI quality inspection, deployment occurs at the edge, and the model training capability is deployed in the cloud. The cloud continuously optimizes the detection model by collecting the real-time production data. According to the inspection standards, the inspection results, such as category information, defect type, and defect location of the inspected products, are fed back to provide early warning and control on-site equipment action. If an image that cannot be recognized by the existing model is encountered in the inspection, the AI inspection model uploads the image to the cloud, thus triggering the cloud-based iterative training process. Once the model update is complete, the new model is deployed to the edge, which performs inspection according to the updated model.

2. Analysis of Technical Requirements

The massive unstructured data in the manufacturing industry provides many scenarios for AI to be implemented in the industry, such as intelligent defect detection, smart recognition and sorting, intelligent dimension detection, intelligent visual guidance, etc. AI-based quality inspection puts higher requirements on data collection, transmission, model training, and algorithm deployment.

(1) High-frequency Transmission of High-definition Pictures and Videos

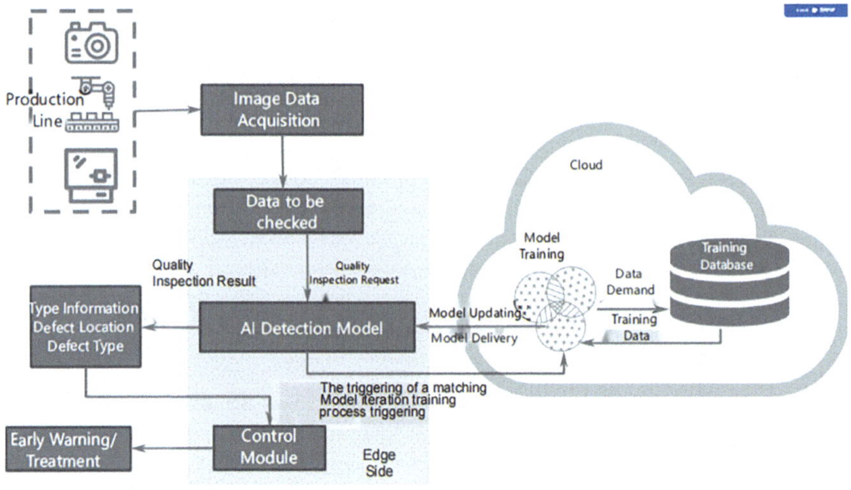

Fig. 6.73 AI quality detection process

Machine vision primarily focuses on uplink transmission and is divided into two subcategories: video and image transmission. In video applications, typical video frames include three types: P-frames and B-frames are less bandwidth-demanding and I-frames are more bandwidth-demanding.

(1) I-frames: keyframes, containing the complete image.
(2) P-frames: predictive frames used to reconstruct the image based on the previous I-frames and P-frames.
(3) B-frames: bi-directional difference frames, which record the difference between this frame and the frames before and after it.

Typical bandwidth requirements for uplink transmission at different resolutions are shown in Table 6.12.

In image-based applications, factors affecting the uplink bandwidth include:

(1) The original image size, influenced by factors such as resolution, pixels and color depth;

Table 6.12 Uplink bandwidth requirements for machine vision video categories

Resolution ratio	Horizontal pixel	Vertical pixel	Average uplink bandwidth	Bandwidth retrograde on peak (I-frame)	1-frame Probability (%)
1080P	1920	1080	4 Mbps	10 Mbps	10
2k	2048	1080	4–6 Mbps	10–15 Mbps	10
4k	4096	2160	10–12 Mbps	20–30 Mbps	10
8k	7680	4320	40–50 Mbps	80–120 Mbps	10

(2) Image compression ratio, with a typical lossless compression ratio around 3:1;
(3) Upload frequency.

The uplink bandwidth requirement for a single camera can be calculated as follows: Original image size × Image compression ratio/Upload frequency.

(2) Real-time Feedback Requirements for AI Detection Results

After AI completes the analysis, different actions are taken according to whether the product has quality issues, such as releasing qualified products or filtering out defective ones. Take a 3C manufacturing enterprise as an example. AI quality inspection can now achieve millisecond-level automatic judgment of defects in single and multi-crystalline polytropism batteries.

(3) Massive Detection Data Storage and Computational Requirements

The detection model is directly related to the accuracy and stability of product inspection, and AI model training requires a large amount of data as a training foundation. In the actual production process, a large number of samples need to be collected for AI learning, and continuous iterative training is needed to optimize the detection model and improve accuracy continually. For example, through deep learning and image recognition algorithms, enabling quality inspection machines to identify defects in solar panels in real time automatically requires learning from approximately 40,000 or more sample images.

3. Value

The value of AI quality inspection includes:

(1) Reducing false alarm rate and enhancing detection accuracy and efficiency in dynamic inspections.
(2) Reducing labor costs.

AGV Scheduling

The production activities of enterprises, such as procurement and sales processes, are often accompanied by the flow of materials. Therefore, while enterprises attach importance to the construction of production automation and intelligence, they are increasingly emphasizing logistics intelligence. Intelligent logistics equipment mainly includes smart storage equipment, sorting equipment, and transportation equipment.

AGV is the key equipment in automated logistics transportation and flexible manufacturing systems, playing an important role in automated logistics warehouse. The hardware of AGV consists of an on-board controller, navigation module, battery module, obstacle detection module, alarm module, charging module, communication module, driving mechanisms, etc. The ground control system of AGV comprises software for task management, vehicle management, traffic management, communication management, and vehicle propulsion. The navigation methods of AGV include electromagnetic guidance, magnetic tape guidance, QR code guidance, laser guidance, and Simultaneous Localization and Mapping (SLAM) guidance, among others. Different navigation methods have very different communication requirements. With the popularity of AGV, more and more manufacturing companies are using flexible combinations of multiple navigation methods to achieve the purpose of flexible distribution.

1. Business Description

In industrial manufacturing and material handling, objects come in various sizes, shapes, and weights. AGV manufacturers cannot design a specific AGV for each type of object. So multiple AGVs must collaborate in the transportation of oversized and overweight items. When transferring oversized and overweight equipment, multiple AGVs must work together to perform tasks such as lifting, starting, turning, and transporting.

AGV collaborative control is shown in Fig. 6.74. After receiving scheduling instructions from the Manufacturing Execution System or Warehouse Management System, the AGV scheduling control system transmits transport tasks, start-stop control, and information gathering to AGVs via a wireless network. When multiple AGVs perform the same task, real-time information synchronization, such as acceleration, key position, and steering angle during operation, is required to accomplish collaborative distribution among multiple AGVs.

Currently, the wireless communication method used for AGVs is Wi-Fi. Wi-Fi operates on unlicensed spectrum communication with carrier frequencies in the 2.4 and 5 GHz bands. Wi-Fi, in practical application, causes issues such as AGVs' inability to receive scheduling control system instructions in real time due to interference and coverage problems, resulting in abnormal stops and affecting the operational efficiency of factory logistics.

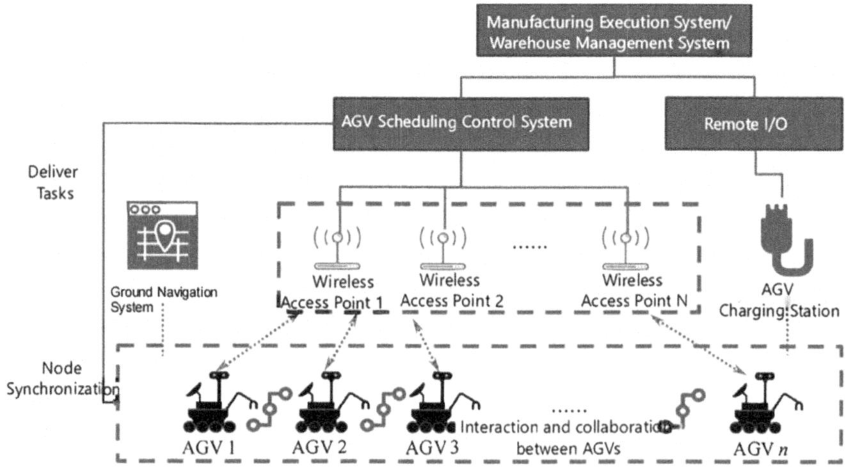

Fig. 6.74 AGV collaborative control

When Wi-Fi is deployed in complex environments, Wi-Fi exhibits poor collaborative communication performance among multiple AGVs. Wi-Fi lacks support for seamless switching, making it challenging to facilitate cooperation across different factory zones. Low latency, high reliability, and excellent switching capability of 5G well support the stability and real-time control of multiple AGVs in collaborative operation.

2. Analysis of Technical Requirements

In terms of the communication requirements for AGV applications from an enterprise perspective, the needs are as follows.

(1) A stable and Reliable Wireless Communication Network

AGV control system needs to collect massive amounts of information from the factory site in real-time through a stable and reliable communication network. The data is processed to generate instructions for AGV execution. At the same time, the complexity of scheduling increases exponentially as the number of AGVs grows.

(2) Stability of Cross-regional and Outdoor Mobile Operation

The scope of mobile operations of AGVs in industrial settings includes not only the factory premises and workshops but also across workshops and factories. The following are the communication network requirements based on the SLAM navigation method.

Bandwidth: The camera adopts 1080P resolution. Considering six cameras in the future along with other sensor data, the usual uplink bandwidth is 40 Mbps.

Latency: AGV travels at a speed of 3 m/s, and obstacle avoidance involves stopping before proceeding. The network latency is less than 50 ms.

(3) AGV Intelligent Control Capability

Combining AGV with edge computing and cloud computing can overcome the limitations of insufficient computing power in individual AGV terminals. However, cloud-based AGVs require communication networks capable of large-scale real-time data interaction. Existing Wi-Fi or 4G communication technologies cannot meet the requirements of cloud-based AGVs, whether in terms of positioning accuracy, network latency, or reliability.

3. Value

The value of AGV scheduling is that intelligent AGVs can improve logistics efficiency, reduce labor costs, and provide full process traceability of information.

Data Collection

Data collection is an important foundation for factory intelligence, and the means of data collection directly determine the degree of integration between OT physical systems and IT information systems. Production data collection is the first step in achieving the transformation towards factory intelligence and serves as the information link between the shop floor production execution and the management layer.

1. Business Description

With the development of technologies such as IoT, big data and artificial intelligence, equipment maintenance and management are moving towards predictive maintenance. Predictive maintenance is an emerging maintenance approach that integrates state monitoring, fault diagnosis, fault prediction, maintenance decision support, and maintenance activities combined with big data analysis.

If the faults of a specific device are to be predicted, its energy consumption data needs to be collected, as illustrated in Fig. 6.75. On the edge side, it consists of sensors, electricity meters, and communication modules, collecting data primarily related to equipment current, voltage, and power consumption. There is one gateway for each floor or building, and the edge side sends the collected data to the gateway via Wi-Fi. The gateway then transmits the energy consumption data to the IoT platform using 5G technology, which is subsequently forwarded to the Management Execution System (MES) for analysis.

2. Analysis of Technical Requirements

Different industries and scenarios have significantly varying network requirements. For example, in the energy consumption monitoring scenario mentioned above, the network requirements are as follows: an uplink bandwidth of 66 Mbps, a downlink bandwidth of 0.2 Mbps, and a latency of 2000 ms. For a data collection point of a CNC machine, there are 50 data collection points, with an average of one data collection per second. If a company has 50 CNC machines, the daily data collection requirement is approximately 216,000,000 data points (number of collection points × number of machines × working hours × collection times per hour). In contrast, the differential protection scenario for the power grid is different. It sends data every 0.833 ms, with a single data packet size of 245 B, requiring a communication bandwidth of 2.36 Mbps (245 B × (1/0.833) × 8). Since power grid faults occur randomly, differential protection for the power grid requires continuous real-time data transmission to determine whether there is a fault in the line, resulting in a continuous uplink bandwidth requirement. An individual terminal's average monthly Dataflow Of Usage (DOU) is approximately 886 GB, demanding high network traffic-carrying capacity.

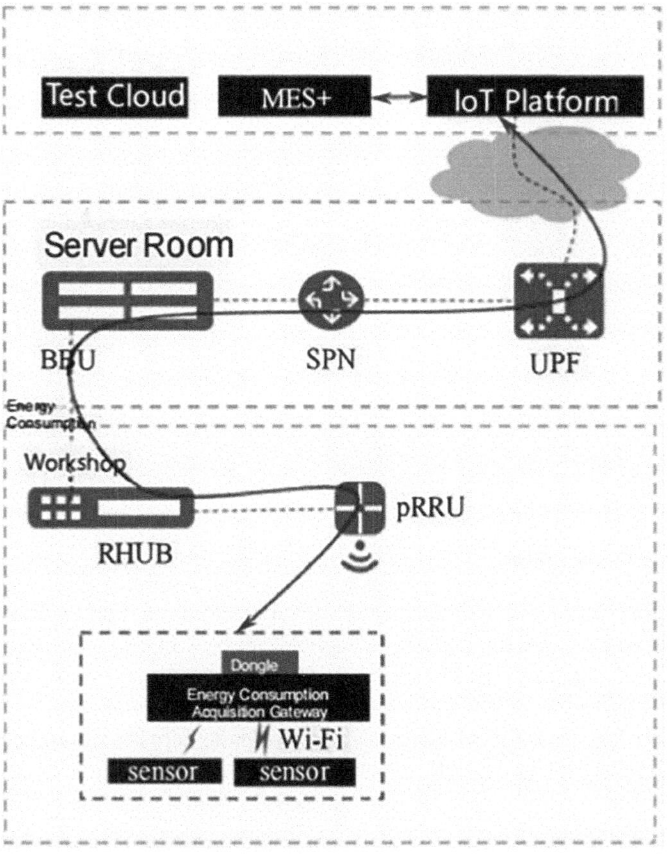

Fig. 6.75 Data collection for energy consumption

3. Value

Value of data collection: Through intelligent monitoring,operation, and maintenance of equipment, production efficiency can be enhanced, equipment downtime due to faults can be reduced, and intelligent scheduling can be facilitated.

AR Assist

Augmented Reality (AR) is an important branch of Virtual Reality (VR) and has been a hot research topic in recent years. AR technology has a wide range of applications in the industrial manufacturing sector, such as remote installation, remote maintenance,

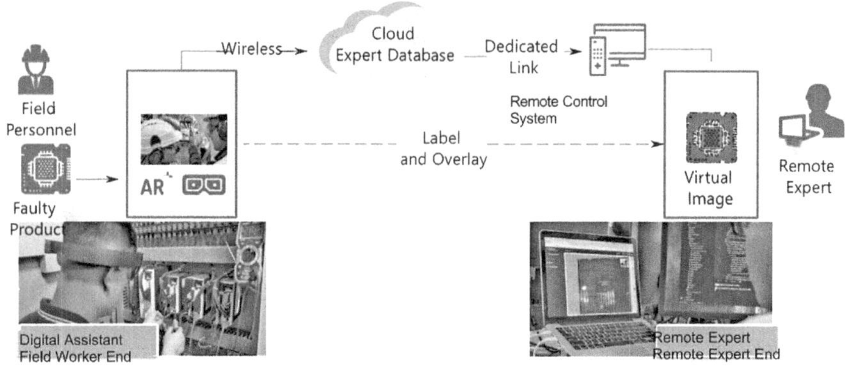

Fig. 6.76 AR remote operation and maintenance

remote collaboration, training guidance, equipment inspection, etc. Due to space constraints, the following mainly focuses on AR remote operation and maintenance.

1. Business Description

AR remote operation and maintenance is shown in Fig. 6.76. Field workers upload local images to the cloud for analysis through AR glasses. In the cloud, query models and knowledge databases are consulted, and maintenance suggestions are overlaid on the real scene, guiding on-site workers in performing maintenance.

AR technology can also be used to request remote support from experts when encountering complex issues. AR glasses are connected to the remote control system through a wireless communication module, and the sound and video images of the scene are transmitted to the remote experts. Remote experts can freeze the screen and use annotation tools to send real-time annotations to AR glasses, overlaying them onto the real scene. In this way, on-site technicians and remote experts can simultaneously view the real scene of the device and the overlaid annotated images, enabling them to guide on-site operations effectively.

In the remote maintenance scenario described above, real-time transmission of high-definition, smooth, and stable video and audio information is required to achieve the effect of diagnosis and maintenance guidance that feels like being nearby, even when remote.

2. Technical Demand Analysis

Based on the application scenarios of HD video and AR in remote operation and maintenance, the requirements for real-time transmission of HD images are as follows.

(1) AR Network Demand Analysis

(1) AR inspection and maintenance: equipment status data is transmitted through IoT sensors, with low bandwidth requirements but with certain requirements for latency.

Table 6.13 AR network demand analysis

Resolution ratio	Code rate	Speed rate(5 times of code rate)	RTT/ms
AR HD (720P)	2 Mbps	10 Mbps	Entry level < 200 ms; Excellent level < 50 ms
AR Ultra HD (1080P)	2.5 Mbps	12.5 Mbps	Entry level < 200 ms; Excellent level < 50 ms
Arl440P (2.5k)	6 Mbps	30 Mbps	Entry level < 200 ms; Excellent level < 50 ms

(2) AR remote collaboration: real-time video sharing and remote collaboration are conducted through live video, placing high demands on both uplink and downlink bandwidth as well as real-time collaboration
(3) AR guidance and training: This involves downlink file/video transmission primarily, making demands on the downlink bandwidth but not requiring high real-time performance.

SinceAR involves real-time transmission and has a higher bitrate, it can withstand greater network fluctuations, providing a better user experience. It is generally believed that the user experience can be excellent when the network bandwidth is five times the bitrate. Table 6.13 summarizes the network requirements for AR at different resolutions.

3. Value

AR remote operation and maintenance provides bidirectional real-time audio and video communication between the first responders on-site and remote experts, reducing operation and maintenance time and lowering operation and maintenance costs. Engineers can free up their hands, access workflow information anytime and anywhere, check equipment status, and enhance work efficiency.

6.4.3 Scenario-Based Solutions

The overall 5G network solution for the 3C manufacturing industry involves network architecture, wireless networking scheme, transport networking scheme, core networking scheme, etc.

Network Architecture

An enterprise's 5G private communication network is an industry-specific network system built based on the 5G public network, and the overall 5G network architecture is shown in Fig. 6.77.

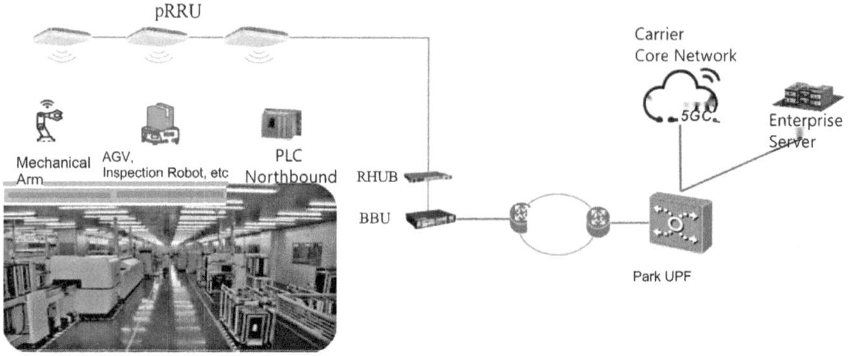

Fig. 6.77 Overall 5G network architecture

The 5G network mainly includes:

(1) 5G core network centrally deployed by operators is generally located in major regions or provincial capital cities;
(2) UPF and transmission equipment deployed locally in the industrial park enable the routing of local services to proprietary servers, ensuring data stays within the park;
(3) Digital indoor small cell base stations deployed locally in the industrial park, include baseband units (BBUs), RF Hub units (RHUBs), and micro Radio Remote Units (pRRUs);
(4) Industry-specific terminals deployed on industrial manufacturing equipment include robotic arms and AGVs communicating with pRRUs via 5G modules or CPEs.

Networking Scheme

1. 5G Wireless Network Networking Scheme

In the early stages of 5G industry applications, the primary focus remains on basic capacity requirements, e.g., video backhaul, picture backhaul, and other services, so the basic network is first designed according to the capacity objectives. Then the network overlay is designed based on low latency, positioning, and other applications.

(1) Capacity Design

The capacity design concept is shown in Fig. 6.78. With the booming development of the 5G Industrial Internet, more industry applications will mature in technology in the future. The types and quantities of deployments in various manufacturing scenarios will also increase. It is recommended that the 5G digital room deployment network construction mode adopt distributed Massive MIMO cells in the early stage and double the capacity by cell splitting in the later stage.

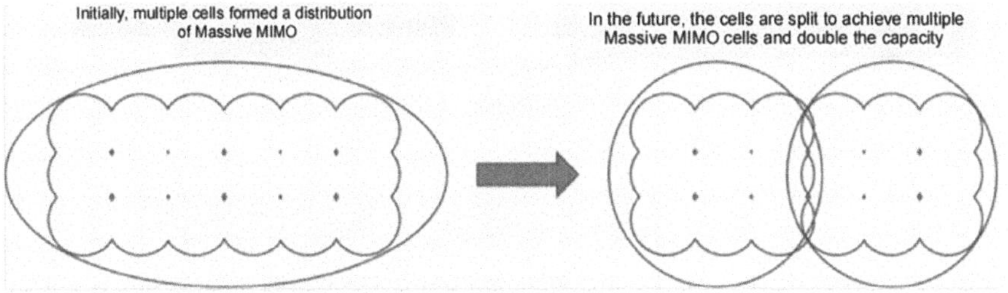

Fig. 6.78 Capacity design schematic

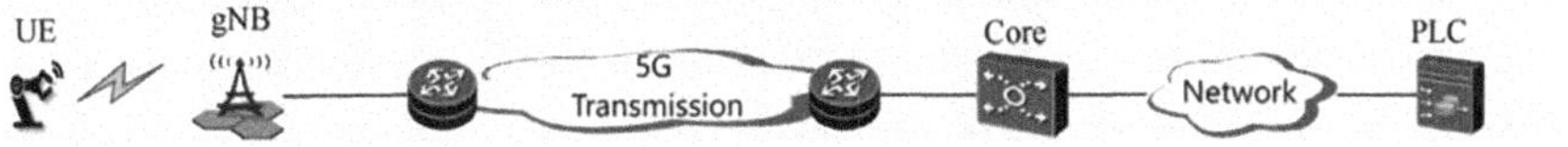

Fig. 6.79 Stable and low latency design

(2) Stable and Low Latency Design

A stable low latency design is shown in Fig. 6.79. The latency in the communication scenario between UE (User Equipment) and cloud services is mainly composed of the following three parts.

(1) Core network processing latency, which is controllable.
(2) Transmission network latency is affected by the number of intermediate node devices in the transmission network and the load of the transmission network. It typically has a low latency, but the traffic burst can significantly impact latency stability. In industrial scenarios, attention needs to be paid to latency fluctuations introduced by the data communication network between gNB and the core network and between the core network and upper-level applications.
(3) Air Interface Latency, which is influenced by network configuration, traffic volume, and channel conditions, leads to significant average latency fluctuations.

Based on business requirements, latency design can be categorized into three levels: basic-level stable low latency, mainly providing a latency capability of 100 ms@99.99%; intermediate-level stable low latency, mainly providing a latency capability of 50 ms@99.99%; advanced-level stable low latency, mainly providing a latency capability of 20 ms@99.99%.

2. 5G Bearer Network Configuration Scheme

The 5G transmission network scheme is shown in Fig. 6.80.

(1) The 5G base station connects to SPN equipment, going through access, common aggregation, and backbone aggregation to reach the park's core node.
(2) N2 interface (base station to AMF) path: base station-urban SPN-IP transport network-regional 5GC core network, end-to-end L3 network, with different networks using a physical port pair-type dual connectivity.

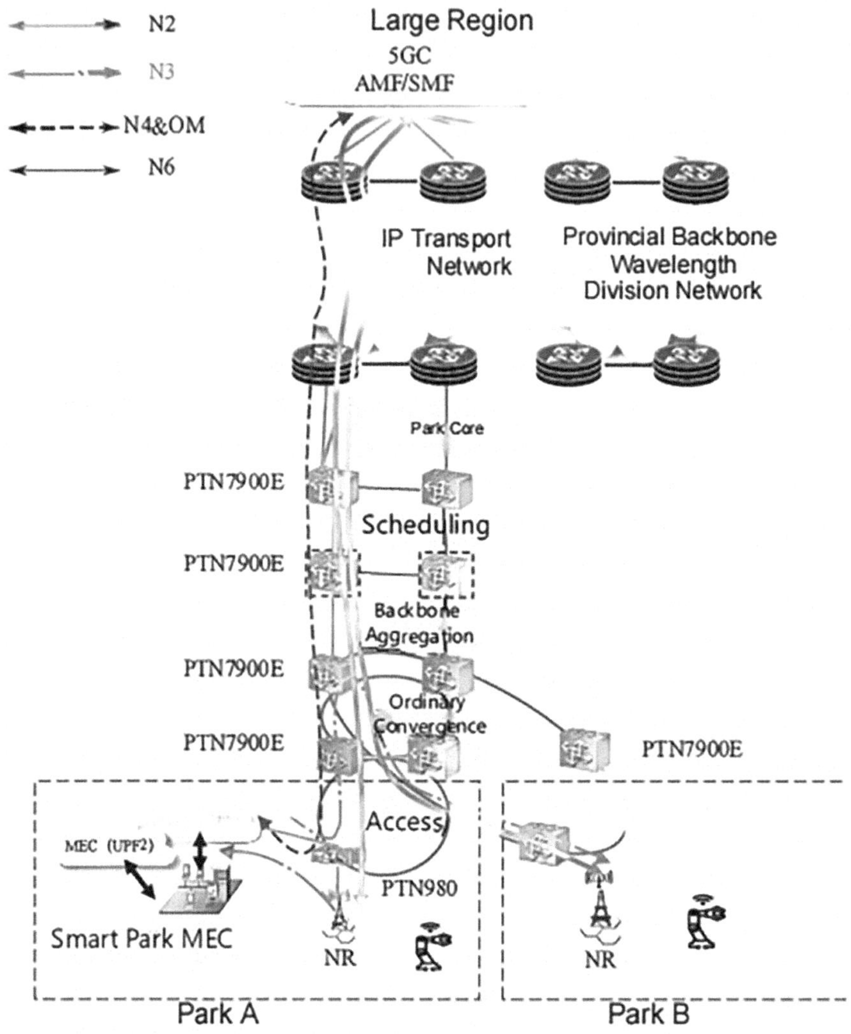

Fig. 6.80 5G transmission network scheme

(3) N3 interface (base station to UPF) path: base station—SPN (local access) — smart park MEC (UPF), traffic from Factory A is forwarded locally. In contrast, traffic from Factory B detours through the common aggregation node before reaching MEC.
(4) N4 (UPF to SMF) & OM interface path: MEC (UPF)—urban SPN—IP transport network—Regional 5GC core network, end-to-end L3 network, with different networks using a physical port pair-type dual connectivity.
(5) N6 interface (UPF to park system data center) path: Direct connection from MEC (UPF) network.

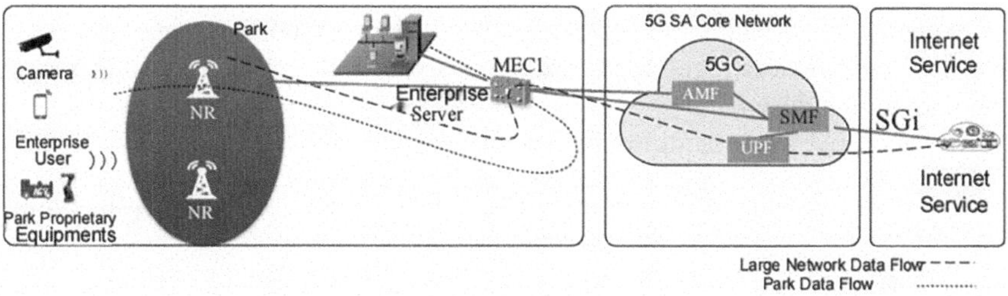

Fig. 6.81 5G core network solution

3. 5G Core Networking Scheme

The 5G core network scheme is shown in Fig. 6.81. The 5G core network control plane is in the central server room, and UPF is deployed in the park.

(1) Enterprise terminals (including AGV carts, robotic arms, surveillance cameras, etc.) access the 5G network through the base stations deployed in the park. Data is exchanged with the enterprise servers through MEC deployed in the park, ensuring a closed-loop and secure data environment.
(2) Ordinary users in the enterprise can access external Internet networks through the base station. At the same time, special terminals (such as inspection equipment or cameras) can be configured to access only the park's private network.
(3) Enterprise users moving to areas outside the park cannot access the park-specific private network services but can still access external Internet networks normally.

6.4.4 Application Case

MD, a large 3C manufacturing enterprise in Guangdong, in collaboration with a telecommunications operator and Huawei, has jointly established a "5G Joint Innovation Lab". Based on real production environments, this laboratory is dedicated to creating research and testing platforms for innovative applications of 5G, cloud, and AI solutions.

Project Background

In 2020, MD and a telecommunication operator and Huawei held an inauguration ceremony for the "5G+ Industrial Internet" application demonstration zone and the 5G Joint Innovation Lab. The three parties planned to focus on solving crucial pain points in the industry and meeting practical production needs based on the actual application scenarios in MD's manufacturing workshops and logistics warehouses. They will work together to construct integrated 5G solutions for the discrete manufacturing industry.

MD Enterprise Manufacturing Park is introduced as follows.

(1) Covering an area of 800 acres, with eight major workshops and nearly 10,000 employees.
(2) Products: microwave ovens, large ovens, small ovens, steam ovens, etc.
(3) There are 59 assembly lines, with an annual production capacity of over 40 million units.
(4) It covers the entire production process, including component manufacturing, product assembly, warehousing and logistics, etc.

Pain Points Analysis

Business pain points are analyzed as follows.

(1) Flexibility: There is a small batch, multi-model production (over 2000 varieties) with frequent production line adjustments (average of 12 times per line per year).
(2) Quality inspection: Manual quality inspection lacks assurance and can lead to batch scrap.
(3) Logistics efficiency: An unstable Wi-Fi network affects logistics efficiency.
(4) Production equipment operation and maintenance: Engineers cannot respond quickly on site when the factory equipment malfunctions.
(5) Wired network: Various production terminals rely on wired networks, which are cumbersome to maintain and incur high costs.

Solutions

The application scenarios of 5G in MD include park security monitoring, AGV application, production data collection, cloud-based PLC, Kuka robot, forklift scheduling, AI quality inspection, MES production dashboard, barcode scanner management, AR remote operation, and maintenance. This section uses forklift scheduling and AI quality inspection as examples to explain service crucial pain points and solutions.

Practice Scenario 1: Forklift Scheduling

(1) Scenario Description

A high-level forklift is used for the placement and retrieval of goods. There are more than 20 high-level forklifts in use. Due to the frequent transportation of goods, each high-level forklift needs to interact with the WMS. A customized PC terminal is installed on the forklift to enable real-time communication with the WMS through the Wi-Fi network and realize real-time synchronization of logistics scheduling information.

(2) Business Pain Points

Packet loss is common in Wi-Fi networks. Issues such as data upload and download failures are prone to occur during access point (AP) transitions. As a result, logistics information cannot be transmitted in a timely and reliable manner, affecting operational efficiency and potential security risks. Additionally, as the forklift system cannot achieve precise indoor positioning, it is challenging to comprehensively understand the location of a forklift, record and analyze the driver's workload, and track the working path. Consequently, it lacks the capability for visual real-time monitoring and management.

(3) Solution

Figure 6.82 shows the solution of forklift management. In the first stage, Wi-Fi will be replaced with 5G to achieve uninterrupted connection, zero packet loss, and smooth operation. In the second stage, once 5G indoor positioning technology is mature, the forklift system will realize visual real-time supervision.

Practice Scenario 2: AI Quality Inspection.

(1) Scenario Description

After the pressing of the microwave oven panel stamping parts, it is necessary to inspect the appearance of the stamped parts. This is done by placing a camera on the front, back, left, right, and top for visual inspection.

Quality inspection hardware deployment: 5 industrial cameras, with a top-line scan 4 K camera, and 2 K cameras in the front, back, left, and right.

Server & AI algorithm: a local industrial control computer and AI software are provided on-site for quality inspection through AI image reasoning and comparison.

Technical requirements include a false positive rate of less than 0.5% and a detection rate over 95%.

(2) Business Pain Points

(1) The false positive rate is relatively high, requiring manual periodic sampling every 30 min.

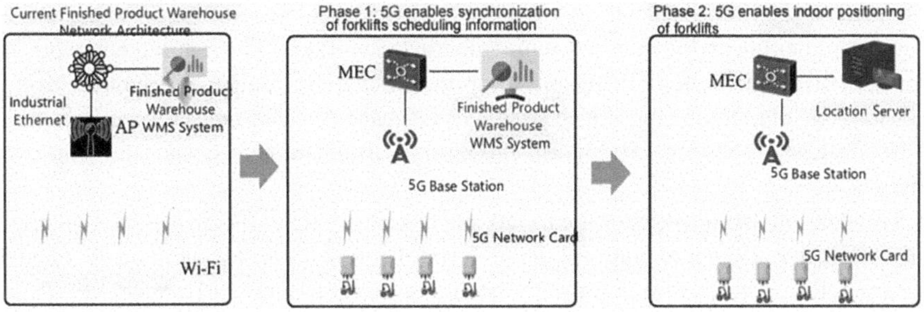

Fig. 6.82 Forklift management solution

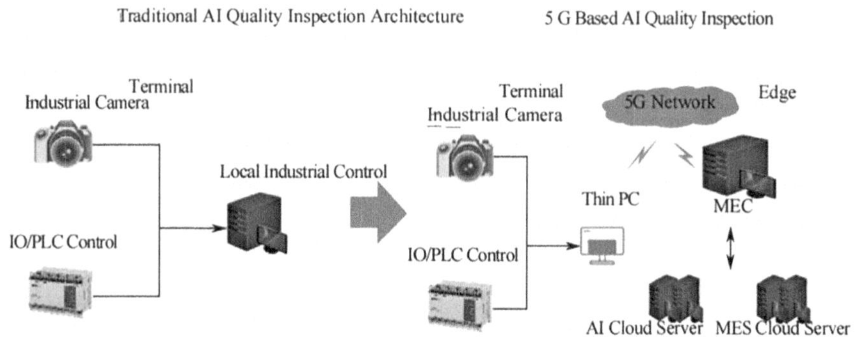

Fig. 6.83 AI quality control solution

(2) During the sampling period, there is a tendency for mass scrapping due to stamping issues.
(3) The cost of local AI computers is high.

(3) Solution

Figure 6.83 shows the AI quality inspection solution. Compared to the local industrial control computer, the MEC cloud server has more powerful computing capabilities, enabling real-time training in the cloud, continuous iteration, and automatic on-site model updates, resulting in more efficient detection.

6.4.5 *Summary (of 6.4)*

China is a major player in the 3C manufacturing industry, with more than 70% of global electronic products being manufactured and assembled by Chinese enterprises. China's 3C manufacturing industry is facing challenges such as the gradual disappearance of the demographic dividend, continuous escalation of labor costs, and a precipitous decline in the number of manufacturing workers. The rise and widespread adoption of 5G technology brings new opportunities for the entire 3C manufacturing industry. 3C manufacturing companies will leverage 5G, AI, and cloud computing technologies to enhance production flexibility further, reduce costs, and improve efficiency.

This section first introduces the trends and business pain points in the 3C manufacturing industry. Then it explains in detail the application scenarios of 5G in 3C manufacturing, the overall solution for the 3C manufacturing industry private network, and industry practical cases.

Through the study of this section, readers should understand solutions involving 5G in the 3C manufacturing industry and become aware of the progress and application scenarios of the industry.

6.5 Summary

This chapter begins by providing background knowledge on four industries: iron and steel, port, coal, and 3C manufacturing. It then analyzes the application scenarios and network requirements of 5G in these industries, offering scenario-specific solutions for each. Finally, it introduces application cases for each industry.

As the 5G arena competition enters the second half, countries are shifting their 5G strategic focus towards applications. In 2021, 5GtoB moved towards large-scale commercialization, illuminating various industries. However, challenges persist in transitioning from exploration to widespread implementation. Questions arise about the urgency and necessity of 5G in each country, the maturity of China's 5GtoB ecosystem, the repositioning of roles in the 5GtoB industry chain, and how the widespread implementation of 5GtoB can be achieved. This chapter aims to stimulate new considerations for the readers on these matters.

Exercise 6
I. True or False

6.1 Currently, China's overall crude steel output production ranks first in the world, and domestic steel enterprises rank at the forefront globally regarding competitiveness and influence. ( )
6.2 The control flow of cranes in ports requires lower network bandwidth than video streams. ( )
6.3 Communication equipment used in shaft and open-pit coal mines should comply with intrinsic safety standards. ( )
6.4 5G, with lower latency, higher speed, and improved overall service experience, is expected to become the cornerstone of the future industrial Internet. ( )

II. Choice Questions

6.5 (Single choice) Which of the following is NOT a pain point faced by remote overhead crane equipment in steel plants? ( )

A. High temperature, noise, dust, etc.
B. Crane movement restricted by the tracks
C. Prolonged driver looking down and fatigue
D. Safety hazards on site

6.6 (Multiple choice) In the end-to-end production process of the iron and steel industry, what are the two main production types? ( )

A. Long process
B. Short process
C. Full process
D. Flexible customization process

6.7 (Single choice) Which of the following is not a 5G application scenario in an iron and steel plant? ( )

A. Remote overhead crane
B. Automatic steel turning
C. Steel plate inspection
D. Vehicle-to-everything (V2X) networking

6.8 (Single choice) Which of the following is correct regarding the operational process of container circulation in ports? ( )

A. Yard crane—Gate—Internal truck—Quay crane
B. Gate—Yard crane—Internal truck—Quay crane
C. Gate—Internal truck—Yard crane—Quay crane
D. Yard crane—Internal truck—Gate—Quay crane

6.9 (Single choice) Regarding the network latency for the assisted driving of AGVs in the port, which of the following is correct? ( )

A. Less than 5 ms
B. Less than 10 ms
C. Less than 50 ms
D. Less than 100 ms

6.10 (Multiple choice) What are the application scenarios of 5G in smart ports? ( )

A. Remote control of cranes
B. OCR intelligent cargo handling
C. AGV-assisted driving
D. VoLTE

6.11 (Single choice) To meet the requirement of ensuring customer data stays within the mining area, which of the following practices is correct? ( )

A. Sink UPF to the coal mine area
B. Sink AMF sinks to the coal mine area
C. Sink UDM sinks to the coal mine area
D. Sink MME sinks to the coal mine area

6.12 (Multiple choice) What are the application scenarios of 5G in the coal industry? ( )

A. Video surveillance
B. Remote control
C. Intelligent inspection
D. Precision positioning

6.13 (Multiple choice) What does the coal industry face the current business crucial pain points? ( )

A. Harsh working environment
B. Difficulty in recruiting workers
C. Low efficiency of manual operation
D. Chimney-style network, high deployment and maintenance cost

6.14 (Multiple choice) Which of the following can be classified as discrete manufacturing? ( )

A. Automobile manufacturing
B. Mold manufacturing
C. Home appliance manufacturing
D. Petroleum production

6.15 (Single choice) In the Smart Factory application scenario, which application does not require MEC configuration? ( )

A. Machine control
B. Machine vision
C. Massive connectivity
D. Remote site

6.16 (Multiple choice) The current industrial Internet is primarily based on wired networks. What are the problems of wired networks? ( )

A. Long wiring cycles
B. Poor scalability
C. Difficult deployment in high-temperature scenariosn
D. Limited bandwidth

III Short Answer Questions

6.17 Briefly describe the overall architecture of a 5G network in a steel plant.
6.18 When CPE is too close to the base station, and RSRP is strong, why does the uplink rate decrease?
6.19 Provide an overview of the wireless-side equipment that should be selected when deploying 5G in an open-pit coal mine and a shaft coal mine.
6.20 What are the navigation methods for AGVs, and what are their advantages and disadvantages?

Chapter 7
5G Wide Area Private Network Scenarios and Solutions

China's 5G+ scenarios hold great promise. On the one hand, there is a strong demand for remote sharing because of the serious imbalance of various social resources. On the other hand, digitization can effectively alleviate the problem of social resource shortage. In the face of China's current development status, digitization has become the key to resolving those challenges. 5G has high bandwidth, low latency, and large communications. Functioning with technologies such as cloud computing, MEC, big data, artificial intelligence, blockchain, and others, 5G is accelerating the transformation and upgrading of social informatization.

This chapter mainly introduces industry insights, scenario and network demand analysis, and scenario-based solutions when applying 5G in traditional industries.

Learning Objectives

- To understand the current needs of traditional industries.
- To master network requirements in 5G+ scenarios.
- To understand scenario-based solutions of 5G+ in various industries.
- To understand relevant cases of 5G+ in various industries.

7.1 Power Private Network Scenario

7.1.1 5G+ Power Industry Insight

The power system is one of the largest scale and the most complex structures in the world, and it is critical to social development and the normal operation of other infrastructures. The classical power system includes five key links: power generation, power transmission, power transformation, power distribution, and power consumption. Specifically, it is a system of power production, transmission, and consumption,

L. Song et al., *5GtoB From Theory to Practice*,
https://doi.org/10.1007/978-981-97-8453-0_7

consisting of power generation by power plants, step-up and step-down transformations by substations, long-distance transmission by transmission lines, power distribution by supply and distribution offices, and power consumption by users. The basic physical law governing the operation of electric power systems is that the amount of electricity generated and consumed must remain equal at every moment because usually, the grid lacks storage capabilities. For the past century, the power system has been using the same basic laws of physics. Even in today's society, where technologies are highly developed and the efficiency of power systems has been dramatically improved, this basic law of physics has not been subverted.

As shown in Fig. 7.1, the common architecture of most existing power systems with the major components mainly includes the power generation at large power plants, the transmission grid that delivers energy at high voltage and over long distances to destinations, and the distribution of electricity to end-users, such as households, business, or industrial facilities through the medium and low voltage distribution networks.

According to the characteristics of the power secondary system, the power system can be further divided into the production control area and the management information area. The production control area is divided into a control area (Security Zone I) and a non-control area (Security Zone II). The information management area is

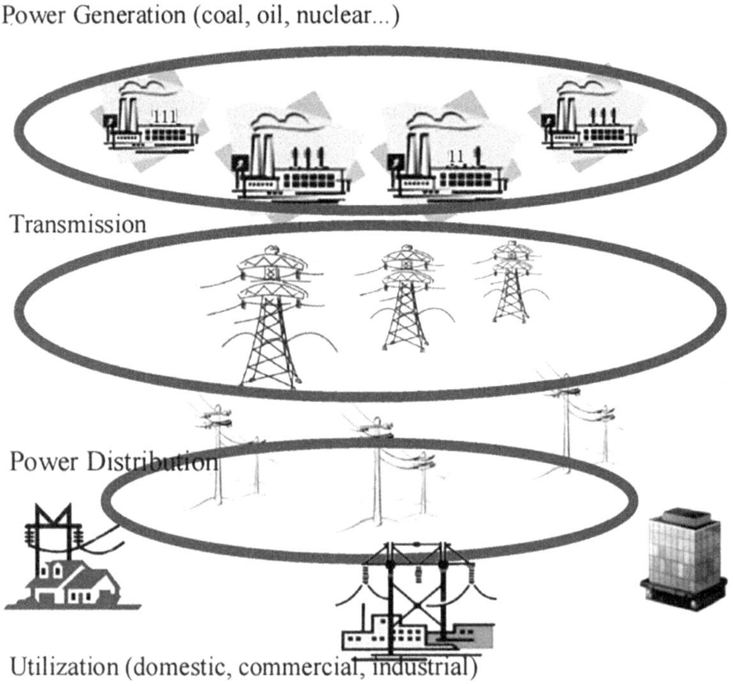

Fig. 7.1 Current scheme of the power system

divided into the production management area (Security Zone III) and the management information area (Security Zone IV). Different security zones have varying security protection requirements, with security Zone I having the highest security level, followed by security Zone II, and others in descending order.

Typical system of Security Zone I: Dispatching Automation System, Substation Automation System, Relay Protection, Safety Automatic Control System, etc.
Typical systems in Security Zone II: Reservoir Dispatching Automation System, Power Metering System, Relay Protection and Fault Recording Information Management System, etc.
Typical systems in Security Zone III: Dispatch Management Information System (DMIS), Lightning Monitoring System, Statistical Reporting System, etc.
Typical systems in Security Zone IV: Management Information System (MIS), Office Automation System (OA), Customer Service System, etc.

Most of the electricity is generated by large power plants. These power plants compensate for imbalances in energy supply and demand by controlling generation through customer feedback on supply and demand. However, due to the large inertia of the generators in large power plants and the relative insensitivity of the feedback-actuating components, the overall power generation system is relatively slow in adjusting to the demand of drastic changes. On the one hand, this inertia benefits the overall system because it constitutes a form of energy storage. On the other hand, it makes controlling large and rapid changes extremely challenging.

In the past, large imbalances between generation and load were mainly associated with faults or interruptions in the power system. Under normal conditions, power systems are assumed to be quasi-static, a fundamental assumption in the design of power control systems. This quasi-static nature relies on slowly changing load capacity and fully controllable generation. In most cases, this assumption is based on the immediate availability of fossil energy (certain fuels). However, some unstable energy sources, mainly renewables, do not possess this property. On the contrary, when they represent a large proportion of the power production, they can lead to huge variations, especially in high permeability.

In the classical view of the power system, the energy flow is unidirectional, from transmission to distribution, from the generator to the load. Generators and loads always maintain a property of "energy injection" and "energy absorption" relative to the grid, while the distribution system is completely passive.

As shown in Fig. 7.2, the power system has undergone profound changes in recent years, which has led to new monitoring and control challenges in its operation and unprecedented coupling with other infrastructures, especially communication and other energy networks. Based on the traditional power system, the new generation power system formed by integrating new energy, new materials, new equipment and advanced sensing technology, information technology, control technology, and energy storage technology has been characterized by high informatization, automation, and interaction. It effectively improves the efficiency and reliability of the grid and reduces economic costs.

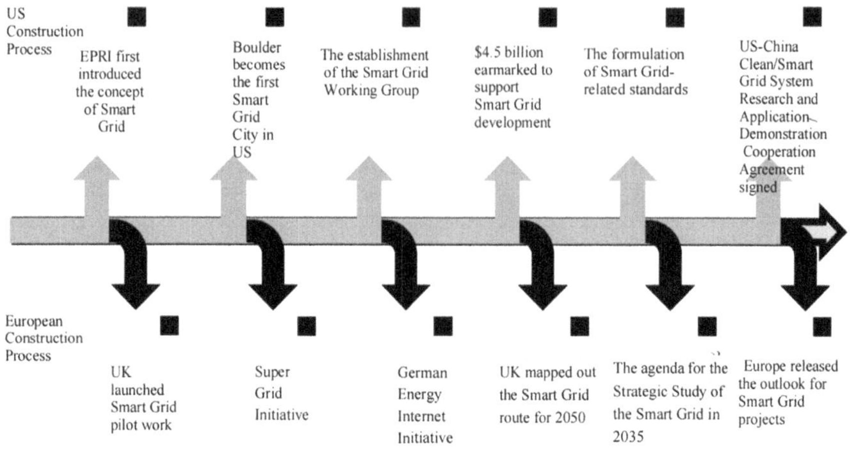

Fig. 7.2 Development trend of the power grid in recent years

The digital grid is a digital transformation of the grid. It makes full use of digital technology to achieve remote monitoring, control, and automated grid operation. The digital grid brings unprecedented opportunities to address changes across the energy spectrum (e.g., distributed generation) and improve the grid's reliability, availability and efficiency. With the rise of artificial intelligence technologies, the concept of a smart grid has emerged. Smart grids are transmission and distribution systems that integrate traditional and modern power engineering technologies, advanced sensing and monitoring technologies, and information and communication technologies with improved performance, providing value-added services to customers (e.g., the Future Energy Alliance).

The difference between the smart grid and digital grid is that the digital grid is only a part of informatization and automation in the smart grid, far less extensive than the smart grid, and only reflects digitization in information acquisition, transmission and control. In summary, the smart grid is a modern power generation and distribution infrastructure that can automatically respond to and manage modern power systems' growing complexity and demands. The functions of a smart grid are as follows.

(1) Support and enable grid integration of renewable energy sources, including solar, hydro, and wind energy.
(2) Provide real-time electricity consumption information to users.
(3) Help corporations of public utilities reduce inspection frequency and power outages.
(4) Differential protection, precise load control, and other functions.

It is estimated that the smart grid is expected to reduce the earth's carbon emissions by 58% in 2030. By applying new information and communication technologies, the level of intelligence and digitization of the grid will be further improved, and the interconnected characteristics of energy will initially appear. In the future, users and long-distance renewable energy sources will be more closely integrated to achieve coordinated control of energy production, transmission, and consumption, promote two-way interaction between energy supply and demand, realize energy sharing, and promote the energy revolution in production and consumption by integrating energy substitution, transmission, trading and dispatching operations.

The smart grid covers five key links: power generation, transmission, transformation, distribution, and consumption. Each link requires the deployment of dozens of smart grid communication services, and the requirements for service level agreement (SLA), such as latency, bandwidth, and security isolation, vary greatly. Compared with public network applications, 5G+ smart grid applications involve the safe and reliable operation of the grid, so there are more stringent requirements for security isolation and SLA guarantees. Smart grid services are widely used in various scenarios, including state-aware, control protection, data acquisition, artificial intelligence assistance and robotic preventive maintenance. Various transmission data types, such as status data, control commands, images, videos have different requirements on network bandwidth, latency, jitter, BER, reliability, etc.

7.1.2 Analysis of 5G Application Scenarios and Network Requirements

There are three main application scenarios of 5G in the electric power industry: control service, collection service, and mobile application service. In the control service, 5G technology will optimize energy allocation to avoid large area blackouts that affect electricity usage in enterprises and households. It will also meet the online monitoring needs of real-time dynamic data in the distribution network. 5G will collect and provide raw power consumption information across the system in the collection service. In the mobile application service, 5G can prevent security accidents and environmental pollution to reduce manual inspection workload. In future mobile inspection, it can also carry out simple hot-line work.

Control Service

The typical requirement of the grid control services for communication with low latency, high reliability, and high security is suitable for URRLC application scenarios in 5G technology systems. Typical technologies include precise load control technology and distribution network differential protection technology.

1. **Precise Load Control Technology**

1) Application Scenarios

Precise load control technology refers to when multi-infeed DC power grids experience multiple DC continuous commutation failures and faults resulting in DC block, large active power shortage and sharp frequency drop in receive-side power grids, according to the magnitude of DC power loss, through precise control of decentralized massive power user interruptible load, realizing friendly interaction between grid, power source and load, achieving transient equilibrium of power supply and demand, supporting large scale optimization of energy allocation. It avoids the occurrence of large-scale power outages, minimizes grid losses, and minimizes the impact on enterprises and residential electricity consumption.

The precision load control system based on the 5G communication network consists of a service terminal, demand response terminal, communication network, and main station system. When the load needs to be excised, the corresponding commands are sent by the co-control master station of the service master station or the service master station itself. After receiving the commands, the load capacity to be excised in each service terminal is obtained by calculating and analyzing the commands in the service master station. Then the corresponding commands and the load capacity to be excised will be sent to each service terminal. After receiving these commands, the corresponding load capacity will be excised by the load control terminal under the control of the service terminal through the corresponding communication interface. In this process, the service terminal is accessed by the building demand response terminal through the local network. In contrast, the demand response terminal is accessed by the main station of the demand response system through the 5G communication network. The specific application scenarios are shown in Fig. 7.3.

2) Communication Requirements

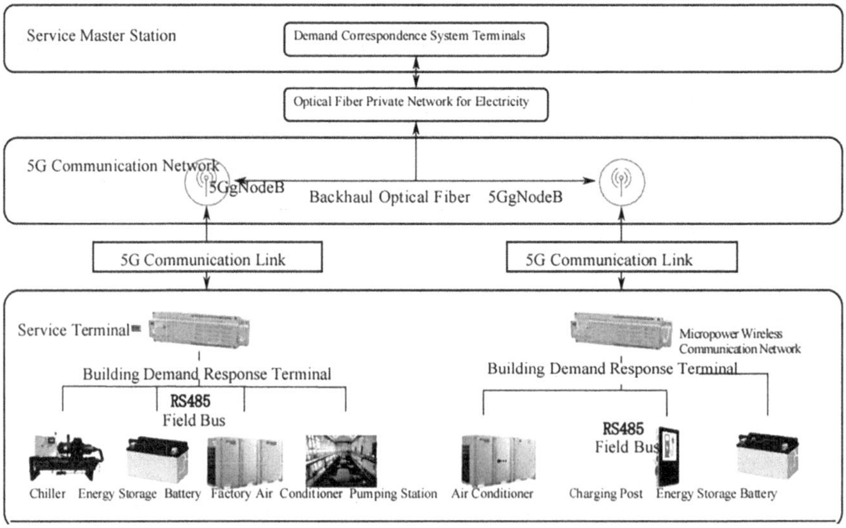

Fig. 7.3 Application scenarios of the technology of precision load control based on 5G communication network

The large power grid's balance of supply and demand needs to be quickly restored in the precision load control system. After DC block faults occur, It needs to be ensured that the grid frequency is restored to its normal value (50 Hz) within about 650 ms. Therefore, after the command of load excision was given, the transmission latency of communication through the channel between the master station and the command terminal cannot exceed 50 ms. The less the latency of the whole process in the precision load control system, the faster the grid faults be repaired. Therefore, a lower latency is better for communication. As for the communication requirements, the emphasis is on latency, availability, security, and reliability. The specific requirements are as follows.

(1) Bandwidth for sending control commands: less than 256 kbps.
(2) Latency: $\leq$ 50 ms from the control master (sub) station to the terminal.
(3) Communication reliability: greater than 99.999%.
(4) Connection density: less than 1000 per 100 square kilometers.
(5) Network slicing: end-to-end hard slicing and exclusive slicing resources.
(6) Security: strong security requirements requiring exclusive use of resources and physical isolation.
(7) Terminal timing accuracy: less than 10 μs.
(8) Communication mode: master–slave mode, permanent online, continuous high-frequency communication.

2. **Distribution Network Differential Protection Technology**

1) Application Scenarios

As a grid technology with mature applications in high-voltage power transmission networks, distribution network differential protection technology can solve many problems caused by distributed power access to the power distribution network. The principle is that the power distribution differential protection terminal compares the current values (vector) at the same moment at two or more ends and determines that a fault has occurred when the current difference exceeds the setting value. Then the circuit breakers or switches will be disconnected, and the action of differential protection will be carried out to realize the precise positioning and isolation of the fault in the power distribution network.

With the large-scale distributed new energy access to the distribution network, the charging load of electric vehicles has increased rapidly, and the interaction between users' supply and demand has become more frequent. The distribution network's source, network, and load show regular random fluctuation and intermittency due to its stronger space–time uncertainties, and the distribution network's bilateral flow and multi-source faults are increasingly prominent. For the systematic characteristics of a distribution network with multiple harmonic waves and strong noise, a high-precision, synchronous phase measurement unit (PMU) is adopted to meet the online monitoring application of real-time dynamic data of the distribution network.

A distribution network automation system generally consists of a distribution master station, a distribution substation (usually settled inside the substation, optional for distribution), a DTU (Distribution Terminal Unit), and communication networks. The work of modeling, real-time monitoring, and clock synchronization to the whole distribution system is in charge of the distribution master station; the distribution substation is a computer subsystem responsible for the collection, processing, and transmitting of the data of terminal unit for distribution network automation system within a residential quarter; Distribution Terminal Unit (DTU), usually installed in places such as regular switching station, small-scale outdoor switching station, RMU (ring main unit), small-scale substation, prefabricated substation, is to achieve the data collecting and calculation of the position signal, voltage, current, active power, reactive power, power factor and electric energy of the switch gear, and then perform a switching-closing operation on switches to achieve the fault recognition, isolation and power restoration for non-faulty zones. In addition, the Distribution Terminal Unit (DTU) includes the Feeder Terminal Unit (FTU) and Distribution Terminal Unit (DTU), etc. The communication network provides channels between the distribution master station and the distribution substation. Through the communication network, a large amount of information and data will be transmitted at high speed. In general, the mode of optical fiber communication is adopted in the distribution network automation system. It is characterized by large capacity, high safety factor, strong stability, etc. As a subsystem of the distribution network automation system, the distribution network differential protection system uses DTU and FTU to collect sampling data and tripping information and transmit them to adjacent terminals. Finally, the system can quickly respond to faults through differential protection algorithms. This system has characteristics such as fast speed, dynamic adaptability, reliable fault detection, etc. The specific application scenarios are shown in Fig. 7.4.

2) Communication Requirements

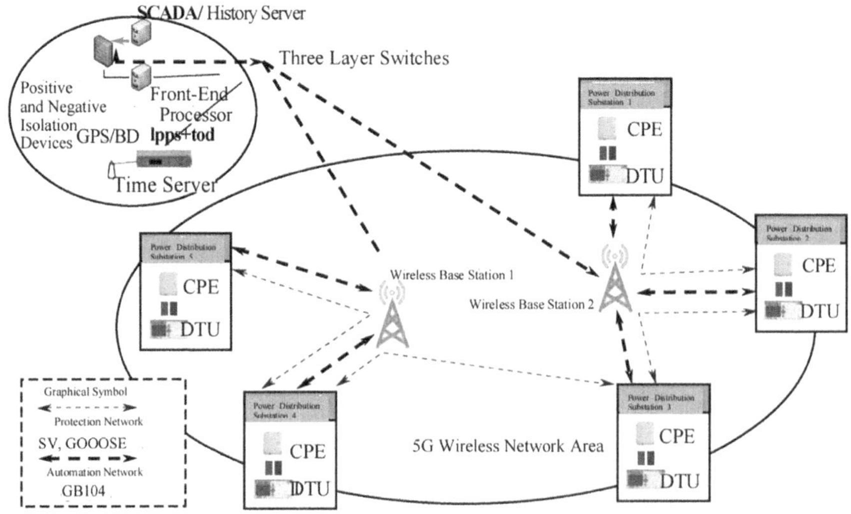

Fig. 7.4 Architecture of differential protection system based on 5G technology

The differential protection system requires the calculating of current difference to be based on the current values at the same time. Two or more correlate differential protection terminals must ensure time synchronization with an accuracy no more than 10 μs. The transmission latency of mutual information must be controlled under 12 ms [the maximum latency for Peer to Peer (P2P)]. Hence, the requirements for communication place great emphasis on latency, availability, and reliability. The specific communication requirements are as follows.

(1) Bandwidth: Less than 10 Mbps.
(2) Latency: Equal or less than 15 ms in service systems, equal or less than 10 ms in communication systems.
(3) Communication reliability: Greater than 99.999%.
(4) Connection density: Less than 1000 per 100 square kilometers.
(5) Network slicing: End-to-end hard slicing and exclusive slicing resources.
(6) Security: Strong security requirements requiring exclusive use of resources and physical isolation.
(7) Timing accuracy: Less than 10 μs.
(8) Communication mode: Multi-party communication, master–slave (automation service) mode, device-to-device, permanent online, continuous high-frequency communication.

Collection Services

The communication requirements of power grid for collection services are characterized by multiple points and wide coverage areas, low latency requirements for communication, difficulties in the coverage of wired communication methods, suitability for massive machine-type communication (mMTC) application scenarios under the 5G technology system, and typical services include the collection of electricity consumption information, etc.

1. **Application Scenarios**

The collection of electricity consumption information can realize the automatic collection of electricity consumption information, abnormal metering monitoring, power quality monitoring, and the analysis and management of electricity consumption, etc. The electricity information collection system consists of a master station, remote and local channels, concentrators, collectors, or electric meters.

The collection system of electricity consumption information is logically divided into three layers: the master station layer, the communication channel layer, and the collection equipment layer. The master station layer has two major servers: the application server and the data server. The application server implements the service logic of various applications in the system, collects electricity information from terminals, and analyzes protocols; the data server is used to realize communication management and scheduling of terminals. The communication channel layer is the link between the master station and the collection equipment, providing wired and wireless communication channels and serving as a link for information interaction between the master

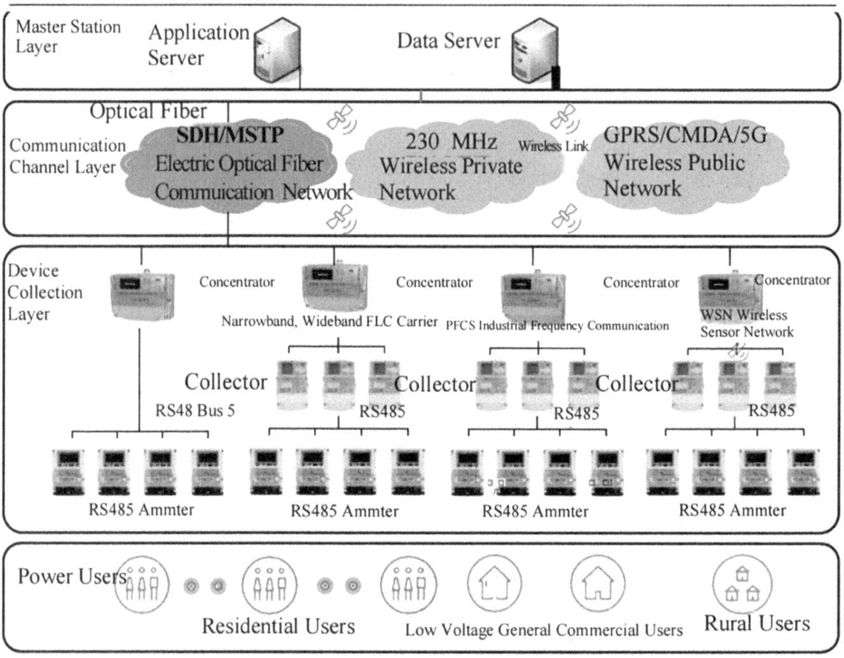

Fig. 7.5 Framework of the collection system of electricity consumption information

station and the terminal. The communication modes mainly include optical fiber private network, wireless public networks, wireless private networks, and medium voltage power line carrie. The collection equipment layer is at the bottom of the electricity information collection system, collecting and providing the original electricity information of the whole system. The framework diagram of the collection system of electricity consumption information is shown in Fig. 7.5.

2. **Communication Requirements**

The collection service of electricity consumption information requires the communication rate (no less than 1.05 kbps) of different users and the transmission rate of load control commands is not less than 2.5 kbps. The specific communication requirements are as follows.

(1) Bandwidth: Uplink less than 2 Mbps and downlink less than 1 Mbps.
(2) Latency: Public or specialized substation testing and low-voltage meter reading need to be less than 3 s, and precise cost control needs to be less than 200 ms.
(3) Communication reliability: Greater than 99.99%.
(4) Connection density: Less than 1000 per 100 square kilometers.
(5) Security: High-security requirements, security encryption authentication, security access area authentication, IoT slicing, and logical isolation.
(6) Communication mode: Permanent online, frequency less than or equal to once every 5 min.

Mobile Application Services

The communication requirements of mobile application services are characterized by high communication bandwidth and security. They are suitable for enhanced mobile broadband (eMBB) application scenarios under the 5G technology system, with typical examples being mobile inspection services.

1. **Application Scenarios**

Mobile inspection services mainly include three scenarios: substation inspection robot/high-voltage line inspection UAV, mobile on-site construction control, and the integrated application of self-assembled networks on urgent accident spots. These services are mainly suitable for low and medium-rate mobile scenarios in the management of power production, and they are capable of collecting and identifying pictures and videos of distribution cabinets and switchgears to extract some information about their operating status and switch resources. In performing this task, these services can effectively oversee the machine room's overall environment, thereby preventing potential accidents and environmental pollution. Additionally, they can significantly reduce the workload associated with manual inspections, eliminating the uncertainty that often arises from manual on-site operations and lowering labor costs. Hence, there is a great improvement in the operation and maintenance efficiency. In the future, simple hot-line operation can be performed for the mobile inspection service.

The following is an example of the introduction of the inspection robot in the substation. The substation inspection robot services mainly target the requirements of comprehensive monitoring, and security inspection of power primary equipment in the substations with the voltage of 110 kV and above. The current inspection robot mainly uses Wi-Fi access, and the video information of the inspection is mostly retained in the station instead of being backhauled to the remote monitoring center in real-time. The main application scenarios of the substation inspection robot services are shown in Fig. 7.6, including the inspection service master station, the substation inspection service platform, and external sensors for detecting the state of the substation.

2. **Communication Requirements**

In the future, substation inspection robots will be mainly equipped with multiple high-definition video cameras or environmental monitoring sensors to backhaul relevant testing data. The data must have the capability to be backhauled to the remote monitoring center in real-time. Inspection robots can sometimes even carry out simple hot-line work, such as the barrier gate switch control. The communication requirements are mainly needed in the multi-channel high-definition video backhaul (Mbps level), and low-latency remote control inspection robot (millisecond level). The future communication requirements for mobile inspection services include:

(1) Bandwidth: A sustainable and stable guarantee of 4–100 Mbps is required depending on the scenario.

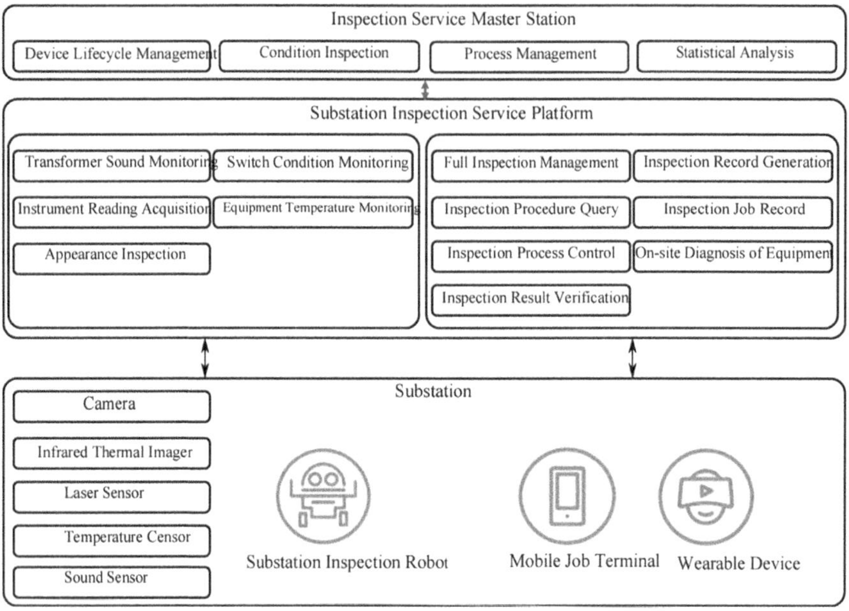

Fig. 7.6 Application scenarios of mobile inspection services

(2) Latency: The latency of multimedia information needs to be less than 200 ms, while the latency of control information needs to be less than 100 ms.
(3) Reliability: A reliability of 99.9% is required for multimedia information, while a reliability of 99.999% is required for control information.
(4) Isolation requirements: Belonging to Grid Zone III services, requirements for security are lower than those for Zone I or Zone II; however, some control functions, such as the remote operation of the control information by inspection robots, belong to Zone I or Zone II.
(5) The number of connections: Concentrated in local areas ranging from 2 to 10.
(6) Mobility: The mobility rate is relatively low, within the 10–120 km/h range.

7.1.3 Scenario-Based Solutions

In power generation, transformation, and transmission phases, fiber optic network coverage is usually available. However, in the phases of power distribution and consumption, the decentralized distribution of a large number of nodes leads to the relatively low coverage of the communication network. It is essentially the "last five kilometers" problem, the main bottleneck for smart grid development. According to the National Energy Administration, 95% of the power failures occur in the last 5 km of the grid.

Fig. 7.7 Typical service scenarios of smart grid

To make smart grid a reality, building a terminal access network in the last 5 km of power distribution and consumption is critical to realize the communication and control to these terminals. To accelerate the pace of development, the National Energy Administration proposed that by 2025, China's terminal access network should have covered 90 percent of the total terminals, achieving the intelligent distributed feeder automation, millisecond-level accurate load control, low-voltage distribution system and distributed power information acquisition, as shown in Fig. 7.7. These scenarios have different requirements for network performance, including latency, bandwidth, massive connectivity, and reliability.

Deploying terminal access networks is difficult due to the high cost of using optical fibers, large connection volume, and wide coverage area. However, 5G can effectively solve these problems.

5G network slicing is ideal for implementing smart grid services. It divides the 5G network into logically isolated networks, each of which can be considered a separate slice. 5G network slicing allows the grid to flexibly customize specific slices with different network functions and service level agreement (SLA) guarantees as needed to meet the diverse network requirements for the various services mentioned above.

Innovation projects have demonstrated that 5G network slicing can meet the diverse needs of terminal access networks, including wide coverage, low latency, high bandwidth, high reliability, high security, and differentiated network capabilities. Moreover, it is more cost-effective and easier to deploy than traditional fiber networks.

5G network slicing has rich functions. Network slicing is a tenant-oriented virtual network that can meet differentiated SLA requirements and be managed independently according to the life cycle. 5G network slicing automatically builds isolated instances based on specific service needs and differentiated SLA requirements. 5G network slicing provides end-to-end network security, enabling SLA, service isolation, and the customization and automation of network functions, as shown in Fig. 7.8.

(1) Isolation Services: Security isolation of different levels can be realized.

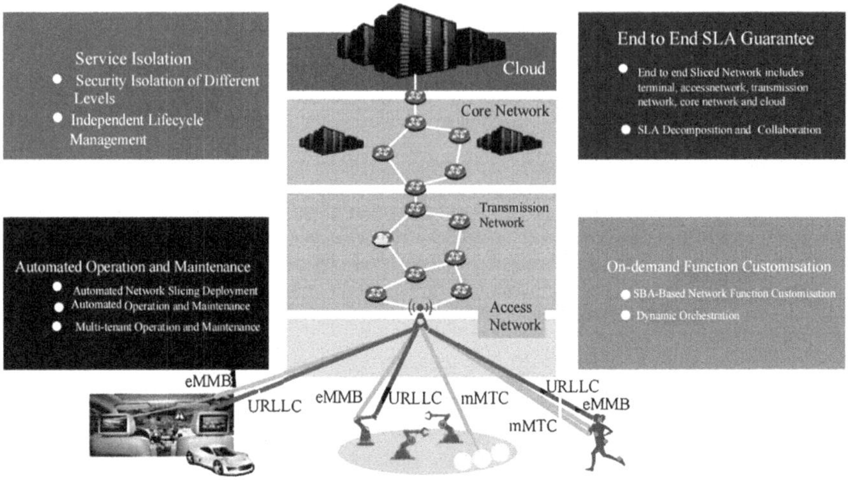

Fig. 7.8 Key capabilities of 5G network slicing

(2) End-to-End SLA Guarantee: End-to-end sliced network includes the terminal, access network, transmission network, core network, and cloud. SLA can realize decomposition and collaboration.
(3) Automated Operation and Maintenance: Using 5G edge computing technology and distributed gateway deployment, local traffic processing and logical computation can be achieved to save bandwidth and reduce latency. This further meets the ultra-low latency requirements for industrial control services on the power grid.
(4) On-Demand Function Customization: It enables the customization of network functions and dynamic scheduling of SBA.

Table 7.1 shows that 5G network slicing can meet the connectivity requirements of core industrial control services on the power grid.

In terms of service characteristics, the typical smart grid service scenarios discussed in this section can be divided into two categories.

(1) Industrial Control Services: Typical examples include the automation of the intelligent distributed feeder and millisecond accurate load control, etc. URLLC is a typical slice designed for such services.
(2) Information Gathering Services: Typical examples are the low-voltage distribution system and the information acquisition of distributed generations, etc. mMTC is a typical slice designed for this type of service.

In addition to the two typical types of slices mentioned above, the grid industry may also need eMBB (typical service scenario: UAV remote inspection) and voice communication (typical service scenario: manual maintenance and inspection).

Regarding service deployment, 5G can provide new industrial control services for power grids and retain the existing information collection services supported by

Table 7.1 5G network slicing meets various demands of smart grid for different scenarios

Service scenario	Communication latecy requirement	Reliability requirement	Bandwidths requirement	Terminal number requirement	Service isolation requirement	Service priority	Slice type
Intelligent distributed feeder automation	High	High	Low	Middle	High	High	URLLC
Millimeter-scale precise loud control	High	High	Middle/low	Middle	High	Middle/High	URLLC
Information acquisition in low voltage distribution system	Low	Middle	Middle	High	Low	Middle	mMTC
Distributed power information acquisition	Middle/High	High	Low	High	Middle	Middle/Low	mMTC (Uplink) + URLLC (Downlink)

the 2G/3G/4G public network. In this way, multiple partitions of the grid can be deployed, managed, and maintained unified, helping customers in this industry to reduce costs effectively.

The 5G network slicing architecture for smart grid applications is shown in Fig. 7.9. The overall architecture of the design and management of the 5G smart grid is as follows: low-voltage distribution system information acquisition, smart distributed feeder automation, and millisecond accurate load control are used to meet the technical specifications required by different service scenarios. They differ from each other in terms of how the distribution network is managed, network latency, whether data is encrypted, forwarding, connection scheduling and routing methods. Sub-domain slice management and end-to-end integrated slice management can meet the service requirements in these scenarios.

The life cycle management of 5G network slicing includes slice design, deployment, and enablement, closed-loop control, operation and maintenance (O&M) monitoring, capacity exposure, etc., as shown in Fig. 7.10.

Both operators and power companies are involved in the operation and maintenance of smart grid slices. Due to the different professional levels of users in this industry and the operators' different O&M processes and maintenance requirements, two types of O&M need to be designed. The differences are shown in Table 7.2.

For operators, FCAPS capabilities are required, namely Fault Management, Configuration Management, Accounting Management, Performance Management, and Security Management. These five basic functions can improve the overall professional abilities of operators' O&M staff and network efficiency. For tenants, an easy-to-use O&M interface is needed to help them achieve the fastest speed, the most natural experience, and value through the network and applications.

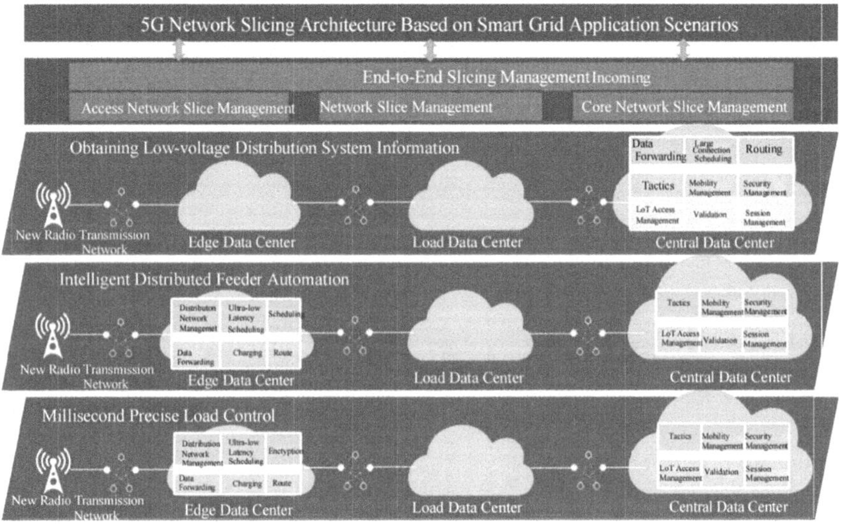

Fig. 7.9 Architecture of 5G network slicing

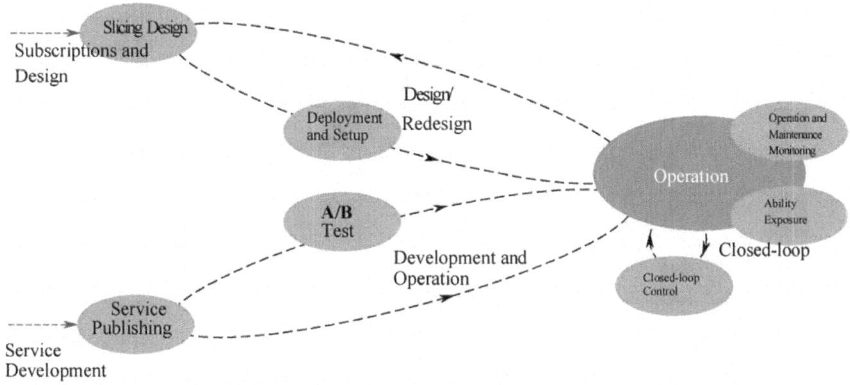

Fig. 7.10 Life cycle management of 5G network slicing

Table 7.2 Two O&M modes for 5G network slicing in smart grid

Project	For operators	For electric companies
GUI	Same habits based on traditional EMS	Easy to use
Goal	Integrated network perception	SLA verification
Data presented	Integrated status and statistics, fixed	Customize key information and variables
Control range	Comprehensive services and resourcing	Restricted service configuration
Slice range	Cross section	Bulit-in Slice

7.1.4 Application Cases

In April 2019, a power supply company developed the world's first power slicing that meets the latest 5G SA specification released by 3GPP. This is the first time in the industry that a power grid slicing test was conducted in a real power grid environment. The success of this test marks the in-depth exploration of 5G vertical industry applications.

The SA power slicing fully leverages the millisecond-level latency of 5G networks and the SLA assurance of network slicing. It strengthens the bi-directional communication between the grid and end-users, ensuring precise management of low-power units on overloaded grid terminals. These benefits help minimize the economic loss and social impact of power failures.

The power supply company conducted end-to-end field testing of 5G SA power slices with electrical terminals. Figure 7.11 shows that after deploying 5G base stations in a plaza and a district, the company conducted indoor and outdoor local, mid-range, far-end, and obstacle-blocking tests. During the power server processing,

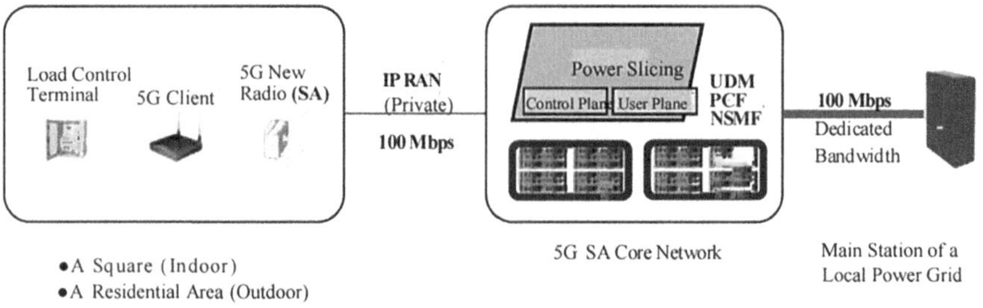

Fig. 7.11 Power slicing as the testing network

network command transmission, and load control terminal processing, an approximately 35 ms end-to-end latency was identified, which may occasionally experience slight fluctuations. In addition, slicing isolation was also fully verified. The slice has been proven to meet the critical requirements for millisecond-level precise management of load processing units on a telecom network.

To adapt to the trend of business development, a power grid company actively promotes digital transformation. However, one of the main bottlenecks in this process is the distribution network communication. The current communication network of this company is the main network communication, mainly self-built plus the distribution network communication, which mainly consists of public networks with self-built networks as a supplement. The main network communication has a perfect optical fiber communication network in which the fiber optic cable achieves the full coverage of plant stations, offices or business halls with a voltage of 35 kV and above; the distribution network communication: there are many distribution terminals with large coverage areas. Currently, the distribution network communication mainly relies on the wireless public network due to the high cost, large investment and difficulties in the operation and maintenance of laying fiber optic cable in the distribution network. Specifically, 99.34% of the metering automation business uses wireless public networks, while 89.08% of the distribution automation business does so.

To support the deep integration of 5G and power grid business, this power grid company has been working with several communication companies since 2018 to build a 5G VPN. They have jointly promoted research on the 5G+ digital power grid and completed technical verification in 2019. Since 2020, they have gradually started commercial use to support the deep integration of 5G and power grid business.

In August 2020, the power grid company pioneered the commercialization of the world's first 5G timing Customer-Premises Equipment (CPE) terminal in a certain place. This groundbreaking technology successfully underwent commercial trial operations, meeting the requirements of power grid control class services with millisecond low latency and microsecond high precision timing. The significance is that the previous power grid control service must be carried by wired means, including some security regulations at the national level, which must also be carried by wired means. It is a breakthrough in using 5G wireless technology to carry the control commands of the power grid.

5G brings all-round upgrades to the power grid, significantly improving the efficiency in production and operation. Firstly, the efficiency of the inspection of equipment in the grid has been greatly improved. A survey form with 1330 inspection items from Shenzhen Power Supply Bureau, which would have taken three days for manual inspection, can now be completed in just 1 h using 5G+ UAV. To inspect an electric transmission line of 500 kV, the traditional manual inspection requires two people to inspect for ten days, but now it can be done in 2 h by using a 5G+ UAV. Secondly, it can achieve zero perception of power failures. 5G's low latency characteristics effectively carry the distribution network differential protection to change the fault isolation time from seconds to milliseconds, thus significantly reducing the recovery time of the power supply when a power failure happens in the distribution network. Currently, typical business applications for transmission, transformation, and distribution have been implemented in the region, and application practices for the 5G+ digital power grid will be launched gradually in provinces and regions under the jurisdiction of the company in the future.

7.1.5 Summary (of 7.1)

The power system is an electricity production and consumption system composed of power plants, transmission and distribution lines, power supply and distribution substations, and end-users. Its function is to convert the primary energy in nature into electrical energy through the power plant and then supply the electrical energy to each user through transmission, substation, and distribution. To achieve this function, the power system also has corresponding information and control systems in each link and at different levels. These measures, regulations, controls, protects, communicates, and dispatches the production process to ensure that the users get safe and high-quality electricity. The power system is one of the world's largest and most complex infrastructures. The classic power system covers five key aspects: power generation, transmission, transformation, distribution, and consumption. Digital power grid refers to the digital transformation of the power grid, which means fully utilizing digital technology to realize remote monitoring, regulation, and automated operation. The construction of a digital power grid is the process of the digitalization, intellectualization, and internetization of the traditional power grid. The digital transformation of the traditional grid follows the network security standards and constructs a corresponding digital twin power grid based on a unified power grid data model. With advanced digital technology platforms, it forms powerful "computing capacity" by using "computing power + data + model + algorithm", relying on the IoT (Internet of Things) and the Internet to connect all parties related to the power grid in perception, analysis, decision-making, services. This gives the power grid company super perception, wise decision-making ability, and rapid execution capabilities. Consequently, the boundaries of the digital power grid are expanded beyond traditional power grids to encompass all aspects of society. This transforms traditional power grids' management, operation, and service modes while driving wide allocation of energy flow, capital flow, logistics, business flow, and talent flow in related industries. Furthermore, it promotes energy revolution and new energy system construction using "electric power + computing capacity", thereby contributing to national economic modernization and building a secure digital power grid system.

A smart grid, as a typical example of vertical industry, poses new challenges for communication networks. The diversity of power grid services requires a flexible and coordinated network, while high reliability demands an isolated network. Furthermore, millisecond ultra-low latency requires a network with optimal performance. 5G network slicing is an on-demand networking approach that allows operators to separate multiple virtual end-to-end networks on a unified infrastructure. In this process, each network slice is logically isolated from RAN (Radio Access Network) to the transport network and then to the core network, catering to various types of applications. In 5G systems, the network will be further abstracted into network slices: the functions of such connectivity services are defined by a number of customized software. These software functions include geographic coverage area, duration, capacity, speed, latency, reliability, security and availability. 5G network slicing can meet the needs of terminal access networks for wide coverage, low latency, high bandwidth, high reliability, high security and differentiated network capabilities. It also has lower cost and easier deployment than traditional fiber-optic networks. 5G network slicing can address and meet various network connectivity requirements from different vertical industries. The 5G Smart Grid Innovation Project has been proven to be a model in terms of the cooperation between global operators and enterprises in the 5G era.

7.2 Police Private Network Scenarios

The complex social environment, heavy police services, limited police force, and the increasing safety requirements of the people have greatly pushed the public security industry to accelerate its transition from traditional policing to smart policing. This section will provide a brief overview of 5G+ policing industry insights, application scenarios, the analysis of network demand, and scenario-based solutions. Through relevant case studies, the section will also systematically demonstrate how emerging technologies such as 5G can be leveraged to achieve mobile smart policing.

7.2.1 Insight into 5G Applications in the Police Industry

At present, the huge improvement in China's economic strength, technological strength and comprehensive national power have led to the increasingly pressing need to realize smart policing. In 2019, the National Public Security Work Conference made it clear that the commission should focus on "building the police politically, strengthening the police by reforms, developing the police with science and technologies, and governing the police strictly", fulfilling the duties and missions entrusted by the Party and the people in the new era. The Fifth Plenary Session of the 19th Central Committee clearly stated that by 2035, China aims to achieve a higher level of security construction, creating a comprehensive, three-dimensional and intelligent social security network and promoting the modernization of the social security prevention and control system. Entering the twenty-first century, China has a large population density, high mobility, and complex service forms. However, the overall number of police officers in public security has not increased, which means that this industry must enhance the efficiency of policing with the help of technologies like 5G and big data. China has many scenes differentiated from each other, covering a wide range of areas. Therefore, without the ability to quickly set up controls over a wide range of areas, it is impossible to complete global space–time retrieval, analysis, and rapid processing. In addition, the front-line officers with limited numbers cannot fulfill increasingly heavy tasks like street patrols and major security. Therefore, to meet the people's pursuit of a safe environment, the public security industry has moved from traditional policing to smart policing and shift from post-investigation to advanced prevention.

Technological innovations like 5G promote smart policing into a mobile and fully connected era, while the mobile scenarios of smart policing also face new demands and challenges.

1. **Daily Patrol**

At present, there are many problems in daily police patrols and they are mainly as follows.

(1) Insufficient Police Resources: Manual monitoring and daily patrols take up a lot of police force. Currently, the proportion of police officers in China is about 12 per 10,000 people, while in Russia, there are 55 officers per 10,000 people, and in the US, about 24 officers per 10,000 people.
(2) High Workload: With a fivefold increase in cases, the police force has only increased by 30%, resulting in heavy tasks and a high workload for grassroots public security agencies.

2. **Surveillance in Key Areas**

The current surveillance technologies have the following shortcomings for key areas, such as airports and train stations.

(1) Video Surveillance: insufficient front-end capacity, incomplete scene acquisition, low accuracy rate and poor mobile video performance
(2) Temporary Checkpoint Deployment: insufficient coverage of fixed cameras, long time and high cost for temporary deployment and construction

3. **Police Enforcement**

Now the requirements for technologies in scenes of police dispatch are as follows.

(1) Strict Requirements for Police Timeliness: Regarding the speed of police dispatch, it requires 3 min for police to be on the scene, and the disposal method needs to be formulated in advance.
(2) On-site Law Enforcement: Most of the existing law enforcement devices cannot backhaul real-time scene footage and, therefore cannot achieve remote analysis and command scheduling.

4. **Mobile Video Backhaul**

Transmitting back videos from a variety of mobile devices suffers of the following challenges:

(1) Police Car/Robot Patrol: poor synergy of temporary command system, insufficient clarity of high-definition video under 4G network.
(2) Unmanned Aerial Vehicle (UAV) Patrols: UAVs with their own backhaul channels cannot patrol a wide range of areas, and unauthorized frequency bands are susceptible to interference.

5G technology advantages can address the needs and challenges listed above well, which brings a new paradigm to the policing industry. As shown in Fig. 7.12, various industries are vigorously promoting the ecological deployment of 5G police terminals. The large bandwidth of 100 Mbps to 1 Gbps in 5G uplink can help cover the process of mobile policing with the application of broadband and realize high-definition video surveillance. The low latency of 5G promotes the application of UAVs in police affairs and the remote control of unmanned vehicles. In addition, 3D fine-grained network positioning makes security monitoring and police positioning more accurate. The security assurance of data, users and architecture also meets the requirements of credible policing, which has given rise to many 5G-based smart police terminals, as shown in Fig. 7.13, and has gradually formed an increasingly mature industrial ecology.

7.2.2 Analysis of 5G Application Scenarios and Network Requirements

5G technology enables the landing of business scenarios. High bandwidth can meet the network demand for image transmission, and high capacity can better build police IoT. MEC and slicing technology can realize timely alarms and fast-speed alarms, as well as high-speed alarms and highly reliable transmission of police data. As shown in Fig. 7.14, 5G can change five major scenarios for the policing industry.

Scenario 1: Video Deployment and Control (Patrol and Control)

5G technology makes the wired network wireless, which can be used to wirelessly backhaul the view of fixed points (including electronic police, checkpoints, road

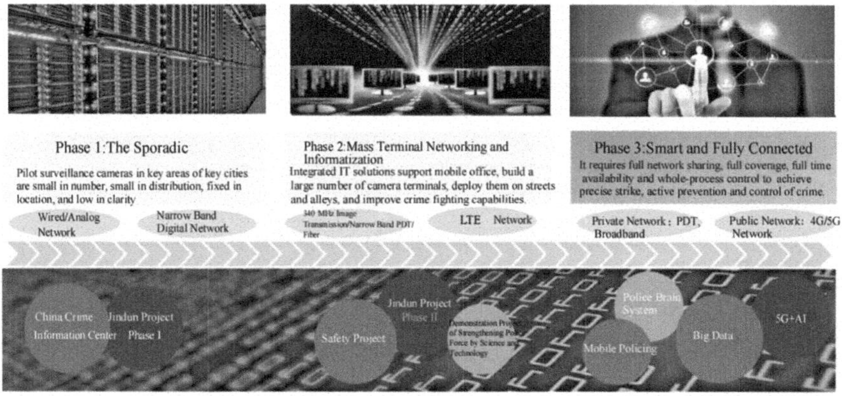

Fig. 7.12 Transition of smart policing in different eras

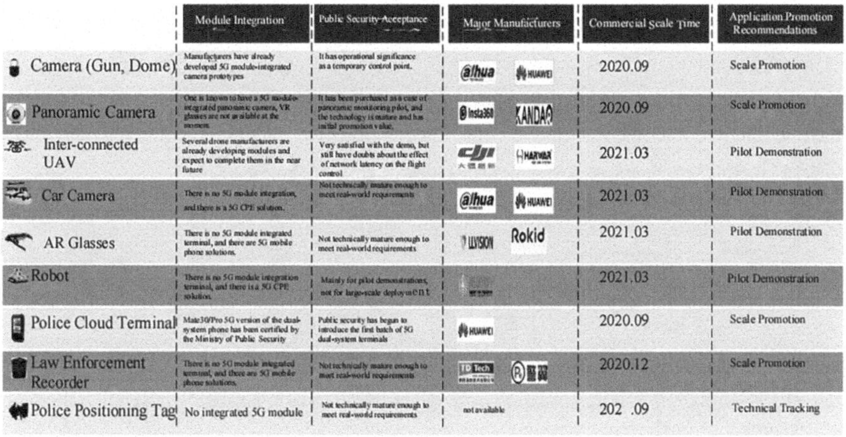

Fig. 7.13 Industrial ecology of 5G police terminals

video surveillance, etc.) in places such as high-speed road sections, where ground-breaking construction is inconvenient and wired optical fiber is not easily accessible. In addition, it can also achieve rapid mobile deployment to meet the needs of temporary supplementary points, such as outfield electronic police, checkpoints, and major event protection. As shown in Fig. 7.15, these pictures depict several patrol and control scenarios. The large uplink bandwidth of 5G allows for higher resolution, a higher recognition success rate and higher video quality. Thus, it can explore the construction of 5G intelligent video surveillance in key areas and monitoring blind zones, which can increase the monitoring coverage and achieve flexible deployment and rapid response.

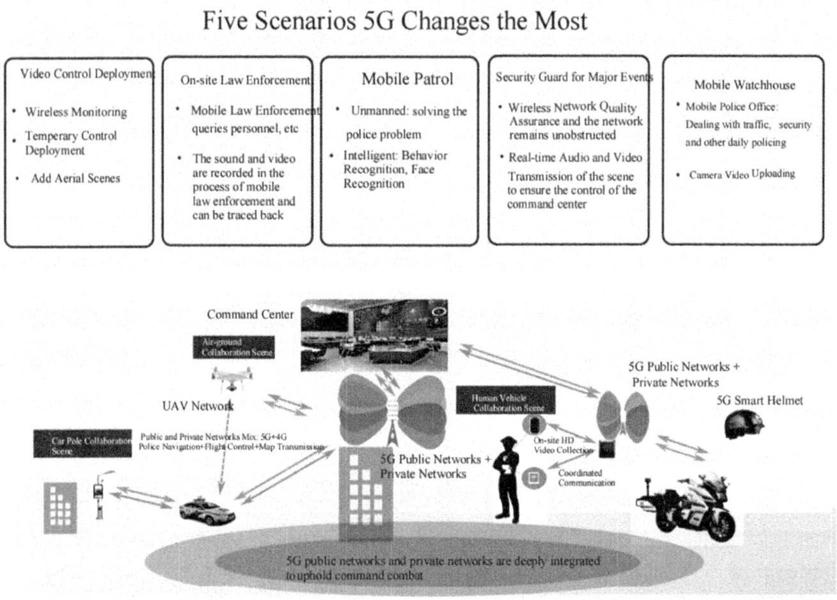

Fig. 7.14 Pictures focusing on five major scenarios of smart policing

Fig. 7.15 5G smart policing—cases of patrol and control scenes

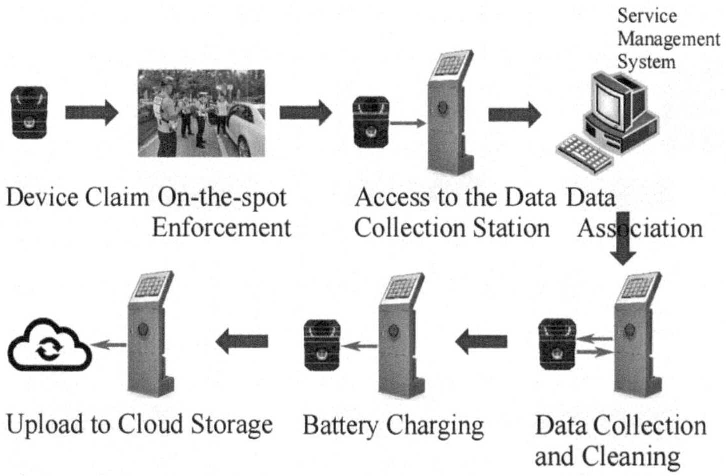

Fig. 7.16 5G smart policing—basic business process of the on-site law enforcement scene

Scenario 2: On-site Law Enforcement (Police Service System, Police Enforcement Apparatus and AR Glasses)

With the integration of 5G technology, it is now possible to achieve real-time backhaul and handling of image and video footage, as well as law enforcement records during traffic police patrols on motorcycles and vehicles. This allows for the immediate investigation and punishment of any illegal vehicles. The secure and synchronous transmission and verification of driving licenses and identity cards is also made possible. Apart from that, videos recording the whole process of law enforcement will be backhauled to the command center. The police AR glasses can also provide the first-person perspective video shooting function to realize the noninductive comparison and alarming for suspicious vehicles and persons. Figure 7.16 shows the basic business process of the on-site law enforcement scene.

Scenario 3: Mobile Patrol (Police Car, Unmanned Vehicle Patrol)

As shown in Fig. 7.17, problems include insufficient police force and difficulties in fixed network deployment for some hotspot areas, such as parks, scenic spots, and intersections. Aiming at these problems, police cars and robots are deployed nearby and 5G networks are used to access high-definition cameras to achieve high-definition video surveillance. For areas with large jurisdictions, dense populations, and many large-scale activities, 5G networks are convenient to deploy. For those areas that are not deployed with fixed networks, efficient and low-cost solutions are available. Patrol robots help police officers conduct regular and fixed-line inspections on jurisdictions, and 5G police cars can realize video backhaul in mobile scenes.

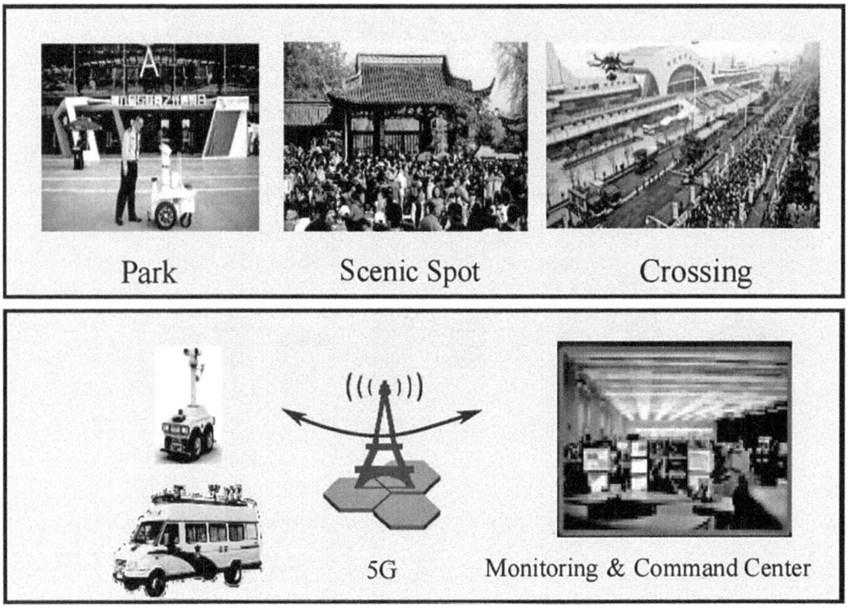

Fig. 7.17 5G intelligent policing—sample pictures of patrol and control scenarios

Scenario 4: Security Guards for Major Events

The utilization of new-type sensing equipment on testing objects, like documents, portraits, vehicles, and signs, can achieve the total-factor real-time dynamic monitoring in key areas of security guards. Police cars using 5G technology can implement the plan automatically according to the route and trace the target object by opening monitoring videos on high or low locations along the route. High-point video surveillance is deployed in a higher position, which can achieve a wide range of accurate monitoring; on the other hand, low-point video surveillance is deployed in a lower position, and it focuses on details like the local screen of close-up shots. Hence, the monitoring probes are coordinated at different heights, with long-range and short-range complementary views, forming a three-dimensional and all-round video surveillance system. In addition, 5G technology can guarantee that police vehicles for security can achieve rapid site deployment. Overall, 5G can realize the visualized deployment of security police forces in major events, full control of major events, and real-time warning of emergencies. Figure 7.18 shows several scenes of guarding for major events.

Scenario 5: Mobile Watchhouse

5G technology can deploy business capabilities to the edge to enable the rapid processing of policing. This can effectively fill the gap of police-public contact and serve as a useful addition to fostering harmonious ties between the police and the public, allowing for efficient handling of specific matters, opinions and suggestions

Large-scale Concert   Holiday Celebration   Diplomatic Activity

Fig. 7.18 5G smart policing—examples of security guard scenarios for major events

Fig. 7.19 5G intelligent policing—examples of mobile watchhouse scenes

raised by the public. As shown in Fig. 7.19, the mobile police office is equipped with a police office system, police communication system, monitoring system, image transmission system, power supply system, lighting system, and simple living facilities, which can allow many people to work simultaneously in it. They can deal with affairs like doing office work, consultation, and field operations. Mobile watchhousess have powerful functions such as an on-the-spot office and command, emergency communication, image transmission, strong sound broadcasting, and patrolling in harsh environments.

7.2.3 Scenario-Based Solutions

The 5G+ policing scenes need to continuously explore the network capability baseline of the 5GtoB wide-area private network and launch standardized 5GtoB commercial products. As shown in Fig. 7.20, some industrial terminals like unmanned police cars, 5G cameras, unmanned aerial vehicles, and panoramic cameras need different atomic applications and the support of atomic capabilities, respectively, which poses higher challenges for network requirements. It also imposes more stringent demands on network design, integrated deployment of services, and services of business assurance, etc.

After continuous exploration and attempts, 5G police terminals and network solutions are now available for preliminary commercial use. As shown in Fig. 7.21, the

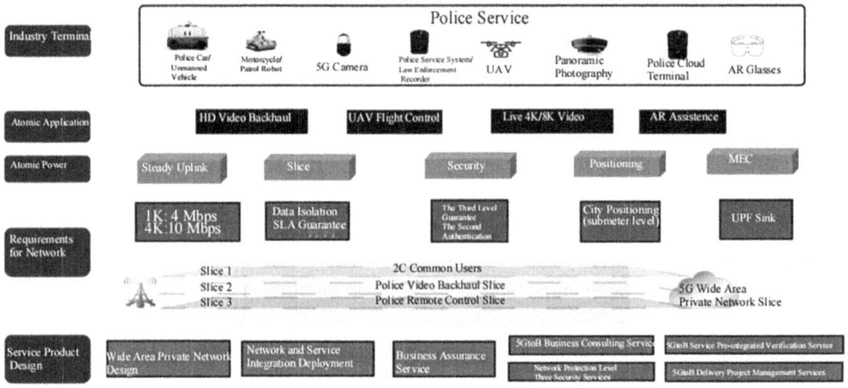

Fig. 7.20 Network appeal for 5G police wide-area private network

network products for 5G policing can be divided into three levels: dedicated channel (5G VPN), priority guarantee (5QI), and resource reservation (RB or frequency cut). The three different product grades of 5G policing private network can be ordered and used on demand. Users can choose a high, medium, or low level of service guarantee according to their needs and incur a one-time charge for the additional expenses required for security equipment, personnel, and network costs during major event protection. In addition, slicing services are required to be ordered and take effect at a fast pace. If the user-level slicing subscription method is adopted, it needs to support online ordering, real-time activation, dynamic effectuation, and other functions; if the shared resource pool method is adopted, it needs to support the managing capability of slicing members, which is convenient for users to add, delete, change, check and other operation in real-time.

Take the solution test of 5G smart police network slicing as an example. The demonstration scenarios can be divided into daily security patrol (Case 1 and Case 2) and major event security (Case 3).

Case 1: Daily security patrol. As shown in Fig. 7.22, 5G policing slicing (5QI = 9 noGBR) can meet the service requirements in networks under the light-load state

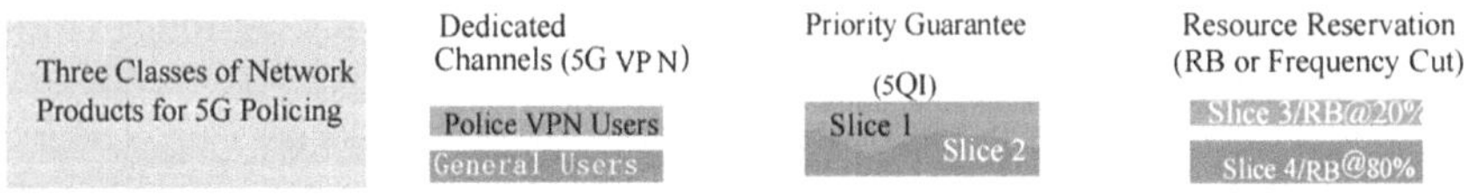

Applicable Scenarios	Daily Monitoring	Routine Enforcement	Major Security
Applicable Terminals	5G Cameras, Patrol Robots Unmanned Aerial Vehicle	5G Police Service System, Law Enforcement Police Vehicles, Mobile Police Booths	5G Police Service System, Law Enforcement Recorder
Resource Occupancy	Low	Middle	High
Price Level	Low	Middle	High

Fig. 7.21 5G product analysis of intelligent police network

Demonstration Scenario	2C Public Users (5QI=9noGBR) Police Users (5QI=9noGBR) Case 1 : Daily Security Patrols	Police Slicing User Business Performance		
		Bandwidth	Video Quality (freezing and blurred)	Video Latency
Network Load Simulation	• Network Light Load (<30%) 2C mobile phone users do not add uplink pressure on the network	10 Mbps	Clear and Smooth	2~3 s
	• Network Mid-load (30% to 60%) A single mobile phone user imitates to create the uplink pressure on the network	7 Mbps	Notably Freezing	>5 s
	• Network Heavy Load (60~90%) Two to three mobile phone users imitate to create the uplink pressure on the network	4 Mbps	Frozen and Blurred	Frozen
	• Network Overload (>90%) Four to five mobile phone users imitate to create the uplink pressure on the network	NA	NA	NA

Fig. 7.22 5G smart police network slicing Case 1 scheme technical requirements

but not the service requirements in networks under the medium-load or heavy-load state.

Case 2: On-demand police enforcement. As shown in Fig. 7.23, 5G police slicing (5QI = 4 GBR) can provide a high-priority bandwidth guarantee for police services in networks under medium-load or heavy-load state. Still, they cannot provide absolute bandwidth guarantee in networks under heavy-load or overloaded state.

Case 3: Event major security. As shown in Fig. 7.24, 5G police slicing (5QI = 4 GBR + RB resource reservation) can provide an absolute bandwidth guarantee in networks under heavy-load or overloaded state to meet the service requirements in such networks.

Demonstration Scenario	2C Service Slice: 5QI=9 noGBR Police Service Slice: 5QI=4 GBR Case 1:Daily Security Patrols	Police Slicing User Business Performance		
		Bandwidth	Video Quality (stuttering and spluttering)	Video Latency
Network Load Simulation	• Network Light Load (<30%) 2C mobile phone users do not add uplink pressure on the network	NA	NA	NA
	• Network Mid-load (30% to 60%) A single mobile phone user imitates to create the uplink pressure on the network	10Mbps	Clear and Smooth	2-3s
	• Network Heavy Load (60~90%) Two to three mobile phone users imitate to create the uplink pressure on the network	9~10Mbps	Clear, Occasionally Freezing	3-4s
	• Network Overload (>90%) Four to five mobile phone users imitate to create the uplink pressure on the network	7 Mbps	Notably Freezing and Blurred	>5s

Fig. 7.23 Technical requirements for Case 2 program of network slices in 5G smart policing

Demonstration Scenario	2C Service Slice: 5QI=9noGBR Police Service Slice: RB Resource Reservation+5QI=4 GBR Case 3: Major Event Security	Police Slicing User Business Performance		
		Bandwidth	Video Quality (freezingand blurred)	Video Latency
Network Load Simulation	• Network Light Load (<30%) 2C mobile phone users do not add uplink pressure on the network	NA	NA	NA
	• Network Mid-Load (30% to 60%) A single mobile phone user imitates to create the uplink pressure on the network	NA	NA	NA
	• Network Heavy Load (60~90%) Two to three mobile phone users imitate to create the uplink pressure on the network	10 Mbps	Clear and Smooth	1~2 s
	• Network Overload (>90%) Four to five mobile phone users imitate to create the uplink pressure on the network	10 Mbps	Clear and Smooth	1~2 s

Fig. 7.24 Technical requirements for Case 3 program of network slices in 5G smart policing

7.2.4 Application Case

1. **A 5G policing information security case: Window of the World—5G smart police office in Shenzhen**

The 5G network meets the three-level equal security requirements of the public security industry, and can realize functions such as the whole network's operation, the awareness of the security situation of the whole network, and AI-based threat analysis and detection to ensure the services of emergency communication assurance. It supports high-priority access to the 5G network for special terminals in critical functional departments, such as the public security bureau and government.

As a central tourist area in Shenzhen, Shahe Precinct is characterized by many tourist attractions, parks, shopping malls, starred hotels, large-scale events, a large flow of people, frequent gatherings, a long coastline, and the coexistence of urban villages and high-end communities. The police service in this district mainly faces three major challenges.

(1) Complex Environment: The district covers more than 40 CBD (Central Business District) hotels, five major scenic spots, and more than 30 schools, and it also has crisscrossing traffic hubs and seafront promenades.
(2) Dense Flow of People: The flow of people in Window of the World Square alone is 245,000 per day, putting enormous pressure on police deployment and management.
(3) Complicated Policing: Policing deployment needs to be implemented along with on-site law enforcement, temporary deployment and control, emergency rescue and handy services for the public.

In general, the current policing is facing some problems like insufficient police force, high pressure on safeguarding social peace and stability, and urgent need to optimize the efficiency of law enforcement. The all-sense smart defense zone covering an area of 515,000 square meters has been established with the aid of 5G technology, encompassing squares, underground bus connection stations, subway

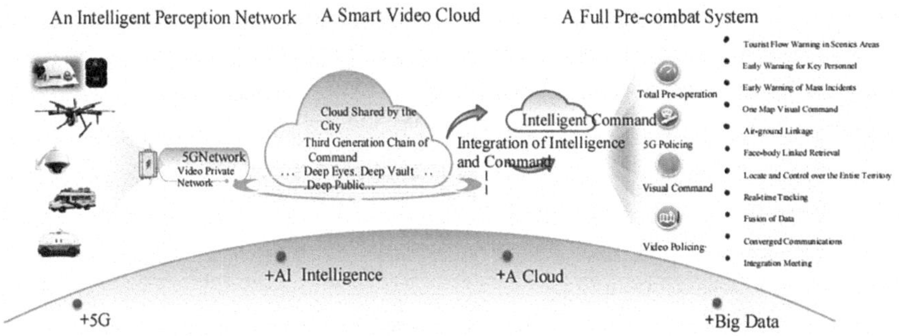

Fig. 7.25 5G helps window of the world to plan a system fully prepared for real practice

stations, and the Yitian holiday business district. This area extends its reach to significant locations such as Happy Valley, Folk Village, Pure Waterfront, Xiangshanli, and more. The highest single-day passenger flows in the past years reached 49,000 passengers at the square, 51,000 at the subway, 83,000 at the bus, and 62,000 at the shopping area. As shown in Fig. 7.25, at present, Shenzhen plans to promote the construction of a police office as the headquarters base based on 5G+ AI + big data + cloud computing and other key technologies. This office is able to take control over the coastline of seafront promenade, the Happy Coast, and the super headquarter base in Shenzhen Bay to realize an intelligent sensing network, an intelligent video cloud, and a set of systems fully prepared for real practice.

7.2.5 Summary (of 7.2)

Smart policing, underpinned by technologies like 5G, requires the whole network sharing, the coverage of the whole area, full-time availability, and full control throughout the whole process to actively prevent and control committing crimes and to actively fight against them after happening. Judging from the progress of smart policing in recent years, 5G, AI, and other emerging technologies are significantly promoting technical and institutional innovation in the public security industry. The construction of policing based on ICT technology is already an inevitable trend. From the perspective of large-scale deployment, the current promotion of smart policing must consider the maturity of terminals in commercial use. Now, 5G cameras can be sold on a large scale, and the technologies of CPE and 5G AR routers used in police cars and mobile kiosks are basically mature, so mobile deployment, mobile watchhouses and mobile police cars can be seen as testing points for the large-scale deployment in the first stage. In addition, efforts must be put into strengthening the expansion of the scope of pilot programs, namely the police mobile wireless system, the 5G law enforcement instrument and the 5G unmanned aerial vehicle. Taking the example of 5G smart policing applied in Window of the World, Shenzhen, at present, 5G helps to bring changes in five aspects of smart policing (as shown in Fig. 7.26), transforming “sweat policing” into “smart policing” to achieve the modernization of governance, imperceptibly in prevention and control, the visualization of police affairs, fully pre-emptive police incidents and integrated disposal. In general, the current 5G+ smart policing projects have begun to bear fruit, forming a major trend where the government has the will, the industry has the technology, customers have investment, and scenarios have mobility. In future work, greater efforts need to be made to promote smart policing.

This section provides readers with an understanding of the current state and future development direction of the 5G+ policing industry, as well as familiarity with application scenarios and network requirements. Additionally, it offers insights into various scenario-based solutions for 5G+ policing.

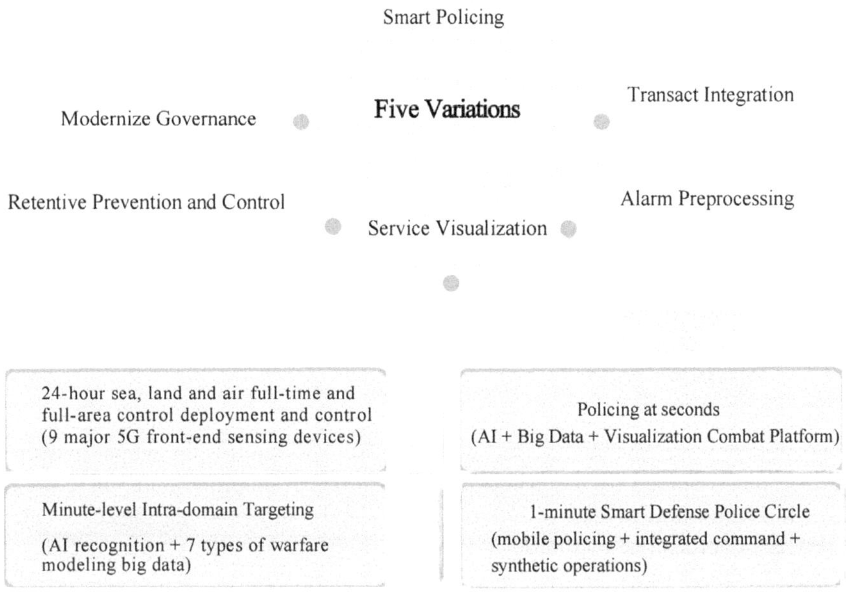

Fig. 7.26 Goals and values of 5G smart policing

7.3 Rail Transit Private Network Scenario

7.3.1 Insight into 5G+ Rail Transit Industry

As an important means of public transportation for urban residents, urban rail transit has outstanding advantages, such as large capacity, intensive efficiency, energy saving and environmental protection. It is an important part of the intelligent transportation system in cities and plays a key role in supporting urban development and residents' lives. The operation level of urban rail transit determines the service quality of urban public transit, and the operation order of it directly affects urban social and economic system.

China's rail transit started its development in the 1980s. With the deepening of reform and opening up, the rapid development of the social economy led to the rapid growth of urban transportation demand. In recent years, China's urban rail transit has developed rapidly, playing an increasingly important role in meeting the travel needs of the people, optimizing urban structure layout, alleviating urban traffic jams, and promoting economic and social development. By the end of 2020, there will be forty cities on the Chinese mainland that will have urban rail transit. With the further acceleration of urbanization, China's urban rail transit construction is expected to usher in a golden development period.

As an important means of public transportation for urban residents, urban rail transit has gradually become networked and intelligent. However, with the increase in rail transit operating mileage and passenger volume, rail transit faces greater challenges in operation, maintenance, security, scheduling, and other aspects. It is an important direction for the future development of urban rail transit to build a smart rail transit operation platform and comprehensively improve the level of rail transit operation service.

The connotation of urban smart rail transit is as follows: It applies emerging information technologies, such as cloud computing, big data, Internet of Things, artificial intelligence, 5G, satellite communication and blockchain, so it can comprehensively perceive, deeply connect, and intelligently integrate physical information of passengers, facilities, equipment, and environment. Through independent evolution, innovative service, operation and construction management modes, the city could build a new generation of safe, convenient, efficient, green, and economical Chinese-style intelligent urban rail transit. Figure 7.27 shows an example of the smart urban rail transit architecture.

As the underlying technology of new infrastructure, the 5G network, together with artificial intelligence, cloud computing, the Internet of Things, and other technologies, constitutes a new generation of information infrastructure, which is used to collect and process huge data resources generated by massive connections, bringing revolutionary upgrades to the entire information infrastructure and providing new basic capabilities for intelligent transformation, mass connection and huge data resources of the rail transit system represented by the subway. In addition, the 5G network based on high bandwidth technology breaks the existing communication

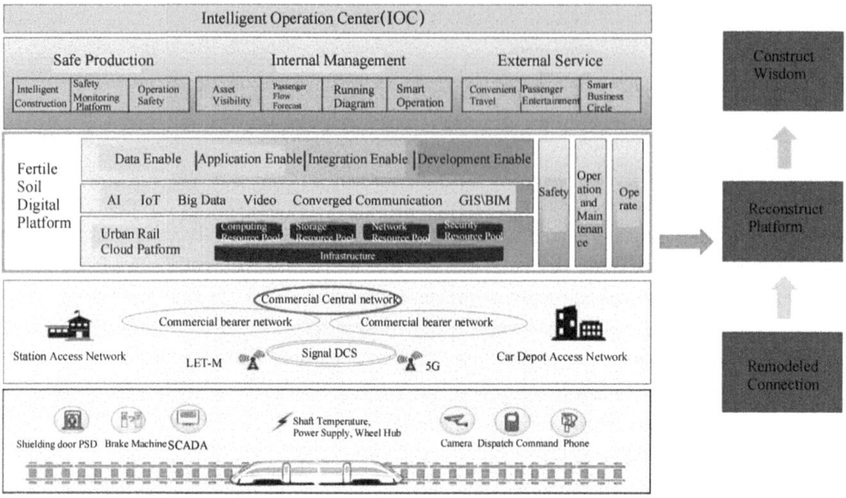

Fig. 7.27 Example of urban smart rail transit architecture

gap between trains and ground, establishing a high-speed pathway between trains and ground, trains and people and trains, real-time collecting monitoring and video data of subway trains, supporting the reconstruction of urban rail cloud platform of rail transit, analyzing and calculating the data, and then pushing it to subway trains. It can provide support for innovative applications, such as intelligent operation and maintenance of subway lines and driverless subway trains.

Therefore, in summary, the operation, service, and guarantee of rail transit have characteristics and challenges of large passenger flow, high degree of data concurrency, cross-station collaboration, high real-time requirements, and complex data analysis. The integration and application of the new generation of information technology, represented by 5G, is an important means for the development of rail transit and the response to the needs of public transport operation guarantee in megacities. It can provide basic capabilities for achieving the goal and vision of urban smart rail transit.

7.3.2 Analysis of 5G Application Scenarios and Network Requirements

1. Application Scenarios of 5G Network in Rail Transit

According to the preliminary research, 5G mainly has the following application scenarios in rail transit.

1) On-board Video Backhaul via 5G

The onboard video of the rail transit train is set in the cab and the carriage of the train, which is mainly used to monitor the running state and driver behavior to ensure safety. The cameras in the carriages are mainly to meet the safety needs of passengers. They can be used to judge the congestion of the carriages, arrange passenger travel, and help passengers find lost items in time. The adoption of 5G network can increase the transmission rate of video data and realize the real-time monitoring of on-board video.

2) Enhance Emergency Communication Capabilities Through 5G

In an emergency, the station needs to build a temporary video communication system, deploy several cameras, and communicate with the ground security personnel in real time. Due to the limited wireless transmission capability of emergency support, voice call is the only option for ensuring real-time communication using walkie-talkies. The ultra-high speed of the 5G network can provide a new bearer channel for emergency communication services, allowing for seamless video integration and communication between inside and outside rail transit stations, across different lines, among people, and between people and objects.

3) Send Back Train Status Data Through 5G

With the intelligent transformation of the train, the train status data also needs to be uploaded to the background in real-time to realize the online monitoring of the train status and ensure the safety of the operation. During train operation, sensors generate a lot of data, which requires synchronous transmission of massive data to realize intelligent operation and maintenance. With the increase of bandwidth resources provided by the 5G network, the speed of the Train Control and Management System (TCMS) can be greatly improved, which can realize real-time upload of other stored data of the train, such as running part, brake, waveform file data of pantograph-catenary, etc. This greatly assists in real-time adjustment of train status.

4) Realize Intelligent Inspection of Rail Running Area Through 5G

With the extension of rail transit lines, the operation time of rail transit is also prolonged, and the operation window time for maintenance is getting shorter every night. To ensure the safety and normal operation of all rail transit lines, a large number of maintenance personnel need to be arranged at night to patrol the rail running area. Suppose the no-load 5G network resources of the tunnel at night are used for intelligent inspection, such as video safety helmet, combined with intelligent inspection robots. In that case, it can replace part of the manual inspection of the track area. The inspection video and data can be backhauled in real-time through 5G. This remote operation can greatly reduce personnel costs and shorten the daily inspection time.

2. Matching Analysis of 5G Key Technologies and Urban Rail Transit Demand

Operators have deployed 5G network coverage in rail transit line tunnels, mainly to provide passengers with services. 5G adopts a TDD system, with a higher ratio of downlink slot than that of uplink slot. The current common ratios are 7:3 and 8:2, which have been solidified. The service model used by rail transit mainly uses uplink data. According to the current train camera ratio, the train needs 48 Mbps uplink bandwidth. For TCMS data, it is recommended to consider 10 Mbps uplink bandwidth. In summary, an average bandwidth of 58 Mbps is necessary.

The 5G network coverage in the tunnel is generally deployed using leaky cables. According to extensive test data, the throughput capacity of the network with a 100 MHz spectrum and a time slot ratio of 7:3 is shown in Table 7.3.

Therefore, the adoption of a 5G network can meet the data service needs of rail transit. There are three construction modes for 5G public networks, as shown in Fig. 7.28.

Wide Area Slicing: Through QoS, RB resource reservation and other means, it achieves logic isolation of services to meet customers' priority needs for specific network rates, latency, and reliability and supports flexible configuration on demand.

Local Area Slicing: Edge computing technology is used to offload data traffic and process local services to meet needs such as data non-display and ultra-low latency and to provide dedicated network services for customers.

Physical Private Network: Through the establishment and sharing of base stations, frequencies, and core networks, dedicated 5G networks are built for enterprises, providing them with high security and high isolation of privileged customized network services.

Table 7.3 Throughput baselines of 5G in different spectral bandwidths of different service models

Spectrum width (Mbps)	Time slot allocation	2T2R Uplink (Mbps)	4T4R Uplink (Mbps)
40	7:3	70	100
60	7:3	108	150
80	7:3	144	200
100	7:3	180	252

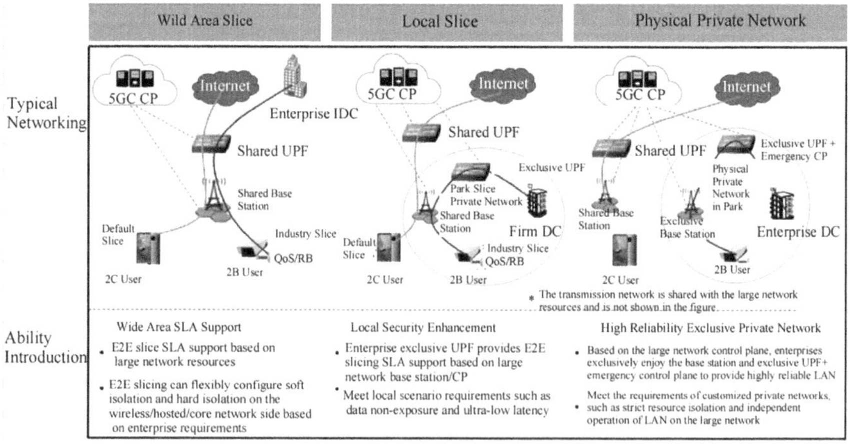

Fig. 7.28 Three private network modes

The preliminary business planning is as follows. The rail transit business is mainly divided into the following five types.

1) Low Bandwidth Load for Production Services

Production data within rail transit need to be strictly isolated. Currently, the existing train control system of rail transit is carried on an LTE-M system, with the channel bandwidth within 20 MHz (1785–1805 MHz). The relevant standards for the train control information carried by the 5G public–private network are not yet mature, and service testing, verification, and innovative exploration will be conducted at this stage.

2) High Bandwidth Load for Production Services

With the development of smart rail transit, smart trains are upgraded. With the demand for various service data types increasing, high-bandwidth data needs to be transmitted. Rail transit service data, including train video data, cluster video fusion system, and video control data of train automatic driving, can be carried by a 5G public network.

3) Broadband Communication for Non-production Services

In addition to production data, there is also a large number of non-production businesses in rail transit. For example, the emergency communication video camera and trackside video camera of the station, as well as the large number of terminals and video transmission required, can be considered to be carried by the 5G public network system.

4) Non-critical Production Business Links

Non-critical production-related service links have minimal bandwidth needs. They can leverage existing operator networks for IoT connectivity, such as station advertising screen control systems, vehicle light sensors, various environmental sensors, etc. The 4G network ecology of operators is mature and has comprehensive coverage. Non-critical production service links can be carried in the 4G network through IoT. This part does not need independent network construction. The capacity demand for this part of the link can be solved by using network service leasing.

5) Intelligent Maintenance Load of the Vehicle Base

With the development of intelligent maintenance services in vehicle bases, new demands are put forward for robot reconditioning and VR/AR remote maintenance. In the future, 5G public networks can be used in vehicle bases to realize intelligent functions, such as remote maintenance and robot inspection, so as to improve maintenance efficiency effectively.

3. Service Requirements of 5G Public and Private Networks for Rail Transit

The 5G test network of Nanjing Maqun Rolling Stock Depot carries six services: rail transit signal, vehicle, communication, urban rail transit Automatic Fare Collection System (AFC), comprehensive monitoring, and escalator. Based on the data obtained from the actual test and combined with the actual situation of rail transit engineering, the following business requirements are formed through analysis and summary.

1) Train Operation Control Service

The train operation control service is responsible for monitoring, controlling, and adjusting the train's running speed and braking mode based on the objective conditions and actual conditions of the train running on railway lines. This service applies to mainline trains, vehicle bases, and parking lots. Two on-board terminals are installed respectively in the front and rear of trains. The concurrency rate of data backhaul by terminals reaches 100%.

(1) Train control services require the system to provide isolated and redundant data channels.
(2) Train control service requires no less than 99.99% system availability.
(3) The train control service requires that communication is the highest priority, and the whole system gives the highest priority to ensure the transmission of this service, which is not affected by the transmission of other services.
(4) The train control service requires that the train's running speed meets the requirements of rail transit.
(5) The probability of single channel one-way transmission delay of train control service is not less than 99%, and the probability of not more than 1 s is not less than 99.92%.
(6) The train control service requires that the packet loss rate shall not exceed 1%, and the probability of communication interruption time not exceeding 2 s shall not be less than 99.99%. Train control service data is periodically sent, and the transmission rate of each uplink and downlink route is no less than 1.5 Mbps.

2) Vehicle Signal Operation and Maintenance Data Service

The onboard signal operation and maintenance data service is backhauling the onboard signal operation and maintenance data to the ground system. When the train is running, it is necessary to backhaul the real-time data. After the train returns to the service warehouse and maintenance depot, the data can be backhauled centralized. The data backhaul must be completed within 5 min. The actual application scenarios include the main line's service warehouse and maintenance depot, vehicle base, and parking lot. The wireless access terminals set at the front and rear of the vehicle are used to send back data, with a concurrency rate of 100%.

(1) The transmission delay of on-board signal operation and maintenance data service not exceeding 150 ms is required to be no less than 99%.
(2) The onboard signal operation and maintenance data service requires the packet loss rate not to exceed 1%.

When the train transmits service data in the running state, the onboard signal operation and maintenance data service requires that the transmission rate of each train is no less than 3 Mbps uplink and downlink. When the static train transmits service data in the service warehouse or maintenance depot, the onboard signal operation and maintenance data service requires that the transmission rate of each train be no less than 50 Mbps uplink and no less than 3 Mbps downlink.

3) Vehicle Running Condition Monitoring Service

Vehicle running condition monitoring service refers to the real-time transmission of train status and fault information collected by the onboard host and subsystems to the ground monitoring center. The actual application scenarios of this service include the main line, vehicle base, and parking lot. The train operation condition monitoring service should meet the following technical requirements:

(1) The probability that the transmission delay of vehicle running condition monitoring service not exceeding 100 ms is not less than 99%.
(2) The packet loss rate of the vehicle running condition monitoring service shall not exceed 1%.
(3) The vehicle running condition monitoring service requires that the transmission rate of each train should be no less than 10 Mbps uplink and no less than 1 Mbps downlink.

4) PIS Video Service

The Passenger Information System (PIS) video service transmits video or image information from the ground to the carriage through broadcast or multicast. The actual application scenario of this service is mainline trains. The PIS video service should meet the following technical requirements:

(1) The probability of PIS video service transmission delay not exceeding 150 ms is not less than 99%.
(2) The packet loss rate of PIS video service should not exceed 1%.
(3) PIS video service requires that the transmission rate of each train is no less than 6 Mbps downlink.

5) Smart Station Integrated Management Service

The integrated management service of the smart station is to summarize the information of station power supply, electromechanical, fire alarm, access control, video surveillance, broadcast, passenger information, ventilation system, water supply and drainage system, air conditioning system, uninterruptible power supply and other information into the comprehensive management platform of the smart station. The actual application scenario of this service is the station. The integrated management service should meet the following technical requirements:

(1) The integrated management service of smart stations requires the system reliability to be no less than 99.9%.
(2) The probability that the probability of the transmission delay of smart station integrated management service not exceeding 150 ms is not less than 99%.
(3) The integrated management service of the smart station requires that the packet loss rate should not exceed 1%.

(4) The integrated management service of the smart station requires the transmission rate of each subsystem to be no less than 1 Mbps uplink and no less than 1 Mbps downlink.

6) Video Surveillance Service

The video surveillance service transmits video surveillance images of train cabs, train carriages, track areas, and other scenes to the control center or ground monitoring station in real time for centralized monitoring. Service application scenarios include the main line, vehicle base or parking lot, with the concurrency rate of 100%. The video surveillance service must meet the following technical requirements:

(1) Mean Time Between Failure (MTBF) required by video surveillance services is over 5,000 h.
(2) The video surveillance service requires the probability of the transmission delay not exceeding 150 ms to be not less than 99%.
(3) The video surveillance service requires that the packet loss rate of the system does not exceed 5%.
(4) The onboard video surveillance service requires the transmission rate of each camera to be no less than 2 Mbps uplink.
(5) The video surveillance service in the rail area requires the transmission rate of each camera to be no less than 4 Mbps uplink.

7) Automatic Fare Collection Service

Automatic fare collection service mainly involves automatic ticket selling and checking systems in the station, including terminal equipment monitoring systems, system operation management systems and transaction data processing systems. The service application scenario is the station. The automatic fare collection service should meet the following technical requirements:

(1) Automatic fare collection service requires that the system's reliability is not less than 99.9%.
(2) Automatic fare collection service requires the probability of the transmission delay not exceeding 100 ms to be not less than 99%.
(3) In automatic fare collection service, the loss rate of the terminal equipment monitoring system and system operation management system shall not exceed 1%, and the loss rate of the transaction data processing system shall not exceed 0.1%.
(4) Automatic fare collection service requires the uplink transmission rate of each group of gates to be not less than 2.5 Mbps and the downlink transmission rate to be not less than 1 Mbps.

8) Escalator Monitoring and Diagnosis Service

The escalator monitoring and diagnosis service mainly involves the escalator in the station, which can automatically detect, diagnose, and send real-time monitoring information to the station's integrated monitoring center. The actual application scenario of this service is a station. Escalator monitoring and diagnosis services shall meet the following technical requirements:

(1) The escalator monitoring and diagnosis service requires that the packet loss rate of the system should not exceed 1%.
(2) Escalator monitoring and diagnosis service requires that the uplink transmission rate of each escalator is no less than 3 Mbps.

9) Video Fusion Dispatching Service

Video fusion dispatching service utilizes the 5G mobile communication system to provide various voice, video and data call communication and management services for emergency command, dispatching, line operation, emergency response and maintenance. The actual application scenarios are in the station (station hall, platform, etc.) and train, using 5G access terminals to transmit data back. The video fusion dispatching service should meet the following technical requirements:

(1) Video fusion dispatching service requires the system reliability to be no less than 99.9%.
(2) Video fusion scheduling service requires the probability of the transmission delay not exceeding 150 ms to be not less than 99%.
(3) Video fusion dispatching service requires that the packet loss rate of the system should not exceed 1%.
(4) Video fusion dispatching service requires that the transmission rate of each terminal should be no less than 2 Mbps uplink and no less than 2 Mbps downlink, respectively.

7.3.3 Scenario-Based Solution

1. Smart Operation Supported by 5G Subway Big Data

The construction of future smart cities must include intelligent and efficient transportation systems as a key element. With the advent of the 5G era, precision positioning, big data collaboration, massive IoT, real-time intelligent analysis, reliability, and security upgrades make it possible to build the smart operation service system of urban rail transit with 5G technology. It also provides conditions for the construction of a harmonious and co-governed smart travel environment. While promoting the vigorous development of public consumption, it can provide communication, computing, and other services for rail transit and other professional fields through network slicing, edge computing, and other technologies, namely, the specialization of public resources and the sharing of transport capacity services, so as to realize the savings and efficiency improvement of traditional industries.

5G technology has greatly improved communication bandwidth, transmission delay, massive IoT, reliability and security of information transmission, and other aspects, bringing new developments for rail transit to improve service process and operation and maintenance management. According to the specific operation process, service demand and social positioning attributes of the subway, the transportation system makes use of the high-precision positioning capability of 5G base stations to realize the digital and intelligent management of subway passenger flow, such as warning and guidance of passenger flow, user travel portrait, queuing time prediction, and traffic transfer linkage. By using big data AI methods to improve the quality of passenger service at stations, expand station service scenarios, and reduce station operating costs, a sound and perfect intelligent service system for rail transit is finally established to ensure the public safety of urban rail transit passenger flow.

1) Establish a "5G+ whole process" service to promote an outbound passenger flow management system

5G will reconstruct the traditional public transport service mode. Based on the open and decoupling three-tier architecture of 5G passenger flow big data perception layer, data layer and application layer, it can sense the change of grid passenger flow in real-time. Data model sharing breaks down various data barriers at the bottom level, and the AI model empowers the iteration and development of applications. It realizes the intelligent management of 5G+ VR / AR / MR + AI passenger flow in the whole process and all elements. The specific links are as follows.

In the link of entering and leaving the station, the passenger flow data can be transmitted back to the dispatching room through 5G for flow forecasting. Holographic images and dynamic signs can encourage passengers to stand on both sides, thereby relieving station congestion, saving passengers' entry and exit time, and improving efficiency.

During the security inspection, the 5G transmission speed is more than 20 times that of 4G, which can support big data linkage, face recognition, and multi-scene recognition. At the same time, a large amount of relevant data can be transmitted to AI identification and cloud identification platforms for matching and comparison, through AI identification and cloud identification, to realize "seconds" entry into the station and improve efficiency. Security staff can also wear AR glasses to achieve visual fusion command and dispatch, face recognition inspection, infrared temperature detection for epidemic prevention and control + dual-stream display of face information, abnormalities alert +, face recording, etc. It can effectively provide convenience for the first on-site control investigation of security check and mobile inspection for security staff.

In the link of card swiping in and out of the station, 5G technology combined with big data can make the card swiping gate fully equipped with real-time interaction, fare calculation, remote operation, cloud collaborative billing, and other functions.

In the process of ticket top-up, 5G subway service robots based on facial recognition, AI and other technologies can be used to realize interact with passengers intelligently and provide suggestions of the best subway routes for subway passengers in combination with 5G.

2) Construction of an immersive waiting experience area of "5G+ Service design"

The smart subway service system and wireless network channel shall be constructed with 5G technology. The integrated platform architecture from the infrastructure layer to the intelligent application layer shall be constructed with multi-point, multi-place, and full function to realize the innovative functions of intelligent perception, intelligent linkage, and intelligent analysis of subway infrastructure, and support the intelligent scenes and applications of operation, service, operation and maintenance. Data-driven improvements shall be made in security, efficiency, effectiveness, and service.

3) Construction of a "5G+ Internet of Things" ride experience process

Using "5G+ IoT" technology and combining user journey and consumer behavior data, key elements and critical links of the subway service system can be identified, and specific applications and solutions for designing the subway service system can be formulated. By conducting passenger flow portrait analysis based on characteristics of subway passenger flow, route distribution, passenger flow distribution area features, origin distribution of migrant workers, as well as analysis of online business preferences, a service system featuring "5G air-ground integration and internet of all things" can be established. Driven by consumer behavior data in the subway industry, this approach enhances the competitiveness of subway enterprises in service differentiation, effectively helps enterprises reconstruct intelligent service systems, and improves the management capability of subway services.

4) Construction of a "5G+ AI" personalized route customization + AI intelligent transfer service system

The subway station is an important node in the urban transportation hub, and it is also the place with the most concentrated human flow. Especially during rush hour in the morning and evening, after a large number of passengers flow out of the subway, there will be a shortage of bus, taxi, and other public transport capacity, resulting in a swarm of outbound passengers stranded, affecting the normal order of the subway and surrounding traffic. One of the main challenges in current station management is how to grasp the flow of people dynamically, formulate passenger flow relief plans in advance, organize public transportation personnel to carry out relief work, and improve passenger service quality.

Through the passenger flow analysis capability of this system platform, combined with the subway schedule information, it establishes a normalized traffic connection induction mechanism with the urban traffic management department to help the management department develop a public transport connection and transfer linkage mechanism aiming at the urban metro transport hub nodes and the tide of passenger flow, and provide big data support for the decision of increasing the subway operation density. This ensures the rapid entry and exit of passengers, avoiding waiting. In combination with VR/AR/MR + AI technology empowerment, platforms such as personalized transfer route customization, holographic transfer guidance, and AI intelligent station interior map under the 5G network will be constructed to build a "5G+ " intelligent transfer ecological chain.

2. Vehicle-ground Communication System Based on 5G

Vehicle-ground communication of the subway is to transmit the onboard data to stations, train depot, and other localities promptly during the high-speed running of the train to monitor passenger conditions, train equipment and running status, tunnel and bow net, etc. However, due to the current technical limitations, the bandwidth of the train-ground wireless network is insufficient, and the transmission rate is unstable. As a result, the onboard data cannot be transmitted in real-time and can only be stored on the hard disk on board. The data can only be copied manually when the train is offline. This not only consumes time and energy but also cannot guarantee reliability and real-time performance, which directly affects the subway scheduling, thus becoming a bottleneck for improving the efficiency of subway operation and maintenance. The application of 5G wireless communication technology will change this situation.

1) Video backhaul

The video backhaul system can take advantage of the low latency and high reliability of the 5G network to transmit the train running images from the ground to the train's onboard display screen in real-time. The system displays the whole process of the train operation to TV sets in the passenger room through high-definition camera video acquisition, wireless transmission of vehicle-ground video, decoding and playing vehicle-mounted video, and finally realizes high-definition and split screen playback. The system can detect the moving position of the train based on the signal of the photoelectric positioning sensor and then switch screens of different stages accordingly, allowing for multi-angle and multi-directional viewing of the process.

2) Automatically upload and update application maintenance data

Compared with other data types, the update priority of application maintenance data is lower. When the network bandwidth is insufficient, the update frequency of application maintenance data is lower. However, the high bandwidth of the 5G network can meet the timely update of application maintenance data and further improve passengers' riding experience.

(1) Video program uploading. Entertainment video programs, such as movies, TV plays and variety shows, can relieve passengers' travel fatigue and improve their ride experience. Entertainment programs can be uploaded in real-time based on 5G technology, and TV video content can be updated in real-time, daily and weekly.
(2) Online resource update. By building an online resource update system, the metro company can update the vehicle information and resources online, reduce labor costs, and improve vehicle operation efficiency. The main functions include setting trains' numbers online, updating trains' displaying and voice broadcast content online, and updating entertainment audio-visual system resources online.

3) On-board 5G public network services

Installing the 5G on-board public network service system, 5G mobile communication base stations can then be set on board. Combined with the characteristics of 5G communication technology, mobile edge computing servers are set on the train to cover the whole carriage. The original sub-base station in the fixed communication room is transformed into a mobile communication base station, which has increased the network technology value and pioneered the business model.

The on-board 5G public network service system of rail transit will absorb off-board users into the train, minimizing the signaling storm problem caused by centralized switching of users. By mounting base stations on board, Massive MIMO antenna technology increases network coverage and capacity while reducing 5G construction investment and maintenance costs.

Due to the characteristics of 5G technology, load-bearing and control are separated, and about 35% of traffic can sink to the onboard edge computing server. It provides an information technology service environment and cloud computing capability close to mobile users. It pushes content distribution close to users, which can provide lower latency, greater bandwidth, and more connectivity capability for games, commerce, video, Cloud VR, smart office, smart education, smart travel, smart shopping, and other businesses. At the same time, it also realizes the application of the Internet of Things based on a 5G network, forming the application scenarios of the innovative connection between people and things and things and things.

(1) Overall structure and composition. Urban rail transit on-board 5G public network service system is composed of on-board 4×4 MIMO antennas, on-board small base station transmission module (vehicle-ground communication module), 5G on-board small base stations (mobile type) and MEC servers, which can realize synchronous coverage of 5G signals in the train, completely overcome the penetration loss of the carriage, and reduce the deployment density of mobile base stations along the rail transit, and ultimately reducing the construction cost of 5G coverage. The overall system architecture is shown in Fig. 7.29.
(2) Module function. Read Table 7.4 for details about its functions.

7.3.4 Application Cases

In 2019, at the China International Import Expo (CIIF) in Shanghai, Shanghai Metro displayed a 5G big data-based passenger flow situation analysis application, which can conduct real-time passenger flow perception and urgent situation warning in subway station halls and platforms. Relying on applications such as HD (high-definition) mobile video, HD intercom, and 5G customer service personnel positioning on the basis of a 5G large bandwidth network, these applications empower subway operating units to launch emergency provisions for mass passenger flow in this flow scenario, including implementing train schedule, running organization, dynamic diverting, and other plans. It can boost linkage and optimize the

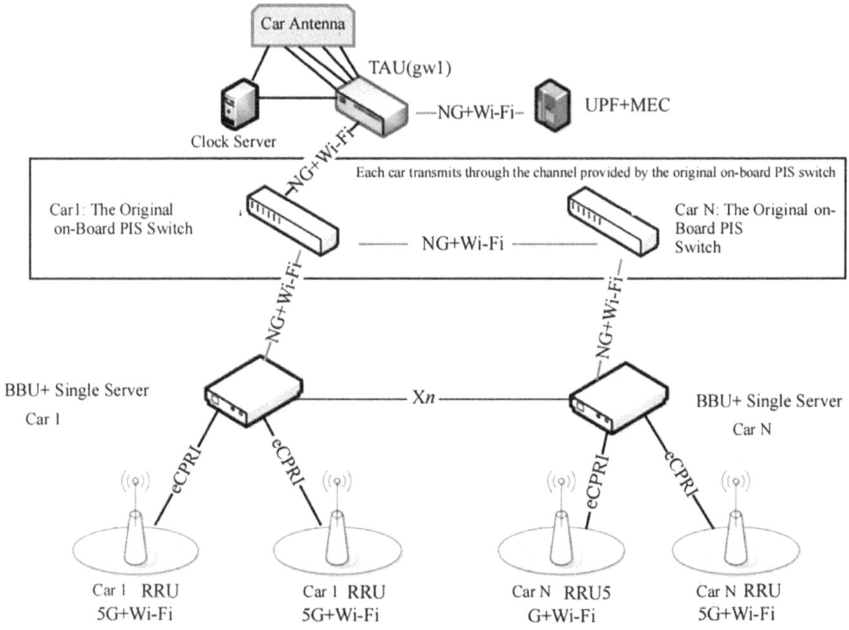

Fig. 7.29 Architecture of the Overall System

transport capacity of relevant regions, serve to divert and guide passengers, and change the previous station management pattern dominated by manual operation and static instructions. It will supply passengers with an all-round experience, which is conducive to the healthy and orderly development of smart subway construction, and help realize the multi-function and efficiency reduction of subway-related staff, promoting the sustainable development of the rail transit industry. The following introduces specific applications and practices of 5G in Shanghai Metro during the CIIE.

1. Use 5G to Analyze the Passenger Flow Situation

5G will reconstruct traditional public transport service formats by integrating multiple data such as subway train signal data, train weighing data, surveillance video data, and mobile high-definition video data, and realizing data transmission through the 5G broadband network. Based on the data center technology, data collection and comprehensive analysis are carried out to achieve situational analysis, use of real-time passenger flow trend changes, and perception of subway station hall and platform. According to various historical data, thermal maps, and diagrams, the distribution and flow direction of passenger flow in each region can be intuitively displayed, which is conducive to improving passenger experience, avoiding traffic gridlock caused by a large volume of passenger flow, and providing decision-making support for security and passenger evacuating work.

The specific application content is as follows.

Table 7.4 Detailed function descriptions

Serial number	Mobile function	Detailed description
1	Car 4 × 4 MIMO Antenna	Responsible for receiving and sending communication signals from operators outside the car, the antenna is installed on the roof of the car to avoid the wireless signal penetration loss of the train box. At the same time, the antenna has a receiving gain, and is used with the remote module to provide a physical link for the 4G/5G base station opening in the car
2	Clock server	Responsible for providing clock and frequency synchronization for 5G small base stations
3	Small base station transmission module (Vehicle-ground communication module)	Responsible for converting 4G/5G wireless signals of operators into digital bandwidth signals to realize the activation of 4G/5G base stations in the car. Users in the car perform services through small base stations, which encapsulate user services through remote backhaul modules and send them back to the background gateway through VPN channels. The gateway unencapsulates user services and then transmits them to the core network for processing. Open the service flow from users to base stations and base stations to core networks
4	5G car small base station (Mobile)	Achieve the coverage of 5G signals of in-car operators, and provide users with 5G services through transmission and access to the core network, achieve accurate and deep coverage, improve the ability of network to support data services, and improve user perception
5	Edge computing survey	Mobile edge computing (MEC) is a new distributed computing method based on mobile communication networks. It is built in the cloud service environment at the base station. By processing services close to users and cooperating with content, application and network, MEC provides low-latency, secure and reliable services to achieve the ultimate user experience, and realizes the core network billing and authentication and data content sinking. Responsible for in-car user communication route selection, data hit processing, etc. It can develop corresponding software plug-ins according to the actual needs, open API interface required by various industries, deeply integrate with direct industries, and maximize the adaptation to the needs of various industry applications

(1) Combined with the real-time passenger flow thermal diagram of the station hall and platform based on Building Information Modeling (BIM), it reflects the congestion of passenger flow. The real-time passenger flow thermal diagram of the station hall and platform is shown in Fig. 7.30.
(2) Use the thermal diagram of the passenger flow of the CIIE venue to reflect the overall passenger flow in the venue.

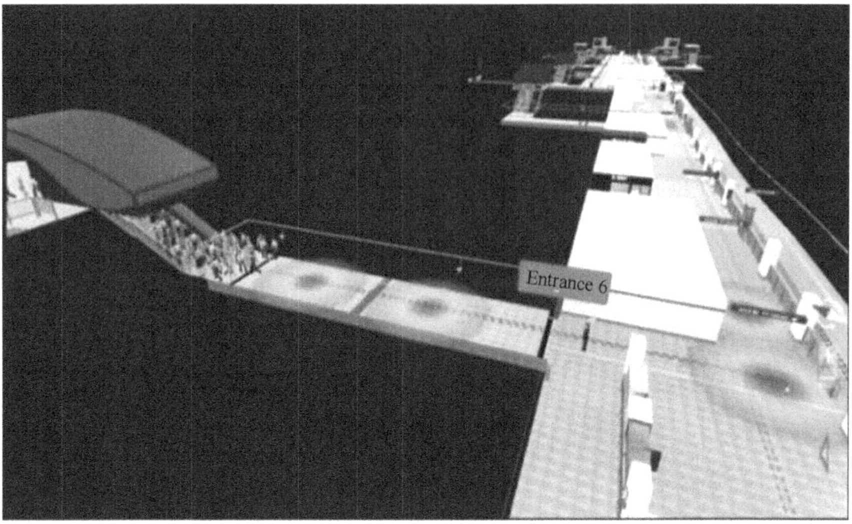

Fig. 7.30 Real-time thermal diagram of passenger flow in the station platform

(3) Use the passenger flow diversion diagram of the CIIE venue to show the current diversion situation.
(4) Based on the normal analysis of historical data, the full-day passenger flow trend prediction is realized, and the passenger flow of the three-line subway (Line 2, Line 17, Line 10) related to the CIIF is predicted and warned for 15 min later.

2. Use 5G HD Video to Evacuate Passengers

According to different passenger flow scenes around multiple stations and main crowd distribution centers (such as venues), timely suggestions for adjusting passenger flow and traffic organization strategies are given to evacuate large passenger flow efficiently, avoid the large areas of passenger backlog and retention, and realize dynamic adjustment and joint command of passenger flow and trains. By mastering the real-time passenger flow, the predicted large passenger flow scene, and the dynamic running condition of trains, combined with AR technology, the subway operating company can implement the emergency plan for large passenger flow and realize the passenger flow evacuation and other services.

The specific application content is as follows.

(1) Use 5G HD video to broadcast the queue situation of passengers outside subway entrances and exits.
(2) The station staff use 5G AR glasses to transmit real-time key area passenger flow information in the station hall and platform.
(3) The commander receives the video backhaul by station personnel using 5G AR glasses on the large screen, observes the implementation and effect of the emergency plan, and gives real-time adjustment instructions.

(4) According to the queuing situation at the entrance and exit of East Xujing Station, the emergency plan should be triggered to inform the venue to guide the passenger flow to Zhuguang Road Station.

3. Realize Immersive Passenger Experience Area with Full Coverage of 5G at Three Stations and Three Lines

A passenger experience area based on a 5G network and cloud services will be established to provide a high-speed Internet download experience, high-definition video live broadcast, and other services. By setting up a passenger 5G experience area on the station concourse and comparing the speed difference between mobile phone downloads using 4G signals and 5G CPE, passengers can experience the large bandwidth and low delay of 5G.

In the practice of the China International Import Expo (CIIE), full coverage of the 5G network for the CIIE venues and three related stations (East Xujing, Zhuguang Road, and Hongqiao Railway Station) and three lines (Line 2, Line 17, and Line 10) has been achieved, including station halls, platforms, key entrances and exits, and other areas.

Specific application contents include:

(1) 5G coverage for CIIE and the three stations and three lines.
(2) Working conditions of 5G equipment in the three stations.
(3) Measurement of 5G network speed in key areas of the three stations.

1. Customer Service Robots Based on 5G

The system is built with multiple intelligent human–computer interaction capabilities such as speech recognition, video inquiry, face recognition and tracking, object recognition, somatosensory interaction, indoor positioning, and navigation vision. On the basis of this system, subway customer service robots will be deployed in 5 G-covered areas. Based on facial recognition, AI, and other technologies, combined with a 5G high-speed network, interactive intelligent service for subway passengers will be realized.

Specific application contents include:

(1) Subway customer service robots obtain Internet data through 5G network and support interactive applications with subway passengers.
(2) The subway customer service robot can obtain face recognition, AI, and other analytical capabilities in the cloud through a 5G network to support intelligent service with subway passengers.

2. Personalized Path Customization Based on 5G

Based on the subway line data, real-time operation data, and historical data, the optimal subway route suggestions and predicted travel time to the Expo venues (and other landmark areas in Shanghai) are provided for subway passengers after analysis and processing. For example, it provides suggestions on the best subway routes from the current station to the CIIE venues, People's Square, East Nanjing Road, Xintiandi,

Lujiazui, Pudong Airport, and other locations, including the required subway fare and approximate travel time.

3. Use 5G to Assist Station Inspection

With the characteristics of large bandwidth and low delay of a 5G network, 5G AR glasses are used to support the inspection work of station staff, send high-definition video information from the first perspective to the command center, and improve operation efficiency. The related functions of 5G AR glasses are shown in Fig. 7.31.

Specific application contents include:

(1) Scan video information (such as two-dimensional code, etc.) backhauled by 5G AR glasses and push AR information after analysis and processing to help staff diagnose and deal with faults.
(2) The video of 5G AR glasses will be transmitted to back-end maintenance experts to guide the on-site diagnosis and handling of difficult problems through real-time intercom.

In addition, 5G auxiliary robots are also used for inspection. The inspection robot is equipped with several groups of high-performance monitoring instruments to carry out full coverage monitoring of the equipment in the station through a mobile platform operating autonomously, greatly reducing the number of installations of fixed sensors and instruments required by "intelligent inspection". There is no need for a large amount of wiring or transformation of the existing switch cabinet. While reducing the comprehensive operation cost, the operation management level is improved. The 5G high-bandwidth network supports the automatic inspection of subway lines, subway tunnels, power supply equipment, and other equipment by robots (Fig. 7.31).

Specific application contents include:

(1) Realize real-time monitoring and data backhaul of robot inspection, improve the efficiency of automatic inspection and background analysis ability.

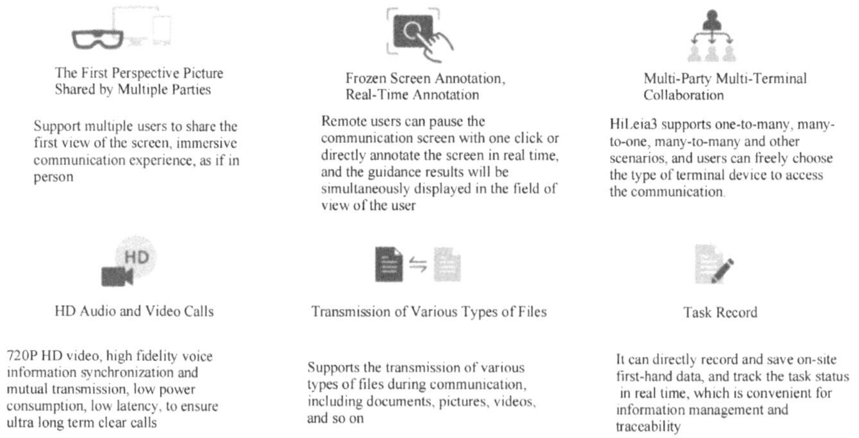

Fig. 7.31 Functions of 5G-AR glasses

(2) According to the data analysis in the system, some intervals need to be inspected with special emphasis.

7. 5G Large Bandwidth Enabling Vehicle-Ground Communication

5G high bandwidth enables vehicle-ground communication to improve the public safety of rail transit. The high-speed channel of vehicle-ground communication is established through a 5G large bandwidth network to realize high-speed dumping and real-time monitoring of train video. The 5 G-based vehicle-ground video transfer scheme relies on 5G wireless network transmission to meet the requirements of vehicle-ground video transfer for vehicle video surveillance. Based on the visual management interface, it supports unified storage management, realizing unified management of transferred videos, reducing management costs, and improving management and control efficiency.

Specific application contents include:

(1) Make an immediate response to various emergencies and illegal behaviors in the carriage through 5G real-time high-definition video surveillance to improve public safety of rail transit.
(2) Download the complete train monitoring video through the 5G network at a high speed in the station for storage and subsequent analysis and processing by the ground video management system.

7.3.5 *Summary (of 7.3)*

In urban rail transit, 5G communication technology is the key to promoting urban rail transit's development as more automated, intelligent, standardized, systematic, normalized, and green. The application of 5G communication technology in urban rail transit promotes the development of the urban rail transit industry to a certain extent. It can effectively enhance urban rail transit communication's flexibility, timeliness, and comprehensiveness. In this regard, under the strategic background of implementing innovation-driven development, it is of certain importance and necessity to understand 5G communication technology, understand the feasibility of its application in urban rail transit, and promote its application in transportation communication systems under continuous research, development, and practice.

5G technology has greatly improved communication bandwidth, transmission delay, massive Internet of things, reliability and security of information transmission, and other aspects, bringing new developments for rail transit to improve service process and operation and maintenance management. Based on the precise positioning capability of 5G base stations and the specific operation process, service demand, and social positioning attributes of rail transit, the system can realize the digital and intelligent management of subway passenger flow, such as passenger flow warning and guidance, user travel portrait, queuing time prediction, and traffic transfer linkage. By means of big data and AI, the quality of passenger service at the station is improved, the service scene of the station is expanded, and the operation cost of the station is gradually reduced. Finally, the intelligent service system of rail transit is established and improved, and the public safety of urban rail transit passenger flow is guaranteed. Based on 5G technology, resource scheduling and system operation efficiency can be optimized to further guarantee the safety and stability of urban rail transit system operation and provide solid foundation support for the green and healthy development of the city.

Under the background of full commercialization of 5G and smart cities entering the next round of fast-track of development, in order to promote the gradual growth and development of the industry, all participants in the rail transit industry need to further cooperate to promote the intelligent upgrading of the industry. In this process, various roles, such as metro operation subjects, government agencies, industry associations and enterprises represented by operators, communication service providers and software service providers, participate together to activate the supporting role of the new generation of information technology in rail transit and its operation services.

Strengthen the system of norms. Further, upgrade the standardization and normalization of 5G network and intelligent system of urban rail transit, make up for the lack of regulations, innovate the construction of intelligent rail transit system, and strengthen the security of an intelligent system.

Build a metro big data system. Formulate standards for constructing large databases for subway operation and service, strengthen multi-source data collection, co-construction and sharing, and smooth the channels for information exchange.

Expand the coverage of smart service platforms. In the process of practice, the coverage of intelligent service platforms in stations, lines, and other public transport is gradually extended to open up horizontal data collaboration. At the same time, based on the smart service platform, the application of the new generation of smart technology in the smart subway is further promoted and encouraged to promote the reform of the innovation form.

Pay attention to information security management. On the one hand, the top-level design of smart subway information security should be strengthened from the initial stage of construction, and an autonomous and controllable information security system should be constructed to strengthen the supervision and management of smart subway information security system to ensure the safe, orderly, efficient and intelligent development of smart subway.

7.4 Education Private Network Scenario

Education is a key driving force for social development and progress and a strong country. A country prospers when its education prospers, and a country is strong when its education is strong. The development of communication technology and information technology has greatly promoted the reform and innovation of education, providing platform support and technical guarantee for the construction of a networked, digitized, personalized and lifelong education system and the construction of a learning society in which "everyone can learn, everywhere, anytime". This section will introduce the insight of the 5G+ education industry, analysis of 5G application scenarios and network requirements, scenario-based solutions, and application cases in detail, and systematically explain how 5G enables smart education.

7.4.1 5G+ Education Industry Insight

Education is the cornerstone and driving force of national development and social progress and plays a vital role in promoting the improvement of national quality, production efficiency, and scientific and technological innovation. As can be seen from Fig. 7.32, countries all over the world attach great importance to education and constantly invest huge funds to implement the strategy of giving high priority to education development. The proportion of public spending on education in the GDP of the 27 EU countries increased from 4.86% in 2000 to 5.03% in 2005, with an average annual growth of 0.7%. China's education expenditure accounted for 3.9% of GDP in 2011. Since 2012, the proportion of education expenditure in GDP has remained above 4% for nine consecutive years, and there is still considerable room for growth in the future. In its *Vision 2021*, the UAE government clearly states that its goal of education is to build a "First-rate Education System".

In addition, as can be seen from Fig. 7.33, the investment in the education informatization market also increases year by year, and the education informatization expenditure of governments at all levels is no less than 8% of the education expenditure. At present, 117 countries and regions have established their own National Research and Education Network (NREN). Among them, 36 European national research network units and the representative units of the National Education Network in the five Nordic countries connect about 10,000 research units and 27,000 primary and secondary schools, with 50 million users.

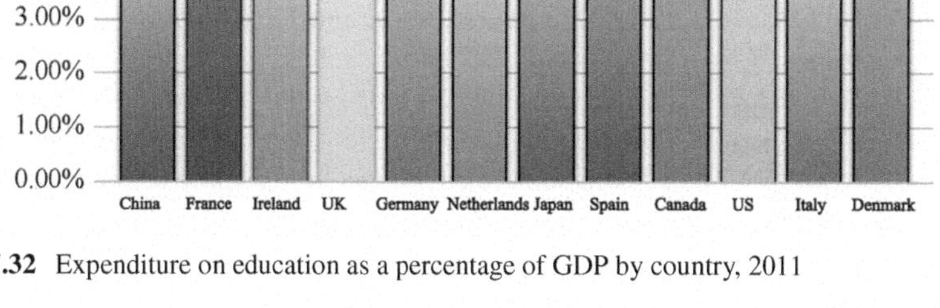

Fig. 7.32 Expenditure on education as a percentage of GDP by country, 2011

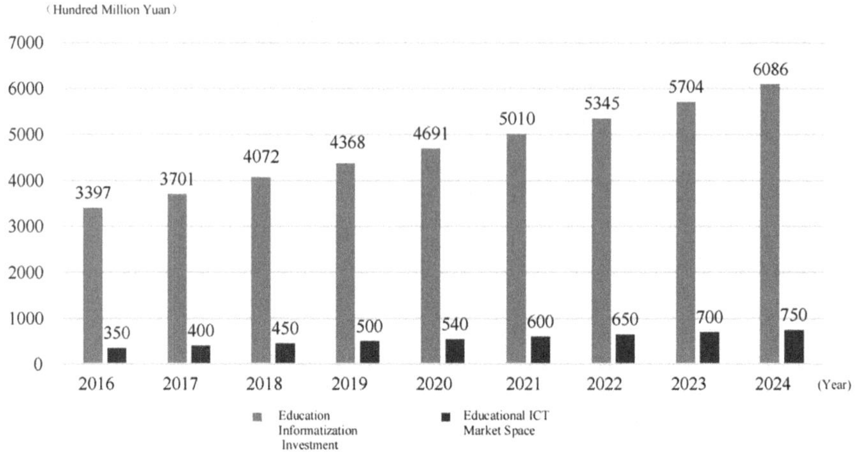

Fig. 7.33 Spatial changes of investment in education informatization and education ICT market in China

Generally speaking, the development of education information has entered the 2.0 era, with a high proportion of general education students and a large market scale, which is suitable for the development of the general education business. The investment granularity of colleges and universities is large, and customized business can be tried. *Education Informatization 2.0 Action Plan* proposed by the Ministry of Education: actively promote "Internet + education", adhere to the core concept of in-depth integration of information technology and education and teaching, adhere to the basic principle of application-driven and mechanism innovation, and establish and improve the sustainable development mechanism of education informatization. To promote the education informatization 2.0 action plan, a vigorous effect must be exerted to develop the new infrastructure innovation application of 5G. In addition, the Ministry of Industry and Information Technology, together with the Cyberspace Administration of the CPC Central Committee, the National Development and Reform Commission, the Ministry of Education, and ten other departments, released the *"Set Sail" Action Plan for 5G Application (2021–2023)* on July 12, 2021, which defines the sustainable development of 5G+ smart education as a key field. Accelerate the research and development of 5G teaching terminal equipment and AR/VR teaching digital content, combine AR/VR, holographic projection, and other technologies to achieve scenario-based interactive teaching and create an immersive classroom. It is to promote the support of 5G technology for education private networks, research and formulate technical specifications for key links of online education such as network, application and terminal based on specific application scenarios. The promotion of 5G in smart classrooms, holographic teaching, campus security, education management, comprehensive evaluation of students and other scenarios should be increased to improve the informatization ability of teaching, management, scientific research, service and other links. The smart development of

5G education has strategic support and financial support. Development scenarios such as AR/VR immersive teaching, 5G+ HD remote interactive teaching, and holographic classrooms are being carried out in full swing. 5G is a good supplement to the campus network. Through 5G wireless access, flexible and unified network carrying can be realized, bringing a smooth experience to school users. 5G is not only an increase in speed but also a comprehensive evolution of 4G. The outbreak of COVID-19 in 2020 is also an important opportunity for education reform. The rapid development of online teaching promotes the development of education informatization and provides a foundation for the use and development of distance education and smart classrooms. In general, the government's education strategy and the development direction of the operator industry are the same, both of which focus on promoting and developing 5G+ smart education.

As can be seen from Fig. 7.34, online education, smart teaching, and safe campuses are the main opportunities and directions for the development of various educational information. 5G+ smart education brings new opportunities to new infrastructure, which can solve the three major pain points in the current education industry resources.

(1) The total amount of educational resources is insufficient and the distribution is uneven: teaching resources (teachers, teaching applications, systems, and equipment) are separated, data is isolated and not shared, and cannot be interconnected. Take junior high school, for example. The ratio of teachers to students in Beijing is 7:73, and that in Jiangxi is 15:85, almost twice the difference in teacher resources.
(2) The teaching is difficult, and the education efficiency is low. The class is not interesting enough and the students can hardly concentrate. Part of the teaching materials are too difficult for students to understand vividly, resulting in low

Fig. 7.34 Development Direction of 5G+ Smart Education

teaching efficiency. Campus management mainly relies on people, lacking individualized teaching and evaluation means, and there is a shortage of criteria for teacher evaluation.

(3) Information asymmetry between home and school leads to potential security risks on campus. Schools and parents have asymmetric access to students' information, and there are hidden dangers to the safety of students. It is difficult to monitor and manage the campus; the information facilities of the campus are outdated, and its operation and maintenance capabilities are weak.

Education Informationization 2.0 Action Plan proposes to vigorously promote the construction of "three accesses and two platforms", namely, broadband network access to schools, high-quality resources access to classes, access to network learning space for everyone, and the construction of a public service platform for education resources and education management. In the past five years, the three accesses and two platforms have driven the construction of education informatization; the construction of the education metropolitan area network paves the way for Education Informatization 3.0. In the future, cloud computing, big data, the Internet of Things, and the mobile Internet, referred to as CBTM, will bring education into the smart era. We should explore a new teaching mode based on information technology (changing single evaluation to comprehensive multi-dimensional evaluation), develop a new model of internet-based education service (integrating online and offline resources to provide rich educational resources at all levels and lifelong learning), and explore a new model of education governance in the information age (promoting the reform of education governance mode based on big data). The implementation of these three new models cannot be separated from the backbone of 5G technology.

7.4.2 Analysis of 5G Application Scenarios and Network Requirements

National documents on new infrastructure point out that current education should focus on the integration, application, and innovation of online education, smart teaching, and safe campuses. All major operators, educational institutions, and enterprises should take the 5G network as the bearer, speed up the construction of HD and even ultra HD video applications, and promote the integration and innovation of 5G+ smart education, as shown in Fig. 7.35. It includes 5G+ smart classroom, 5G+ AR/VR classroom, 5G+ interactive teaching, 5G+ holographic classroom, 5G+ teaching quality analysis, 5G+ student health assessment, 5G+ distance teaching, 5G+ safe campus, etc. In addition, 5G technology can also effectively meet the current ICT status and demands.

(1) The network is equipped with wide coverage and large bandwidth operation capability.
(2) With the support of AI, scenarios, such as smart classrooms and intelligent security monitoring, can be implemented.

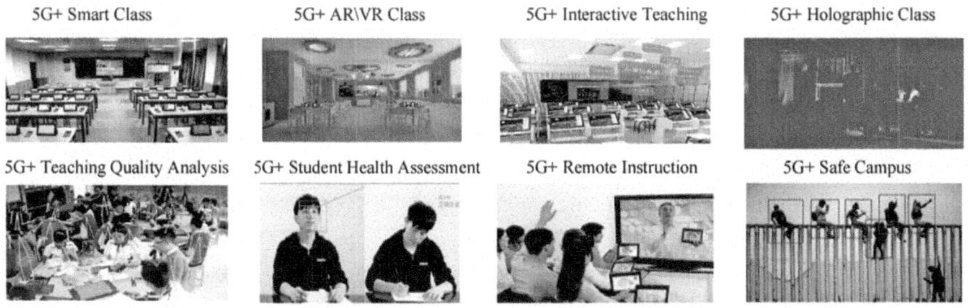

Fig. 7.35 Application scenario of 5G+ smart education

(3) Double guarantee of computing power and platform, application sharing technologies, such as edge computing and Independent Software Vendor (ISV) will help 5G+ smart education.
(4) Protect privacy and security and build a private education network isolated from other networks.

The following describes typical 5G application scenarios.

1. 5G+ Smart Classroom

The emergence of 5G technology makes classroom teaching interactions smoother. 5G technology, through deep integration with artificial intelligence, big data, virtual reality, and other technologies, enables data-driven teaching decisions, real-time assessment feedback, three-dimensional interactions, and intelligent resource delivery. Figure 7.36 is a schematic diagram of a smart classroom scene, which is of two features: the first is high density, that is, the space between students is less than 1 m, which is denser than other dense places such as offices; The second is high concurrency, that is, when teachers send small videos and pictures during teaching, all students need to receive them within 1 s. This puts forward extremely high requirements for network access and transmission. First of all, the smart classroom has a high demand for network bandwidth. As shown in Fig. 7.37, if 50 4 K videos are concurrently transmitted simultaneously, the downstream bandwidth capacity requirement is 50 × 20 Mbps = 1 Gbps. Secondly, smart classroom also has high requirements for network stability. If there are 50 students in the class, a 1-min delay in network transmission would result in a 50-min delay for the entire class.

As shown in Fig. 7.38, currently, operators are working with their education partners to build smart classrooms to form a complete business value chain and jointly promote the informatization and intelligent development of regional education. In this business value chain, the municipal/district Education Commission, as the main investor, provides financial support for the construction of smart education in schools; education content partners provide information course content; education business partners provide course assistance, teaching evaluation, AI behavior analysis, and other smart education auxiliary tools; operators provide high-quality network services and education application integration services. Communication

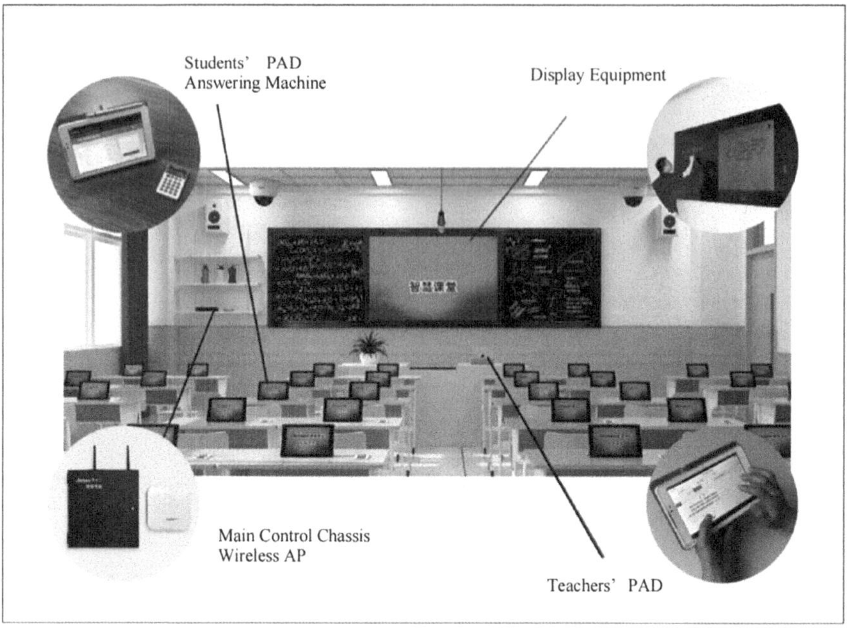

Fig. 7.36 Schematic of smart classroom scenario

manufacturers provide 5G/MEC technical support and CPE/ terminal equipment supply.

2. 5G+ AR/VR Education

The two core advantages of 5G, such as large bandwidth and low delay, enable it to upload virtualized teaching content to the cloud for rendering, reconstruction, and other processing. Finally, the processed content is transmitted to various terminal devices in real-time through a 5G network in the form of audio and video streams to realize AR/VR cloud applications and cloud platforms. Table 7.5 lists the basic requirements of VR education for network transmission.

Through the construction of a virtual laboratory, three-dimensional interactive classroom, and other virtual reality application platforms, students' interest and interactive experience can be enhanced. The knowledge retention rate of immersion learning is five to ten times higher than that of lecturing. AR/VR courses in primary and secondary schools have taken on a small scale, with a total of more than 1,000 courses covering all grades of primary, middle, and high schools. Figure 7.39 shows the application scenarios of 5G+ AR/VR education.

3. 5G+ Interactive Teaching

5G interactive teaching refers to 5G processing of various hardware and software modules in the traditional classroom, and 5G network with high speed, large bandwidth, and low delay as the carrier can avoid the problems of lag and poor scalability

Fig. 7.37 Example of large bandwidth requirements for 5G+ smart classroom

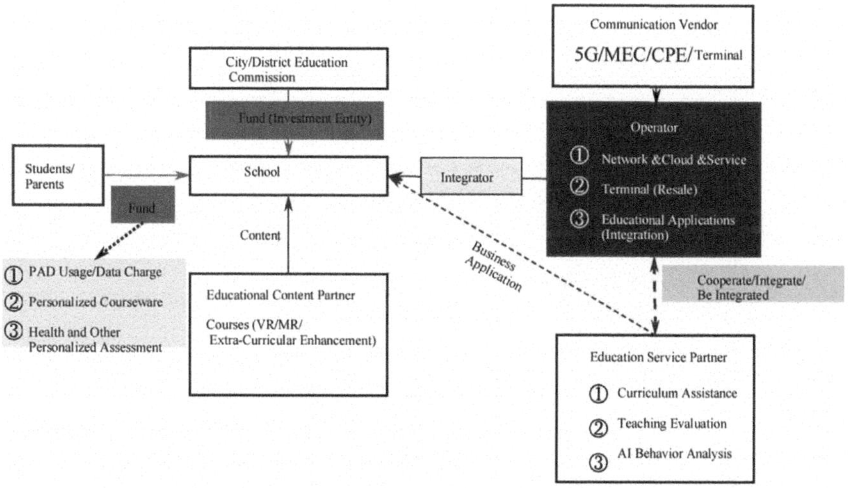

Fig. 7.38 5G+ smart classroom: cooperation mode among participants

Table 7.5 Basic requirements for network transmission in VR immersive teaching scenarios

Scene requirement	Network requirement	
	Bandwidth	Latency
VR Glasses (Monocular resolution 1600 × 1600 pixel)	50 Mbps/student @DL	Network < 40 ms

Hangzhou Caishi Education 5G+VR Air Classroom

China Science and Technology Museum Cloud AR/VR Platform Interactive Experience

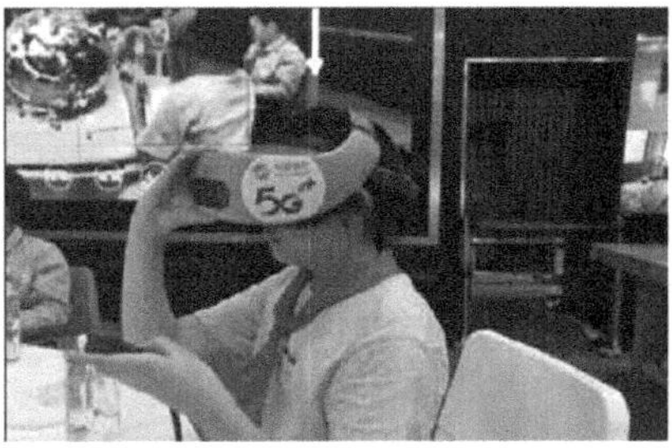

Fig. 7.39 Examples of application scenario of 5G+ AR/VR education

of traditional wired network, Wi-Fi network, Bluetooth and other carriers. Thus, teachers and students can have a faster and smoother teaching and learning experience. Table 7.6 shows the basic network requirements of the interactive teaching class based on the ISV test conducted by the carrier and an AI enterprise.

Figure 7.40 shows the basic configuration of an interactive classroom, where teachers use interactive large screens to teach and interact with students through smart tablets. Through 5G+ interactive teaching, the efficiency and effect of teaching can be improved. Students can timely interact with teachers to answer questions and use richer multimedia teaching courseware. For schools, such interactive teaching and intelligent application-assisted learning can enhance the popularity of schools and facilitate teaching management.

4. 5G+ Holographic Classroom

Table 7.6 Basic requirements for network transmission in interactive teaching classes

Scene requirement	Network requirement	
	Bandwidth	Latency
5G large screen + 5G Pad (50 pieces)	50 Mbps @DL	Network < 100 ms
	100 Mbps @DL	

Fig. 7.40 Example of application scenario of 5G+ interactive teaching

5G+ holographic classroom has realized the remote distribution of high-quality educational resources and has explored a new solution to solve the uneven distribution of educational resources in China. The emergence of holographic classrooms has expanded the coverage of elite schools and enabled celebrities to address the speeches from thousands of miles away, replacing the flow of people with the flow of information.

Table 7.7 shows the basic network transmission requirements in remote holographic classrooms. Based on the transmission rate of more than 100 megabits of 5G and the delay as low as tens of milliseconds, low delay propagation of AR/VR applications and audio and video streams relying on large bandwidth content can be realized further to achieve extensive coverage of 5G+ holographic projection technology, so that teachers and students can enjoy one-to-many, many-to-one and many-to-many holographic classroom remote teaching modes. Figure 7.41 shows an example of a 5G+ holographic classroom application scenario. A teacher in Fujian Province gives a remote class through a live broadcast device and realizes holographic interaction in a middle school class in Wuhan through holographic output.

Table 7.7 Basic requirements for network transmission in remote holographic classroom

Service name	Communication requirement		
	Uplink speed rate	Downlink speed rate	Latency
Remote holographic class	80 Mbps	80 Mbps	Network < 100 ms

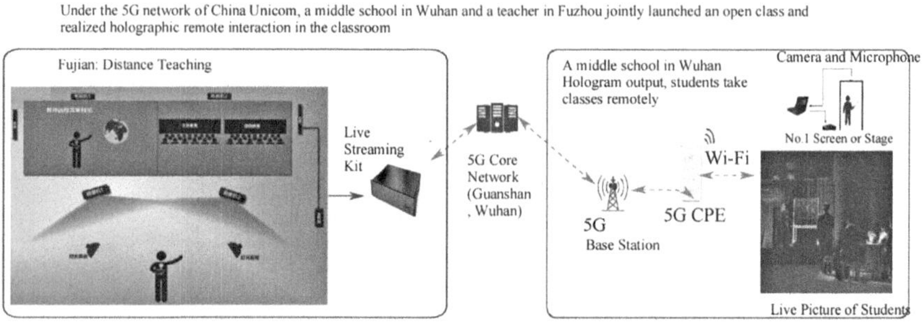

Fig. 7.41 Example of application scenario of 5G+ holographic classroom

5. 5G+ Teaching Quality Analysis

Limited by bandwidth, transmission speed, and other factors, teaching quality analysis in traditional scenarios can only complete data collection in a single form. However, 5G network provides a new possibility for teaching quality analysis with its characteristics of high bandwidth and low delay. Table 7.8 shows the basic requirements of digital intelligent teaching evaluation for network transmission. Figure 7.42 shows an application scenario of 5G+ teaching quality analysis. By installing high-definition cameras in the classroom, collecting and analyzing the status of students and teachers with the help of AI, the intelligent assessment of teaching quality is realized, as well as the collection and assessment of students' learning status, and a one-click personality chemistry analysis report is generated.

6. 5G+ Student Health Assessment

Based on the characteristics of low delay, high speed of 5G, and low energy consumption of the Internet of Things, Narrow Band Internet of Things (NB-IoT) technology can be used to achieve wide coverage of Internet of Things in the whole school, edge computing can realize data summary and distribution. Network slicing technology can ensure data security. Then, a complete student health evaluation system can be realized. Table 7.9 lists the basic requirements for network transmission in the student health assessment scenario. As shown in Fig. 7.43, based on the data collected by the AI camera deployed in the classroom, students' adverse health risks can be monitored through sitting posture analysis, myopia risk analysis, and other means, and students' school data can be pushed to parents so that the school can better care and manage students.

7. 5G+ Distance Teaching

Table 7.8 Basic requirements of network transmission for digital intelligent teaching evaluation

Scene requirement	Network requirement	
	Bandwidth	Latency
HD Camera upload in real time (2–4)	15–30 Mbps @ UL	None

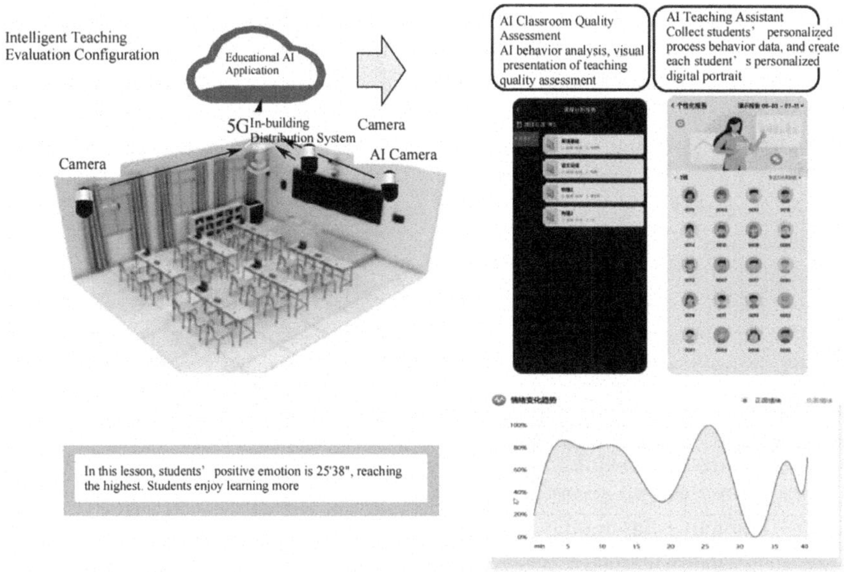

Fig. 7.42 Example of application scenario of 5G+ teaching quality analysis

Table 7.9 Basic requirements for network transmission in the student health assessment scenario

Scene requirement	Network requiremen	
	Bandwidth	Latency
HD Camera upload in real time (2–4)	15–30 Mbps @ UL	None

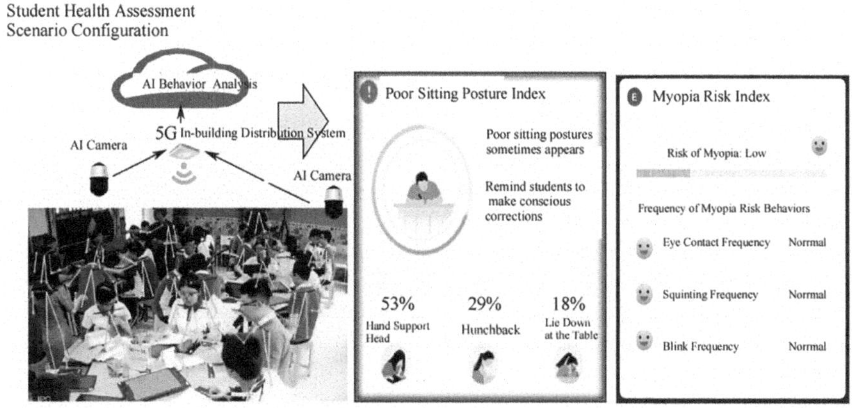

Fig. 7.43 Example of application scenario of 5G+ student health assessment

There are more than 200,000 rural schools in China, with widespread problems, such as a small number of teachers in rural teaching centers, uneven curriculum offerings, poor teaching quality in remote areas and uneven educational resources. The emergence of distance education can greatly supplement teachers and courses for rural schools so that students can get a better and richer education, which is of great significance to the development of education in the country. Therefore, distance education is an inevitable trend and requirement for the development of general education.

Given the problems such as delay and instability in the existing wired and Wi-Fi networks used for distance education, 5G network can provide services such as HD video live broadcast and remote interaction with low delay due to its characteristics of low delay, large bandwidth, and high reliability, and truly realize immersive distance education. Table 7.10 shows the basic network requirements for distance learning. Figure 7.44 shows the architecture of a typical remote synchronous classroom solution. The architecture includes a main lecture classroom, distributed remote teaching point, interactive teaching platform, and cloud server, etc., which can realize a "dual-teacher" interactive classroom with the participation of multiple people and multiple classrooms. Meanwhile, dual-teacher classrooms support the mixed deployment of public cloud and private cloud, which can not only meet the security requirements of private cloud but also realize the sharing of educational resources with famous schools outside the region through public cloud.

8. 5G+ Safe Campus

The country and society are very concerned about campus security issues. Security conflicts between students, teachers, and undesirable personnel from outside the school occur occasionally, while traditional technical means mainly rely on physical security measures, lacking a safety warning mechanism, making it challenging to maintain 24-h supervision. 5G+ campus security can make up for the shortcomings of traditional technology and realize the normalization, intellectualization, and humanization of campus security and hidden danger investigation.

Through the large bandwidth access of 5G, the real-time switching and viewing of HD surveillance video in every corner of the campus can be realized. Intelligent analysis of surveillance video and recognition of various abnormal events can be realized through the combination of edge MEC platform and AI. With the integration of 5G network and cloud-based robots, the campus can patrol 24 h without blind corners, saving security manpower and increasing security guarantee. Figure 7.45 shows several examples of 5G+ safe campus application scenarios.

Table 7.10 Requirements for network transmission in a remote classroom

Service name	Communication requirement		
	Uplink speed rate	Downlink speed rate	Latency
"Dual-teacher" interactive classroom based on 4 K HD	40 Mbps	40 Mbps	100 ms

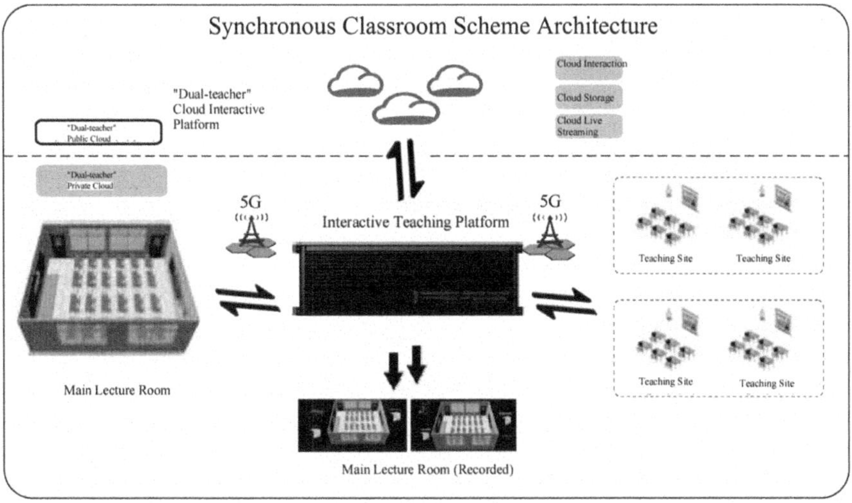

Fig. 7.44 Example of application scenario of 5G+ remote education

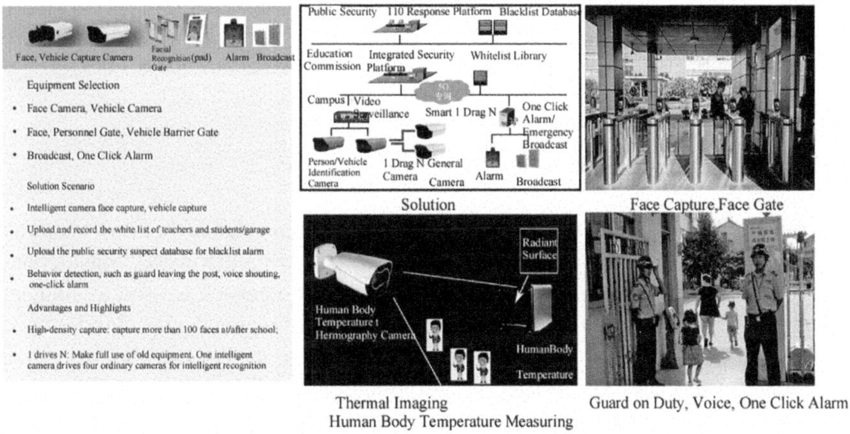

Fig. 7.45 Examples of application scenario of 5G+ secure campus

7.4.3 Scenario-Based Solution

With the further promotion of the "5G+ education" model, the education industry is also showing a new luster with the support of 5G. Compared with 4G, the performance improvement of 5G network equipment, the acceleration of network speed, and the reduction of network delay make 5G have the distinct characteristics of high speed and large capacity. To harness the full potential of 5G in education, it is essential to integrate all kinds of intelligent technologies and, implement them in a variety of

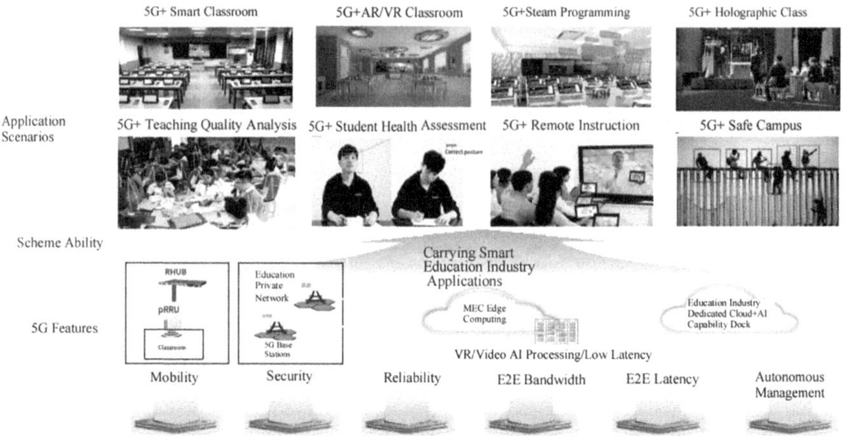

Fig. 7.46 Solution based on 5G+ smart education scenario

potential application scenarios, and provide intelligent services for teachers, students and administrators on- and off-campus. Figure 7.46 shows the solution based on the 5G+ smart education scenario. Next, the corresponding scenario-based solutions are given with some specific application scenarios.

5G Education Private Network

Traditional education networks will limit learning within schools, but now with the help of 5G, the 5G education network can cover all scenarios inside, outside and inter-schools. As shown in Fig. 7.48, a core backbone network is used to connect on-campus, off-campus, inter-campus, and families, and regional edge cloud is used to realize streaming media, VR distribution, and video analysis to finally realize on-campus smart classroom, VR/AR classroom, multimedia teaching, health monitoring, safe campus and AI evaluation, as well as 4 K and holographic live broadcasting between schools (branch schools and lower-level schools), online courses for off-campus and at-home learning, multimedia teaching. The overall 5G+ cloud intelligent education network plan can be divided into the following five specific plans.

(1) Wireless deployment scheme: In the indoor scene, pRRU is deployed in the classroom to meet the 5G access of cameras, students' pads, large screens, and other terminals. In outdoor scenarios, 5G access is provided through macro stations.
(2) Transport network deployment scheme: Share with the bearer network of the large network and isolate services through FlexE slices.
(3) Core network deployment scheme: MEC is deployed in the teaching group/ education commission location and shared by multiple schools. Multiple schools

share teaching resources based on the same platform, which facilitates the rapid promotion and application of teaching resources.

(4) MEC platform deployment: The exclusive MEC platform is deployed at the edge of the network, and the data does not go out to the campus/school district, realizing local traffic unloading and meeting the requirements of VR/security and other low-delay scenarios.
(5) Application deployment scheme: Innovative applications such as VR teaching/dual-teacher classroom/AI video are deployed based on the MEC platform, while traditional office applications can directly connect with existing servers. AI applications are based on cloud edge collaborative architecture. The algorithm is trained on the public cloud, and the inference is performed on edge MEC. After analyzing the results, AI synchronizes with the cloud from MEC.

Figure 7.47 shows an example of a scenario-based 5G education private network. Such a comprehensive 5G education private network can fully enhance the program competitiveness of the education industry so as to build a 5G+ cloud education industry ecology of "production, education, and research". Specifically, the goals can be achieved in the following different scenarios.

1. 5G+ Smart Classroom

With the help of 5G, virtual simulation, VR/MR, AI, and other technologies, the classroom can be more interesting and provide an immersive experience while offering personalized teaching, teacher assessment, student status assessment, and other functions.

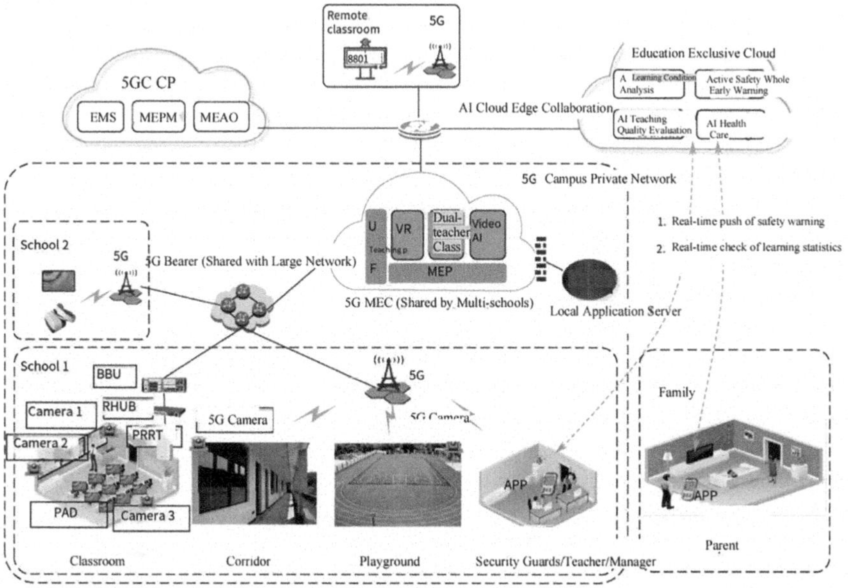

Fig. 7.47 Example of scenario solution for a 5G education private network

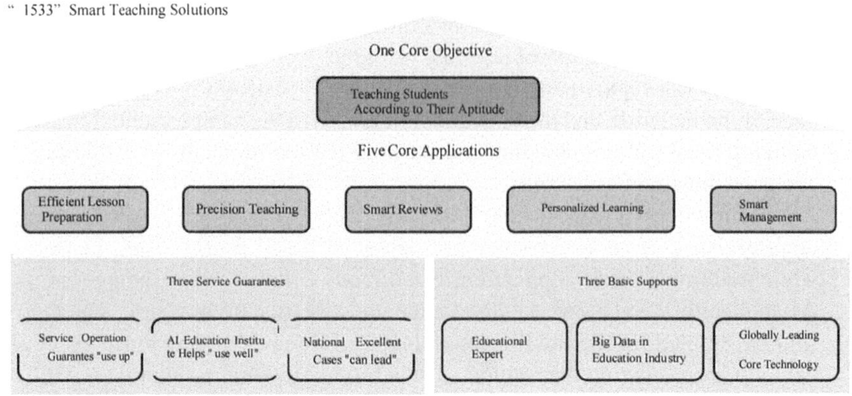

Fig. 7.48 Example of scenario solution for 5G smart classroom

5G smart classroom should meet the following technical requirements:

(1) Classroom VR/MR teaching experience.
(2) Smart classroom management and video analysis.
(3) AI for learning analysis and assessment.
(4) Intelligent analysis of knowledge points.

2. 5G+ Interactive Teaching

Teachers and students can realize ultra-high-definition real-time interaction and campus high-definition video conferences through 5G tablets and other terminal devices and based on a three-dimensional teaching platform. In addition, by pushing high-definition multimedia content, teaching can be visualized to improve teaching quality and efficiency.

Promoting 5G interactive teaching requires cooperation with platforms and providers.

(1) Fast construction of platform-level broadcasting channels.
(2) Introduce online partners—Service providers (SP).
(3) Promote content resources of educational resources—Content providers (CP).

3. 5G+ Safe Campus

Real-time monitoring and forensics will be carried out through 5G cameras and inspection robots, and 4 K high-definition monitoring images will be sent back and analyzed in real-time through a 5G network, providing visible, controllable, and manageable campus-integrated intelligent security.

5G Safe campus should meet the following technical requirements:

(1) Proactively warns of emergencies and intervenes in real-time by AI.
(2) Abnormal behavior recognition and temperature monitoring.
(3) Use existing network monitoring facilities and protect investments.

(4) Drones and robots should be used to perform a 360° three-dimensional inspection.

With 5G+ cloud + AI as the starting point, enterprises will be gathered in the industry to build a cooperative ecology, jointly create educational products welcomed by customers, and respond to the requirements of the Ministry of Education of "anywhere, anytime, co-education of home and school" to provide inclusive educational services. Next, the corresponding solutions are introduced from the above three specific campus scenarios.

5G+ Smart Classroom

Figure 7.48 shows a typical scenario solution of a 5G smart classroom. In the scene of a smart classroom, the "1533" smart teaching solution can effectively improve the quality of a class. Among them, "1" refers to a core goal, namely, individualized education; "5" refers to five core applications, namely efficient lesson preparation, precise teaching, intelligent evaluation, personalized learning, and intelligent management; The first "3" refers to three service guarantees, namely, service operation guarantees "using it", AI Education Academy guarantees "using it well", and national excellent cases "take the lead"; The second "3" refers to three basic supports, namely education industry experts, education industry big data, and global leading core technology. Therefore, taking precision teaching as the core, it is possible to systematically solve the core pain points in the teaching process and achieve the following four goals of smart teaching.

1. Efficient Lesson Preparation

In traditional scenarios, teachers cannot fully grasp the learning situation and use targeted teaching design. Meanwhile, it is difficult to search massive teaching resources systematically. In the 5G scenario, the visual report of the learning situation can collect the interactive homework results of students in class in real-time, providing scientific and accurate support for teachers. The knowledge map designed for lesson preparation can accurately mark and intelligently recommend teaching resources to solve the problem of finding massive materials.

2. Precise Teaching

In the traditional scenario, when the teaching content is relatively abstract, different students have different degrees of acceptance, and it is difficult for teachers to judge the degree of students' mastery. In the 5G scenario, real-time teaching interaction between teachers and students can be realized through 5G tablets, including answering questions and assignments in class, etc., and three-dimensional classroom interaction can be achieved by pushing high-definition multimedia content to visualize teaching content.

3. Intelligent Review

In traditional scenarios, teachers have a heavy burden of evaluating homework, which requires a lot of time and energy, and the results cannot be digitized. In a 5G scenario, OCR technology can be used to realize intelligent evaluation of homework by machine, greatly reducing the time for teachers to evaluate homework.

4. Personalized Homework

In traditional scenarios, there are often problems such as heavy homework burden, which is one task for all, and mostly ineffective training. In the 5G scenario, personalized homework can be generated based on students' learning feedback, and error problem-intensive training and personalized improvement schemes can greatly reduce the homework burden.

5G+ Interactive Teaching

Figure 7.49 shows a typical 5G interactive teaching scenario-based solution. In the scene of interactive teaching, teachers usually use interactive large-screen to teach interactively with students' smart tablets, supplemented by online lesson preparation, multi-screen interactive teaching, after-class personalized tutoring, key playback, and other teaching methods, to achieve fine-grained precision teaching and bring new teaching experience. For students, it is convenient to interact with teachers in time to answer questions and use richer multimedia teaching courseware to realize intelligent application-assisted teaching. It can simplify teaching management, enhance the school's visibility, and so on. Usually, the 5G+ interactive teaching solution consists of two parts: an intelligent terminal and a 5G room.

1. Intelligent Terminal

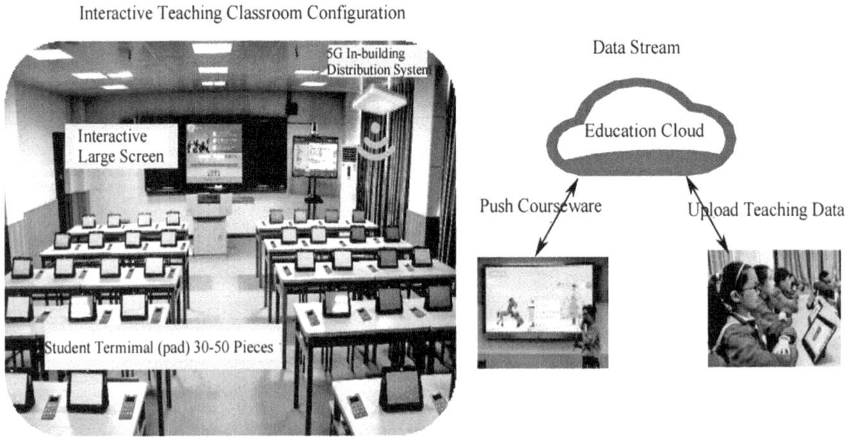

Fig. 7.49 Example of scenario-based 5G interactive teaching solution

The interaction between teachers and students is inseparable from various intelligent terminals. Interactive White Board can realize free writing through gesture control, writing/saving at any time, and free comments. It supports wireless screen projection of notebook, pads, mobile phones, and other types of terminals to realize free sharing of ideas between teachers and students. A highly interactive classroom can finally be achieved with a control panel, a modular intelligent management host, an electronic class board, an interactive teaching coordinator, and a video conference terminal.

2. 5G Indoor Division

Based on the indoor division of 5G, its gigabit per second (Gbps) capacity enables fully connected smart classrooms, which has three advantages.

(1) Easy Deployment: A single BBU can support hundreds of pRRU, while only one BBU needs to be deployed in one teaching building.
(2) Large Bandwidth: 1.4 Gbps downstream and 109 Mbps uplink can meet the speed of more than 20 Mbps per student in the whole class, plus access to large screens and cameras. Meanwhile, uplink and downlink rates can be adjusted by changing the proportion of time slots to meet the bandwidth requirements of VR scenes.
(3) Energy saving and environmentally friendly. Its emission power is low, and there are no concerns about radiation. When people leave, AI can detect that and turn off lights to save energy.

5G+ Safe Campus

In recent years, the community has been very concerned about campus security, and the shortcomings of traditional security technologies have been noticed. In the campus scenario, only through the combination of 5G and AI can 360°movable monitoring be achieved to meet the needs of a safe campus. Figure 7.50 shows a typical 5G safe campus scenario-oriented solution. In general, the solution of this scenario focuses on the following four areas.

1. **Campus entrance and exit security, human-vehicle blocklist/allowlist comparison, alarm linkage**

Use facial and vehicle recognition devices to capture images of individuals and vehicles, upload the allowlist of teachers and students on campus for access and record, and upload the database of public security suspects for blocklist alerts. Detect the behaviors of guards, such as leaving their posts, and realize voice announcements and one-click alarms.

2. **Conduct boundary crossing and intrusion detection in the campus perimeter and record the whole structure's data**

Infrared intelligent perimeter behavior analysis cameras are deployed around the school to link the alarm of transgression and intrusions. Fully structured cameras are deployed on roads around the campus for big data recording of people/vehicles.

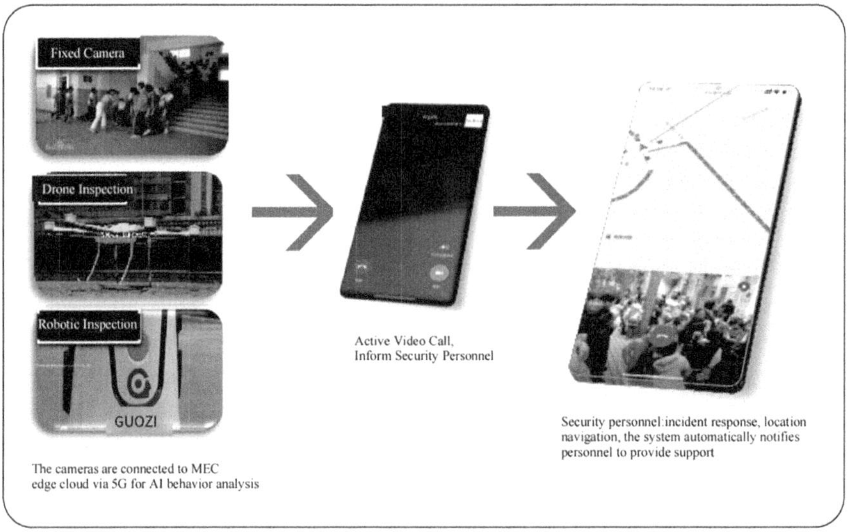

Fig. 7.50 Example of scenario-based 5G campus security solution

3. **Bright kitchen stove, wide dynamic image, intelligent behavior monitoring recording**

Detection of chefs' behavior, such as not wearing chef's hat, will be made public in key cooking areas, and dynamic patrols will be conducted in the dining area of the canteen.

4. **Realize epidemic prevention + inspection robot automation and remote control**

The intelligent robot can perform 360° all-weather ground inspection, and the comprehensive intelligent security platform can detect autonomous obstacles, circumnavigating, autonomous charging, fire warnings, abnormal alarm linkage, and intelligent environment perception. At the same time, in response to the epidemic, intelligent robots can measure human body temperature, disinfect and spray automatically, detect whether people are wearing masks, and conduct safety inspections.

7.4.4 Application Cases

5G, as a technology with great potential, has been the market's focus. A large number of resources and capital investments have brought 5 G-related applications into a stage of rapid development, and education, as one of the application scenarios, has already seen many successful application cases.

1. **Active Security**

As shown in Fig. 7.51, a middle school in Nanjing and a communication operator in Nanjing implement an AI active security video analysis platform by deploying MEC nodes (UPF + MEP) and 5G 4 K cameras. By collecting 30 channels of video streams, it realizes proactive security warnings for crowd gathering, intrusion, suspicious lingering, fighting, crossing barriers, leaving posts, etc. The middle school has successfully realized active security based on 5G+ AI through deploying the innovative project.

The "5G + AI active security" solution has also been deployed in other provinces and cities, as shown in Fig. 7.52.

2. **Smart Classroom**

As shown in Fig. 7.53, an operator and an AI enterprise jointly deploy local smart classroom services and public cloud smart classroom services in Tianjin through the service networking of CMNet, PTN, IPRAN transport network, and 5G MEC, realizing interactive teaching anytime and anywhere and intelligent assisted teaching, bringing a brand new teaching experience. In addition, the 5G network realizes the unification of networking technology, and schools no longer need to deploy multiple networks, greatly simplifying network management, operation, and maintenance, while bringing faster, better, smoother network experience to school users.

3. **Smart Lesson Evaluation**

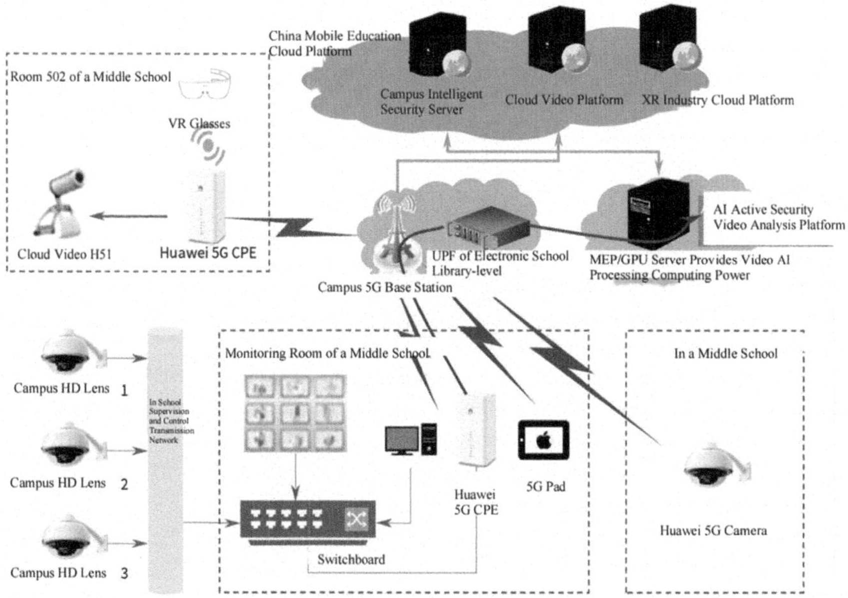

Fig. 7.51 Example of 5G + AI active security for a middle school in Nanjing

Fig. 7.52 Example of "5G + AI active security" in a primary school in optics valley district, Wuhan

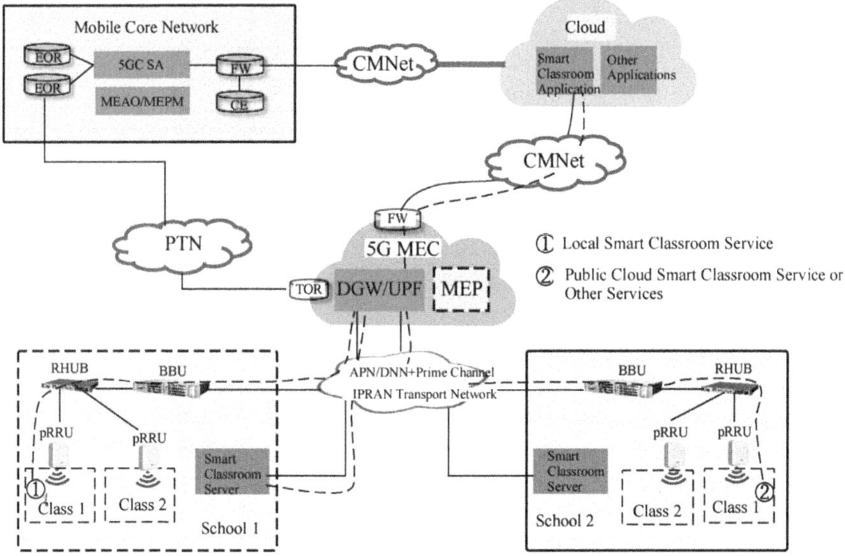

Fig. 7.53 Networking diagram of 5G smart classroom service

A middle school in Beijing has jointly implemented "5G + smart classroom evaluation" with operators and AI technology companies. As shown in Fig. 7.54, the AI classroom quality assessment can visually present the teaching quality assessment through AI behavior analysis. In contrast, the AI teaching assistant can collect the behavior data of students' personalized process and make a personality digital portrait for each student. Therefore, digital evaluation can be enabled through 5G+ AI to provide school classroom teaching assistants and promote specialized teaching in schools.

4. **VR Teaching**

A certain operator group in Suzhou realized the 5G+ video education library application as shown in Fig. 7.55 through the pan-teaching service combining 5G and

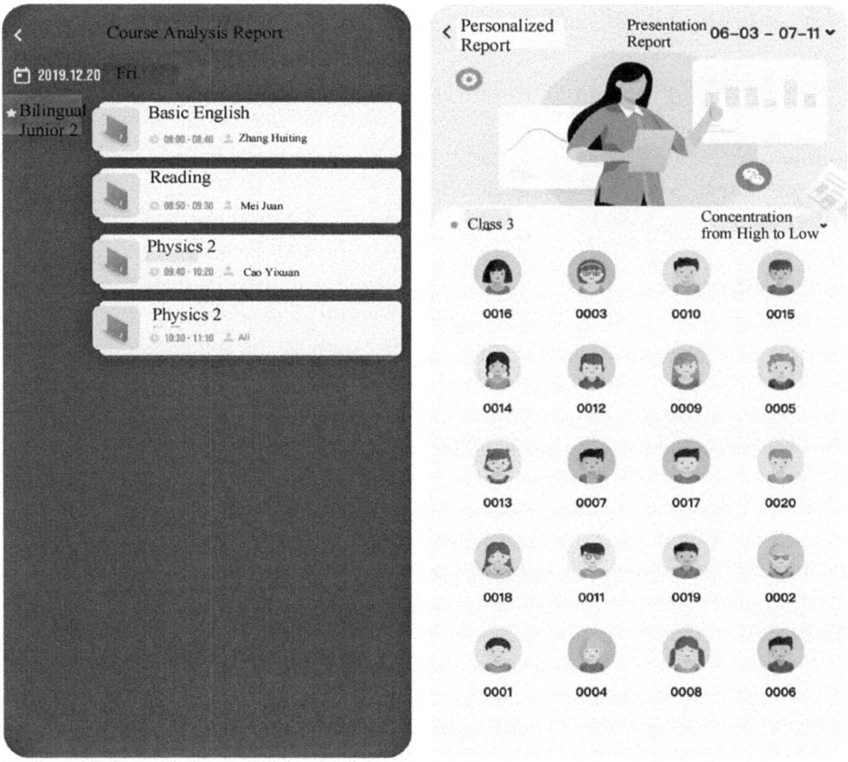

Fig. 7.54 Example of "5G + Smart Classroom Evaluation" in a middle school in Beijing

VR. This project was implemented in Suzhou Video Education Library and three benchmark middle schools. Through the construction of a 5G+ education private network, VR lab, VR music master class, and other applications, this project has been rated as a 5G leading project of a certain operator group. It was awarded the "National 5G Smart Education Demonstration Zone".

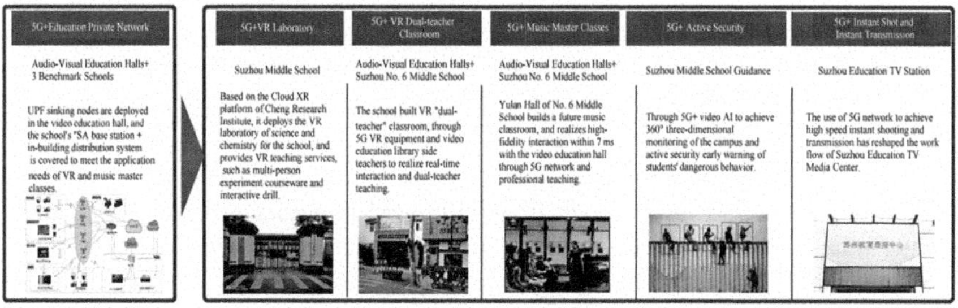

Fig. 7.55 Example of 5G + video education library application

7.4.5 *Summary (of 7.4)*

Education, a highly prioritized field by the nation, has consistently received significant capital investments to promote its development. Nowadays, with the tide of the information age, the defects in the traditional education mode have become increasingly prominent. As a good supplement to campus network, 5G can solve problems existing in traditional Wi-Fi, such as small capacity, unstable coverage, high latency, and poor mobility, to fully empower smart education and further promote education development in the new era. From the smart education industry chain perspective, the upstream construction subjects are divided into software vendors, hardware vendors, and content providers. The three main bodies support the construction of smart education with products from four dimensions: campus IT infrastructure, interactive teaching hardware equipment, information platforms and software, and online content resources. Through the integration of software, hardware, and content, the smart education industry is constituted of smart campus, smart classroom teaching, and online education. Taking these three businesses as the market entry point and centering on the new infrastructure of 5G+ MEC + AI, the core products of 5G smart education will be created, the industry standards will be drawn, and the new 5GtoB industrial market will be incubated to break out of the old competitive market and create a new track.

Through the study of this section, readers can understand the status quo and future development direction of the 5G+ education industry, get familiar with the application scenarios and network requirements of 5G+ education, and understand various scenario-based solutions of 5G+ education.

7.5 Medical Private Network Scenario

7.5.1 *5G+ Medical Industry Insight*

From the perspective of each medical unit, digital medical care and office work are essential for the development of the medical industry. It is one of the development directions of digital hospitals to break through the data barrier between various medical links and realize data interconnection. For example, in digital hospitals, patients' examination results and data will be shared directly with the terminals of relevant units and medical staff, through which the responsible doctors can automatically issue treatment plans. With the support of the digital networks, wearable human body detection devices can be popularized in future hospitals to monitor patients' physical signs in real time and transmit data to medical staff terminals to realize round-the-clock effective monitoring and avoid many accidents such as delayed rescues. As part of the digital platform, the medical guidance robot can save human resources and achieve more efficient and accurate pre-diagnosis guidance services.

Further analysis shows that based on China's condition, the imminent aging population poses a significant challenge to its healthcare system. Along with the accelerated process of industrialization and urbanization and the continuous improvement of economic development, Chinese residents' demands for medical services are also constantly increasing. The spectrum of diseases of Chinese residents is changing constantly. The current medical system is far from meeting the needs of the society in the future.

To sum up, at present, Chinese society is in urgent need of more efficient and accurate medical and health services. Remote and digital healthcare are multipliers of the existing medical talents and medical resources. They will dominate the development of the new medical system in the future. It will be the inevitable path for the development of the Chinese medical industry in the future. Internet, artificial intelligence, big data, and other digital technologies as effective technical means to improve the level of medical and health services have received much attention, and all these will produce massive data transmission demand, such as real-time high-definition video transmission, which requires communication means with super bandwidth. At the same time, medical services also need communication services with stable service quality and very low communication delays because of their particularity.

The theoretical downward speed of 5G is 20 Gbps. Regarding the design concept, 5G focuses on the development of the Internet of everything. Compared with 4G, which focuses on mobile broadband business, 5G has more business scenarios, and the network capability is not limited to uplink and downlink speed. It can simultaneously ensure high speed, low delay, large connection, and reliable service quality. In terms of physical implementation, 5G introduces new wireless transmission technology based on the point-to-point physical transmission of traditional communication technology, realizes the construction of multi-user regional networks, and greatly improves the transmission performance of communication networks. 5G has a peak rate and lower delay several times than 4G, which can meet the needs of remote ultrasonic diagnosis and real-time transmission and sharing of high-definition image data in the mobile environment. 5G can meet the support ability of super-large connection and ensure access to multi-terminal medical equipment devices. Meanwhile, the millisecond delay capability of 5G can support doctors to assist remotely in manipulating robotic arms to carry out various examinations. The features of 5G are well suited to the needs of the medical industry.

Therefore, 5G is an indispensable technology for future healthcare. With the rapid development of 5G, remote consultation, remote ultrasound, intelligent medical robots, and other technologies have gradually become reality, proving their value in clinical practice. In the future, they will definitely change the current medical system structure so that effective medical resources can serve society more efficiently.

7.5.2 Scenario and Network Requirement Analysis

The medical scene can be divided into three parts: in-hospital, inter-hospital and out-of-hospital. Three aspects are introduced: in-hospital data collection, monitoring and analysis, inter-hospital video-assisted diagnosis and control, and out-of-hospital AI + mobile diagnosis and treatment.

(1) In-hospital data collection, monitoring, and analysis: Regarding the hospital scene, 5G mainly helps realize the interconnection of image/test equipment, and supports doctors' remote ward rounds, remote monitoring, newborn visits, and other services through the integration of equipment data. Regarding security and privacy requirements, medical services need to be isolated from public online services, so that medical data cannot be released from hospitals. In terms of transmission capacity, cloud medical services require large bandwidth support. For example, a CT or nuclear magnetic picture (about 800 MB) can be downloaded within 3 s. Medical services' precision and real-time performance also require the network to have low latency. For example, the interactive delay of audio and video services should be less than 50 ms.

Although hospitals' business requirements are high,their social and commercial value is huge. Take an ordinary Grade-III A hospital as an example; the number of daily outpatients can generally reach tens of thousands of people, and 90% need medical equipment inspection. It is a long-standing appeal of medical institutions to save the limited medical resources of hospitals through cloud services. Robot AI can improve triage efficiency and false triage caused by a lack of professional standards by replacing artificial doctors (3 ~ 4 persons/Grade-III A hospital). Moreover, robots work 7 × 24 h a day, effectively solving the problem of limited workforce.

(2) Inter-hospital video-assisted diagnosis and control: Cloud interconnection between hospitals can realize regional medical cooperation and help high-quality medical resources sink to the grassroots level. Through telemedicine means, such as remote consultation and remote surgery, patients in primary medical institutions can also enjoy high-quality medical resources and improve the rehabilitation of grassroots patients. Inter-hospital communication often requires the provision of a dedicated communication channel, and it also puts forward high requirements for delay and bandwidth. Generally, it is required to ensure that the bandwidth of audio and video services is about 30 Mbps and the delay is less than 50 ms. The service delay is less than 20 ms for remote ultrasonic and other control classes.

At present, Chinese Grade III A hospitals usually cooperate with a large number of middle and primary medical institutions in the medical business, thus forming a large-scale medical commonwealth, opening up all nodes in the medical commonwealth, realizing the sharing of medical resources, and changing the current proportion of medical resources in China, which will have far-reaching social influence and commercial value.

(3) Out-of-hospital AI + mobile diagnosis and treatment: the scenario outside the hospital mainly involves mobile first aid and critical hospital transfer, which is characterized by flexibility. The transport network is required to provide temporary special pipes for mobile rescue in time to guarantee SLA. Generally, the network needs to meet the performance of audio and video + mobile medical devices' data transmission bandwidth, which is about 400 Mbps, and the delay is less than 50 ms.

At present, there are a large number of ambulances. If intelligent and cloud-based mobile ambulances can be realized and first aid can be achieved, many delayed cases will be avoided and the success rate of first aid will be improved.

Medical institutions have relatively complex network requirements. Apart from the and inter-hospital differences in the three scenarios of in-hospital, inter-hospital, and out-of-hospital, the network demand can also be divided into long-term occupation and temporary emergency demand. But no matter which scenario, high security and high reliability of the network must be ensured. The needs of medical institutions for network mainly include:

(1) Data security isolation. The network security isolation between medical private network and public ones can meet the demands of medical industry users for the performance and security of 5G medical private networks.
(2) Dynamic adaptation of network bandwidth. According to the different bandwidth requirements of regional traffic at different times, dynamic bandwidth allocation is set up, the idle bandwidth of idle departments is extracted and allocated to departments with large traffic demand, and resources are accurately allocated to maximize resource utilization efficiency.
(3) Dynamic allocation of network performance. It provides different security service levels, such as bandwidth, delay, jitter, packet loss rate, security, and reliability. The network management platform enables SLA to be manageable, controllable, visible, and committed.
(4) High deployability and replicability. It can be rapidly deployed and implemented in various medical scenarios.
(5) Reliable networks carry complex application scenarios. Three scenarios of pre-hospital emergency care, telemedicine, and hospital management are fully enabled to provide a safe, reliable, and flexible network guarantee.

5G medical private network is a 5G network for customers in the medical industry. That is, customers in the medical industry can enjoy different degrees of exclusive construction, exclusive sharing, exclusive transportation, and exclusive maintenance. Network slicing technology allocates end-to-end (wireless, transmission, and core) network resources for industry users and realizes network isolation. By sinking 5G network physical resources, customized network deployment is carried out according to industry applications to realize the unloading, shunting, and isolation of local service data. The capacity requirements of the 5G medical private network are as follows.

(1) Wide coverage. Differentiated deployment according to the actual situation of the park, indoor and outdoor, and macro and micro collaboration are considered to ensure seamless coverage.
(2) High speed. Support up to 1 Gbps user experience rate, and peak user rate peaks at tens of Gbps.
(3) High capacity. Independent/shared carrier frequencies for the large-scale network; Support ultra-dense networking, up to 10 to 100 times the capacity of LTE; The flow density can reach Tbps/km^2 at least;
(4) Higher flexibility. Flexible allocation of slot ratios supports asymmetric and differentiated configurations of uplink and downlink rates.
(5) Massive connection. eMBB supports a connection density of 200 ~ 2500/km^2; mMTC supports high-density and mass device connections (at least 106 connections/km^2).
(6) Seamless switching. Fine-tuned optimization for continuous switching and smooth transitions in various scenarios (indoor/outdoor, private network, WAN, etc.).

7.5.3 Scenario-Based Solution

The 5G smart medical solution involves in-hospital, inter-hospital and out-of-hospital scenarios. Three major medical scenarios require 5G network, such as data staying within the hospital, reliable connection, fiber-like performance, and accurate positioning. Given the above requirements, a comprehensive network scheme of 5G smart medical is designed.

1. **Overall Networking Diagram**

Figure 7.56 shows the overall network structure of a 5G smart medical network.

Medical business data needs to be isolated from public Internet services for the in-hospital scenario. In this case, medical data is only transmitted within the hospital. In the inter-hospital scenario, medical services are isolated from the public network, and special channels must be provided. For the out-of-hospital scenario, medical services should carry temporary dedicated channels for mobile ambulances to ensure SLA.

In the overall networking of 5G smart medical network, Grade III A hospitals and lower-level hospitals are connected through exclusive channels through the core network's medical section and medical platform. Grassroots offices, institutions, and ambulances are interconnected through the core network medical section and medical platform through the optimal access channel.

Figure 7.57 is a schematic diagram of medical slices.

(1) Wireless (exclusive/superior): For the in-hospital scenario, consider hard isolation with RB resource reservation; for the inter-hospital and out-of-hospital scenarios, consider soft isolation with 5QI priority scheduling.

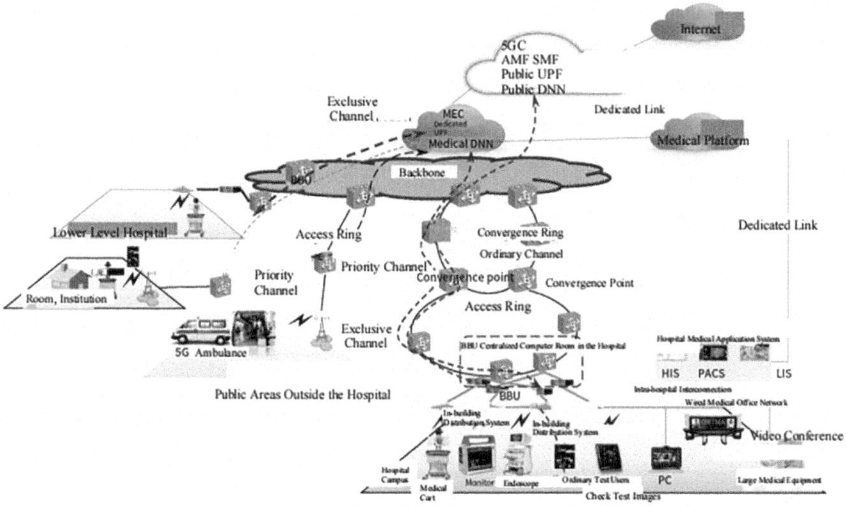

Fig. 7.56 Overall networking of 5G smart medical network

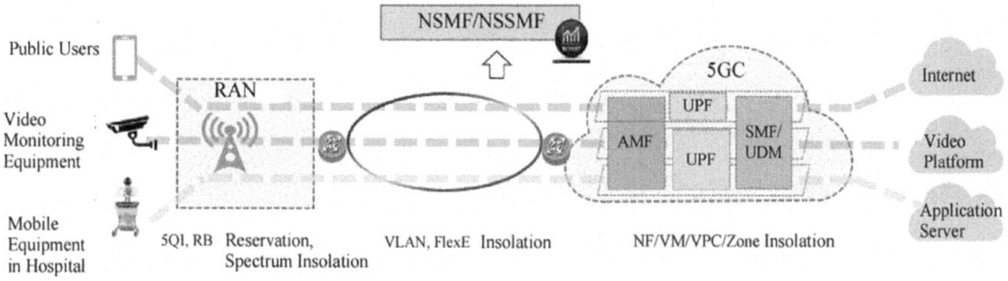

Fig. 7.57 Schematic diagram of medical slices

(2) Transmission (exclusive): Consider the physical isolation of toB and toC users based on FlexE; Superpose channelized sub-interfaces to realize small particle slicing.
(3) Core network (exclusive): Modules (SMF, UPF/MEC) strongly related to the conversation are considered as independent construction of hospitals; Modules unrelated to the session itself (AMF, SDM, PCF) are considered to reuse large network resources to reduce management difficulty and save investment costs so that users can get the supreme enjoyment.

2. **Access Network Scheme**

Suitable base stations for deployment are selected according to medical scenarios: macro stations are used for outdoor scenarios, and digital indoor distribution systems are used for indoor applications to ensure that the signal strength of receiving base stations in business application scenarios meets business requirements. The scale

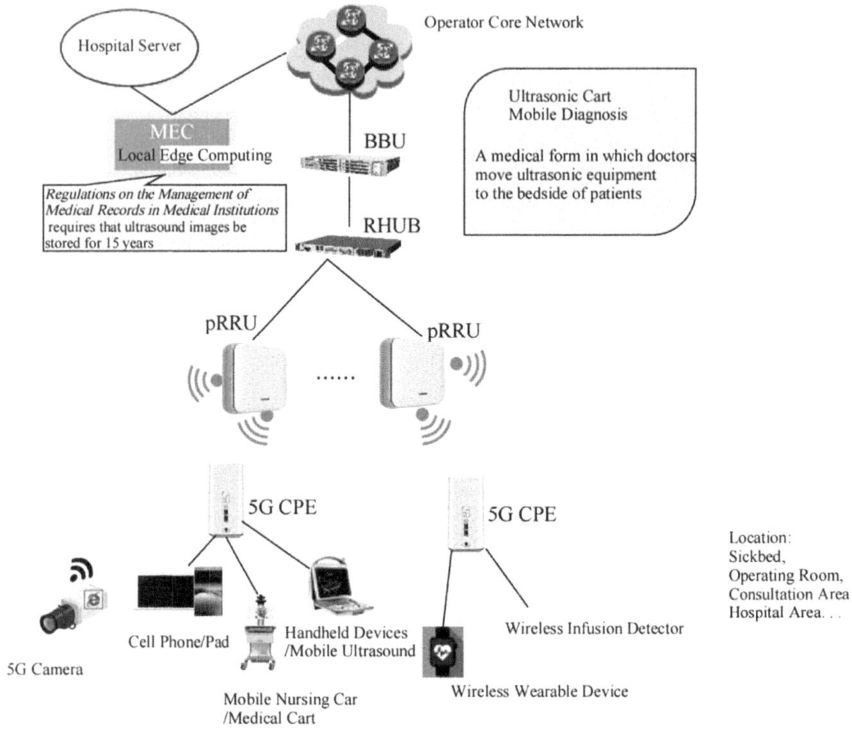

Fig. 7.58 Access network scheme: inside a hospital

of the base station should be planned according to the hospital's coverage area and bandwidth demand. The base station in the park connects with the operator's core network through the municipal transmission network, and the core network interacts with the base station in the park to manage and control 5G equipment in the park. Figures 7.58 and 7.59 show internal and external access networks.

3. **Transmission Scheme**

Figure 7.60 shows the load-bearing requirements of the optimal and exclusive channels in all medical scenarios.

(1) Exclusive channel: A hardware-isolated hard pipeline that provides full-time dedicated broadband to carry payable services with high latency requirements.
(2) Optimal channel: A hard pipe providing temporary elastic bandwidth, carrying payable services with temporary speed increase needs.
(3) Ordinary channel: Providing the best possible ordinary pipeline, carrying the ordinary business without payment.

To guarantee the definite delay requirements similar to those in remote B-ultrasound, FlexE hard slicing technology is introduced to slice a physical network

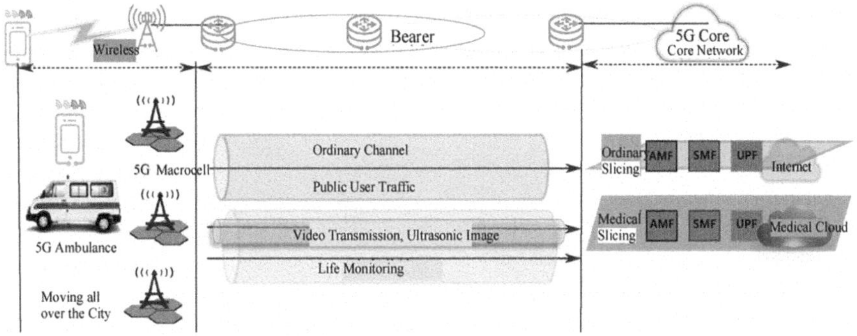

Fig. 7.59 Access network scheme: outside a hospital

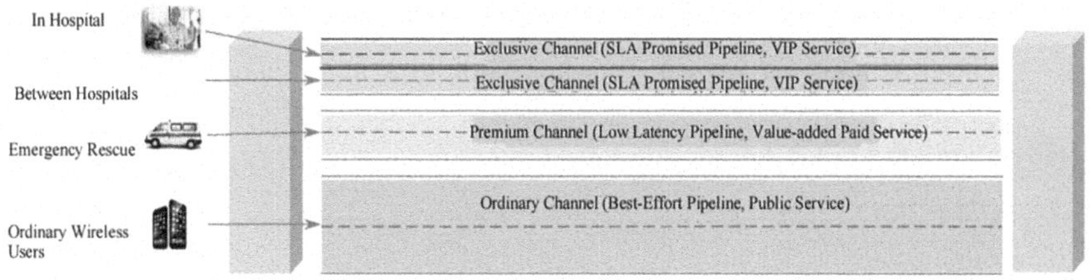

Fig. 7.60 Features and bearer requirements of each channel

into multiple logical slicing networks. Relevant medical services are carried out to ensure that hospital services are not affected by other traffic on the network and meet the delay and bandwidth requirements of relevant medical services. End-to-end FlexE-based network slices from the access layer to the aggregation core ensure the delay and bandwidth requirements in related medical services and improve the work efficiency of medical staff.

This section describes how to divide hard isolation slices into FlexE ports on the line side of the device.

(1) Divide the line-side FlexE ports on the access, aggregation, and core devices into three sub-ports, as shown in Fig. 7.61.
(2) The bandwidth of the prime slice and exclusive slice is 1 GE, and that of a common slice is 3 GE.
(3) The bandwidth of the optimal slice and exclusive slice on the core NPE interconnection link is 2 GE, respectively.
(4) Each slice port is configured with an independent IP address and added to the IGP instance.

VRF deployment and binding relationship with tunnels are shown in the following.

(1) Deploy L3VPN instance VRF2B for vertical industries and L3VPN instance VRF2C for common Internet users on the core network and in the hospital building, as shown in Fig. 7.62.

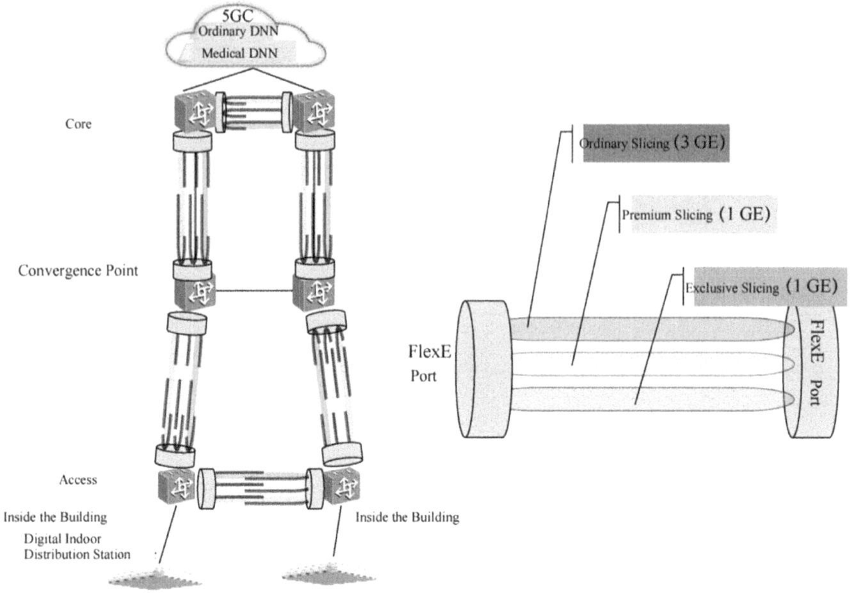

Fig. 7.61 Schematic Diagram of Hard Isolated Slices of a FlexE Port

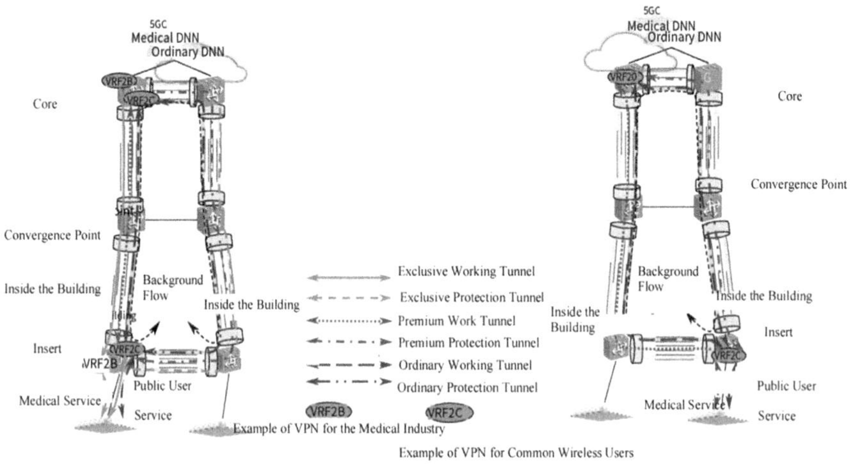

Fig. 7.62 VRF deployment diagram

(2) Deploy L3VPN instance VRF2C for common Internet users within the hospital.
(3) VRF2B is bound to an exclusive tunnel (working & protection), and VRF2C is bound to an optimal tunnel (working & protection) and a common tunnel (working & protection).

4. **Core Network and MEC**

User Plane Function (UPF) is an important part of the 3GPP 5G core network system architecture. It is mainly responsible for the routing and forwarding user plane packets in the 5G core network. UPF is pivotal in 5G edge computing and network slicing technology for low delay and large bandwidth. UPF is the connection anchor between the 5G network and Multi-Access Edge Computing (MEC). UPF must forward all core network data before flowing to external networks. MEC is the hallmark capability of the 5G service application.

Figure 7.63 shows a schematic diagram of the core network side.

The core network can be divided into the following two parts.

(1) Control plane and management plane: Responsible for accessing carrier 5GC, including AMF, SMF, UDM/PCF.
(2) User plane: Responsible for UPF at the edge of the hospital (new) to ensure that the hospital's data is kept locally.

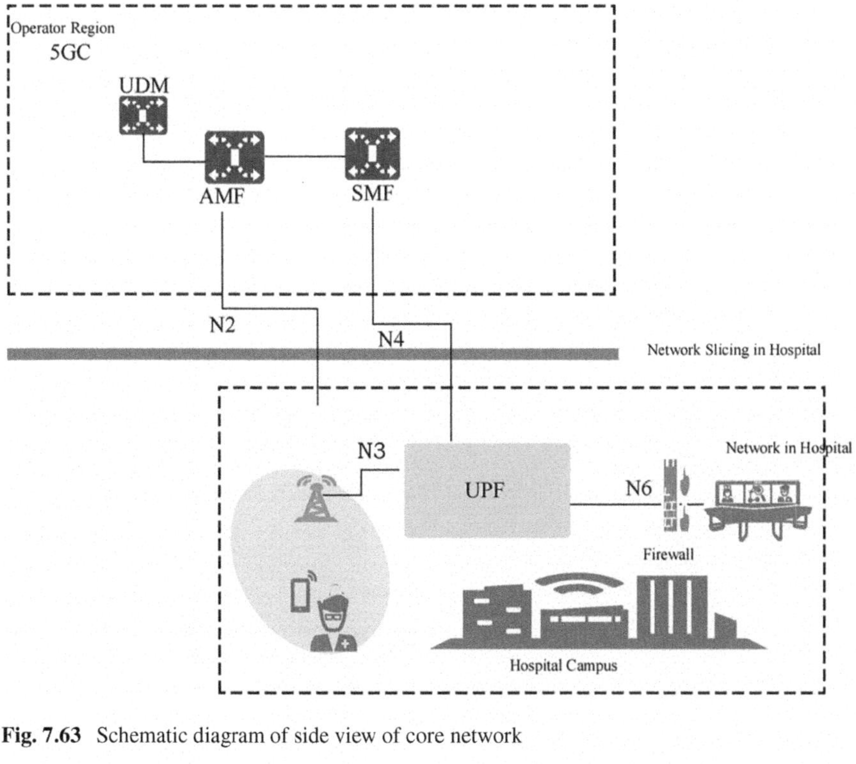

Fig. 7.63 Schematic diagram of side view of core network

In this scheme, it is necessary to plan slices of the hospital network, plan independent DNNS, select a UPF slicing network for hospital users to conduct service based on DNNS, as well as uninstall local traffic of the service. The following describes its service process.

(1) Registration: Users apply for 5G registration requests.
(2) Subscribed NSSAI (sNSSAI) is obtained from UDM for subscribed information of toB users by AMF, and sNSSAI (slice ID) is sent to the terminal to complete the 5G registration.
(3) Activation: The user carries a new DNN and sNSSAI (slice ID) to initiate activation. AMF interacts with NRF and obtains SMF information based on information such as sNSSAI.
(4) AMF selects SMFS in the edge UPF network slice.
(5) SMF selects UPF as UPF Anchor through DNN.

The above 5G+ MEC scheme has the following advantages.

1). Large bandwidth and low delay.

(1) The deployment of a 5G network improves the uplink and downlink bandwidths of the network, bringing convenience to nurses' ward rounds, mobile film reading, and case uploading.
(2) The MEC is deployed by integrating into the hospital area to reduce network transmission distance and delay.

2). Good Mobility

The newly built 5G network in the hospital area can cover outpatient buildings, inpatient buildings, elevators, and underground parking lots to ensure the continuity of network coverage, as shown in Fig. 7.64.

3). Low maintenance cost

(1) To build LAN in a hospital, special lines and network planning are required, and special personnel are required to maintain the network. Locating network faults is inefficient. It takes 2 to 3 h to identify and resolve a problem.

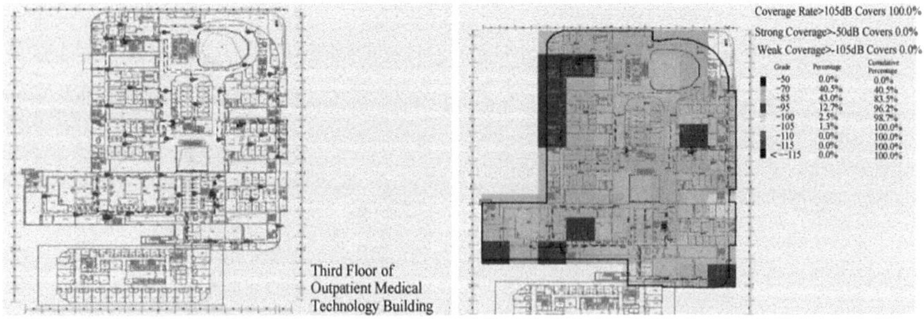

Fig. 7.64 Full coverage of 5G network

(2) the operator shall build 5G+ MEC, and the network maintenance shall be undertaken by the operator, while the hospital only needs to maintain the private cloud platform.

4). Data Isolation

MEC gateway is deployed in the hospital area, and based on the destination IP offloading technology, the internal access of the hospital can be directly closed in the hospital, and the data will not be uploaded to the public network to ensure data security.

5. **Network Security**

Figure 7.65 shows that the network security solution focuses on deploying firewalls between MECs, SPNS, and enterprise servers. MEC uses the cloud-based hardware platform E9000H. Firewalls can be configured in enclosures or boxes. The forwarding delay of firewalls in enclosures is lower than 80 μs, and that of boxes is lower than 15 μs. The delay does not affect services. The specific deployment scheme is as follows.

(1) Deploy a dual-system hot backup firewall between E9000H and SPNS to protect the traffic of N3, N4, and OAM, as shown in dotted box ①. The signaling and control traffic is limited. Box switches can be used.

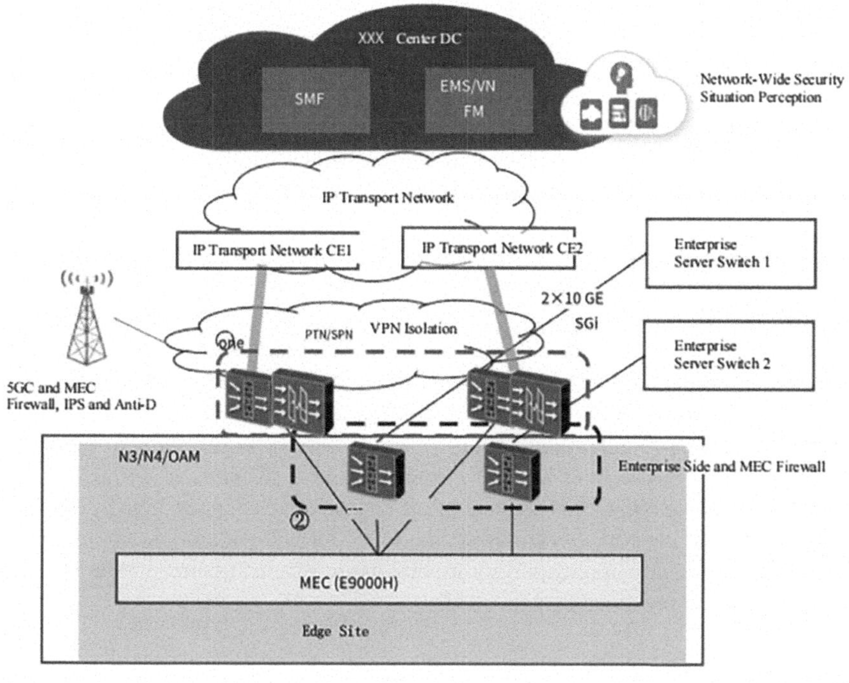

Fig. 7.65 Network security deployment plan

(2) IPS and anti-D protection core networks are deployed between E9000H and SPN devices.
(3) The Huawei HiSec platform is deployed in a DC to provide unified management of all MEC security devices, simplifying operation and maintenance.
(4) MEC security device sends security events and specific traffic to the Huawei HiSec platform for abnormal behavior analysis. It establishes an active defense security system to improve the defense capability of the entire solution.
(5) Deploy a dual-system hot backup firewall between the switches on E9000H and the enterprise server to protect N6 traffic on the user plane between MEC and the enterprise service server, as shown in dotted box ②. To ensure scalability, a framed firewall is recommended.

7.5.4 Application Cases

The following is a 5G+ MEC smart medical case of a city and a region medical alliance. Since 2019, a regional medical consortium, company A, company B, and other units have jointly carried out 5G+ smart medical strategic cooperation in Shenzhen. By deploying a dedicated medical network for the healthcare consortium, efficient and secure information exchange has been achieved among all regional medical institutions, including 7 hospitals and and 83 community rehabilitation centers. It takes the lead in completing applications, such as 5G remote first aid, 5G remote consultation, 5G mobile diagnosis and treatment, 5G community health first aid guidance, and 5G smart wards. This has resulted in the fast innovation and upgrading of remoteness, mobility, and informationization for medical consortium services in this area. During the pandemic, the medical private network based on 5G+ MEC has effectively implemented hierarchical diagnosis and treatment through bedside consultation, remote consultation, and community health emergency services to help precise epidemic prevention and control.

The medical industry mainly faces three pain points, as shown in Table 7.11.

Given the above three pain points, the 5G smart medical project in this region takes 5G MEC and slice technology as the base to deploy the regional 5G health private network, covering three scenarios in hospitals, inter-hospitals, and out-of-hospitals, which can meet the requirements, such as keeping medical business data within the hospital, high bandwidth, low delay, flexible access and so on.

The 5G smart medical project in this region focuses on establishing a 5G health private network, researching and developing 5G medical industry terminals, and innovating and applying 5G business in all scenarios. Under the high-speed and stable 5G private network environment, multi-platform collaboration is realized to create 5G full scenario application for mobile diagnosis and treatment, hierarchical diagnosis and treatment, and smart wards.

Table 7.11 Challenges faced by medical informatization

Higher communications security requirements for new services	Risk of data leakage	Difficulty in sharing frozen data assets
Meidcal video and image type data puts new requirements on network bandwidth, transmission quality, transmission speed rate and reliability. At the same time, 3GPP/ITU and other international standard and industry organizations have launched research on smart medical scenarios and applications, and 20+ new services will be stationed in the new medical ecosystem in the future, and new applications derived from technologies have put forward higher requirements for communication networks	Medical information systems collect a large amount of patients' health information, such as electronic medical records, medical images, etc., and there is a high risk of leakage and tampering of patients' personal privacy and medical data	Due to the lack of uniformity in the standards of medical information systems of various medical institutions, there is a lack of clinical information sharing and exchange between departments within hospitals and across hospitals, a lack of data integration in hospitals, a lack of circulation and difficulty in sharing health data between hospitals and within and outside hospitals, and a low utilization rate of medical data resources

1. **Network: regional-level 5G health private network covering medical services and public health**

1) Regional-level 5G health private network

The regional-level 5G health private network covers seven hospitals and 83 community rehabilitation centers of the medical consortium in the region and innovatively designs regional-level 5G health private network and MEC co-construction and sharing.

2) Dedicated public network.

In the medical consortium project, based on the "5G Dedicated Public Network" architecture, 5G+ MEC + slice technology is adopted to build a virtual private network.

(1) The "private network" mode considers the communication needs of medical industry applications and hospital public users.
(2) It has the advantages of low cost, high performance, wide coverage, safety, and reliability.

3) MEC

MEC is a key technology that is enabling 5G health private network solutions. Integrating MEC can keep medical data in the hospital and guarantee real-time interaction of low-delay services.

4) Independent SA networking & end-to-end independent slicing.

By December 2020, Company A has built 15,000 5G base stations to achieve full coverage of 5G SA independent network, creating the basic conditions for applying end-to-end network slicing technology. 5G SA end-to-end slice ensures stable bandwidth and delay of medical services during peak communication periods and provides a stable network environment for medical services.

2. **Terminal: A 5G communication module is integrated into the medical cart to solve the problem of lack of 5G medical terminals**

Medical carts (mobile terminals) are implanted into 5G modules. The 5G medical terminal is a key link in the implementation of 5G applications, especially mobile medical terminals 5G communication module needs to be highly integrated with terminal equipment and integrated design. Industry terminals integrated with 5G modules usually need a long development, verification, integration, and certification process, and currently, 5G modules have problems of large size, large power consumption and low industry suitability. In the face of the above challenges, Hospital C's 5G project team has carried out a series of innovative designs and explorations. It is the first in China to integrate 5G communication modules on the motherboard of the chassis of the medical cart (5G mobile doctor cart, 5G mobile nurse cart, 5G bedside consultation cart, 5G remote video ward round cart, 5G intelligent rescue cart), which solves the problem of the lack of 5G medical terminals with strong replicability. It has an industry demonstration effect and milestone significance. In the first phase, 5G medical cart modules were implanted, and the first batch of 36 carts were transformed. Figure 7.66 shows 5G smart medical devices and terminals. The average download rate of 5G is 600 Mbps, which is smooth inside and outside the room. The loading of more than 500 CT image sequences can be completed within 10 ~ 20 s, and the image sequences can be viewed smoothly.

3. **Business application: innovative application in all scenarios within, inside and outside of the hospital**

By the wide coverage of 5G and the seamless connection of the 5G health private network to the hospital Intranet, the first phase of the project of Hospital C has realized the collaboration of a 5G home remote care platform, intelligent regional emergency platform, remote consultation platform, intelligent ward interaction platform, and other platforms.

1) Mobile Medical Care

Through 5G network, 5G PAD can be used for daily routine ward rounds, and 5G mobile doctor cars and 5G mobile nurse car can quickly run various information systems on mobile medical workstations, making mobile offices more rapid and convenient for medical staff.

2) Intelligent Rescue Car

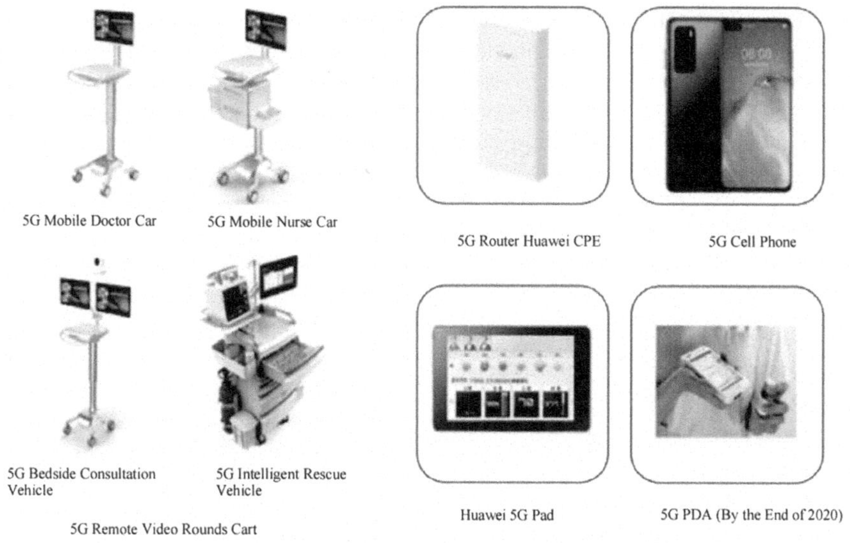

Fig. 7.66 5G Smart Medical Devices and Terminals

When a patient needs emergency rescue in a community rehabilitation center, the 5G intelligent rescue vehicle system of regional interconnection supports one-click intelligent rescue. Suppose the on-site rescue doctor needs expert guidance from a superior medical institution. In that case, he can directly initiate remote rescue guidance on the rescue interface to realize the second-level linkage rescue between hospital and social health and give full play to the nearby first-aid function of the community rehabilitation center.

3) Pre-hospital First Aid

Through the 5G private health network, the network outside the hospital can be connected to realize the "admission upon boarding" service. When patients board the ambulance, the registration and medical records processes begin. The vital signs, electrocardiogram, high-definition video, vehicle location and other information of the patients in the ambulance can be transmitted to the emergency command center of the hospital in real-time. This achieves the principle of "patient's information arrives before the patient does." Emergency doctors can promptly understand the patient's condition, prepare for in-hospital emergency care in advance, and integrate pre-hospital data with the in-hospital pre-check triage system. This preparation includes setting up green emergency channels, reserving beds, and more, aiming to reduce in-hospital examination and handover time and ensuring that patients receive timely treatment.

4) Remote Consultation

Experts can carry out mobile consultation, ICU/special patient bedside consultation and inter-hospital remote consultation anytime and anywhere through 5G mobile phones, 5G PAD and consultation carts, breaking time and place restrictions, taking care of difficult patients, and improving diagnosis accuracy and guidance efficiency. On August 12, 2021, hospital D was connected to conduct bedside consultation in ICU successfully. The audio and video communication at both ends of the bedside consultation was clear and smooth, without delay, and the data, such as medical records, medical orders, images, and tests, were displayed synchronously on the same screen with zero waiting.

5) Smart Ward

5G smart ward provides an integrated solution for auxiliary diagnosis and treatment, intelligent nursing, and ward management based on the connection of 5G health's private network to the hospital's intranet. 5G smart ward realizes real-time collection, backhaul, and monitoring of patient signs and indicators, infusion monitoring, and other data, helping the hospital's lean management and efficiency improvement.

By July 2021, the 5G smart medical project of the medical consortium in the region has been running for eight months in Hospital C. In the future, continuous promotion and exploration will be carried out for applications such as 5G full access to medical terminals, public network dedicated slice management system, and accurate positioning for the Internet of Things.

7.5.5 Summary (of 7.5)

In the process of realizing modernization and informatization of hospitals, the following three levels of problems of the 5G smart medical network should be considered: the first level is the informatization of medical business, realizing the network, paperless and film-free office; the second level is the management of information resources to realize the integration of information and application and give full play to the advantages of informatization; the third level, starting from the service, is the one that activates the demand of medical informatization, activates the time–space barrier, fully circulates and shares information, continues to innovate and meets the continuous development of medical services.

The 5G smart medical network can be divided into three business scenarios: in-hospital, inter-hospital, and out-of-hospital.

In-hospital data collection, monitoring, and analysis

The interconnection of image/test equipment can support intelligent medical treatment, doctors' ward rounds, patient monitoring, newborn visits and other services, and all kinds of information and data can be shared through the medical cloud.

Inter-hospital video-aided diagnosis and control

Inter-regional medical cooperation and integration of quality medical resources can be realized through mobile Internet. Remote ultrasound, endoscopy and video consultation can be conducted between branch hospitals and primary hospitals.

AI+ mobile diagnosis out-of-hospital

During mobile first aid and critical hospital transfer, audio and video equipment, multi-parameter monitor and electrocardiograph data can be transmitted to the first aid center using a 5G network to realize first aid pre-processing.

The smart medical network will be built in three categories of 5G medical scenarios: in-hospital, inter-hospital, and out-of-hospital. In the scenario-based solution and the 5G overall medical network design scheme, in-hospital, out-of-hospital and inter-hospital indirect network access shall be constructed. For data transmission, exclusive, optimal, and common channels are designed according to the level of requirements, hard isolation slices are divided into FLexE ports on the device line side, and VRF technology is introduced. Plan hospital network slices and independent DNNS at the core network side. Combine 5G and MEC and deploy network security solutions on the above network.

Based on the above plan, the 5G+ MEC smart medical consortium in a city and a region has realized three major innovations in network, terminal, and industry from technology to business, including the establishment of a 5G health private network, the research and development of a 5G medical industry terminals, and the full range of applications of 5G smart medical in hospitals, inter-hospitals, and out-of-hospitals.

Exercise 7

I. True or False

7.1. Transmission line inspection UAV that adopts 5G communication can extend the inspection range to thousands of meters away. ()

7.2. Smart grid will use both wireless and wired networks for distribution. ()

7.3. 5G cannot replace existing fiber optic infrastructure in distribution automation. ()

7.4. After implementing intelligent distributed power distribution automation, the troubleshooting level reaches the minute level. ()

7.5. The business model used for rail transit is mainly uplink data. For training TCMS data, it is suggested that the uplink bandwidth of 10 Mbps be considered. ()

7.6. Business data of production within rail transit does not need to be strictly isolated. ()

7.7. In non-production services of rail transit with low bandwidth demand, many non-critical links can be carried in 4G networks in the form of Internet of Things cards. ()

7.8. In the future, 5G public network can be used to realize intelligent functions such as remote maintenance and robot inspection of subway vehicleto improve the efficiency of maintenance operations effectively. ()

II. Choice Questions

7.9 (Single choice) Which of the following technologies enables a carrier to split multiple virtual end-to-end networks in hardware infrastructure so that each network slice is logically isolated in terms of devices, access networks, transmission networks, and core networks, adapts to various types of services, and meets different requirements of users? ().

A. Network slicing technology
B. Edge computing technology
C. SB technology
D. NFVSDN Virtualization technology

7.10 (Multiple choice) What are the three slices in 5G? ().

A. eMBB
B. URLLC
C. mMTC
D. MEC

7.11 (Single choice) In 5G core network slicing technology, which network element is mainly used to realize the selection of slicing? ().

A. NSSF
B. MANO
C. NEF
D. NRF

7.12 (Multiple choice) What scenarios would 5G bring to policing with breakthrough changes?

A. video deployment
B. on-site law enforcement
C. mobile patrol
D. major event security and emergency
E. mobile watchhouse

7.13 (Multiple choice) What are the three levels of network products for 5G policing? ().

A. Dedicated Channel (5G VPN)
B. Priority Protection (5QI)
C. Intelligent Sensing Network (AI)
D. Resource Reservation (RB or cut frequency)

7.14 (Multiple choice) Which of the following are the needs and challenges currently faced by mobile policing scenarios? ().

A. Insufficient police resources and high work intensity
B. The video surveillance collection scenario is incomplete, and the accuracy is low
C. On-site law enforcement cannot backhaul scene images in real-time, and remote analysis, command, and dispatch cannot be realized

D. The return channel of UAV cannot be patrolled in a large range, and the unauthorized frequency band is vulnerable to interference

7.15 (Single choice) Which system is 5G mainly used? ().

A. TDD
B. FDD
C. UMTS
D. GSM

7.16 (Multiple choice) Which of the following are not part of the education industry's current pain points? ().

A. The total amount of educational resources is insufficient, and the distribution is uneven
B. High teaching difficulty and low educational efficiency
C. Information asymmetry between home and school; campus security risks
D. High cost of education

7.17 (Multiple choice) Which of the following are 5G+ Smart education application scenarios? ().

A. 5G+ Smart Classroom
B. 5G+ AR/VR Education
C. 5G+ Interactive Teaching
D. 5G+ Holographic Classroom
E. 5G+ Teaching Quality Analysis
F. 5G+ Student Health Assessment
G. 5G+ Remote Teaching
H. 5G+ Safe Campus

III. Short Answer Questions

7.18 Why is the smart grid an inevitable trend of power grid development?
7.19 What is 5G Network Slicing Technology? What is the specific application of 5G network slice in a smart grid?
7.20 Why should 5G and other new-generation information technologies be used in urban rail transit scenarios?
7.21 What are the main application scenarios of the 5G network in rail transit? How exactly is it applied?
7.22 How many construction modes exist in the current 5G public–private public network? Please give a brief introduction to each mode.
7.23 Please describe the out-of-hospital, inter-hospital, and in-hospital access network solutions in the 5G smart medical solutions.
7.24 What are the advantages of MEC technology?
7.25 What technologies are used in the 5G program of the medical consortium? What problems have been solved?

Chapter 8
Summary and Outlook

This chapter briefly reviews the evolution of mobile communication networks over the generations and provides an outlook on the evolutionary direction and technical route of 5G to 5.5G. Finally, it introduces the current status of the 5G industry and the development and direction of all branches in the industry.

Learning Objectives

- To review the evolution of mobile communication networks over the generations.
- To look to the evolution direction of 5G to 5.5G.
- To understand the industrial situation and development direction of 5G.

8.1 Iterative Evolution of Communication Technologies

The development of mobile communication technology is a small step in 10 years and a giant leap in 20 years. 1G, 3G, and 5G are relatively big technological developments, while 2G and 4G are technical improvements and optimizations of 1G and 3G.

From a technical perspective, with continuous efforts, China has gone through a process of "1G blank, 2G following, 3G breakthrough, 4G running", and then 5G has taken the lead.

8.1.1 5G Full Connectivity Era

In the 5G era, China has shown leadership in 5G technology, standards, industry, applications, etc. Moreover, communications will begin to penetrate deeper into the industry in the 5G era, to achieve "5G changes society".

L. Song et al., *5GtoB From Theory to Practice*,
https://doi.org/10.1007/978-981-97-8453-0_8

5G is the network infrastructure that realizes the interconnection of people, machines, and objects. From 1G to 2G, the transition from analog communication to digital communication was realized, and mobile communication entered millions of households; from 2G to 3G and 4G, the transformation from voice service to data service has been achieved, with the transmission rate increased hundreds of times, promoting the popularization and prosperity of mobile Internet applications. As a new type of mobile communication network, 5G can be used in many aspects of work and life with its slicing function and high reliability, making a "qualitative" leap in wireless communication. The strong connectivity of 5G enables various industries to carry out more technological innovation, product innovation and business model innovation based on big data and artificial intelligence, promoting high-quality development of more vertical industries. 5G technology will penetrate all sectors of the economy and become a key new infrastructure to support the digitization, networking and intelligent transformation of the economy and society.

5G technology features mainly include the following three aspects: Enhanced Mobile Broadband (eMBB), Ultra-high Reliability and Low Latency Communication (URLLC), and Massive Machine Type Communication (mMTC), as shown in Fig. 8.1. Enhanced Mobile Broadband is mainly oriented to the explosive growth of mobile internet traffic and provides a better application experience for mobile internet users. Ultra-high Reliability and Low Latency Communications mainly meet the needs of industrial control, telemedicine, autonomous driving, and other vertical industry applications with high requirements for latency and reliability. Massive machine-type communication is mainly oriented to the needs of applications targeting sensing and data collection, such as smart cities, intelligent IoT, environmental monitoring etc. 5G+ can be applied to a wide range of industries, which will improve industry efficiency and reduce operating costs. 5G will drive industry digitization and boost global GDP growth.

8.1.2 From 5G Triangle to 5.5G Hexagon

The transformation from 5G to 6G cannot be achieved overnight but will require an evolution to 5.5G in the next 3 to 5 years. In April 2021, 3GPP defined the evolution from 5G to 5G-Advanced.

To more fully meet the differentiated needs of various industries, the industry should consider expanding the 5G scenario "triangle" into a 5.5G scenario "hexagon" in the next 10 years, as shown in Fig. 8.2, by adding Uplink Centric Broadband Communication (UCBC), Real-Time Broadband Communication (RTBC), and Harmonized Communication and Sensing (HCS), thus supporting the transformation from Internet of Things to intelligent net of things.

To achieve the 5.5G industry vision, 5G must achieve more than ten times capability enhancements in uplink and downlink experience rates, high-precision positioning, and connection density. New capabilities of 5G, such as sensing, need to be added to meet the needs of more diverse and complex scenarios.

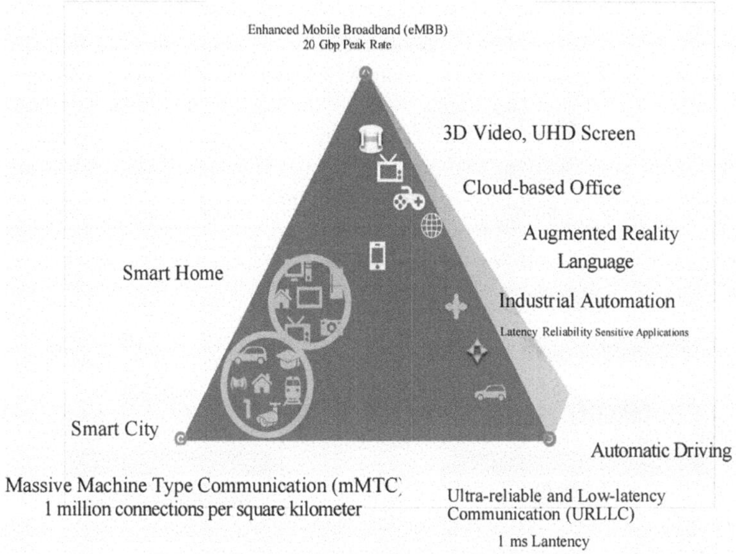

Fig. 8.1 5G high-rate, low-latency, and wide-connection scenarios

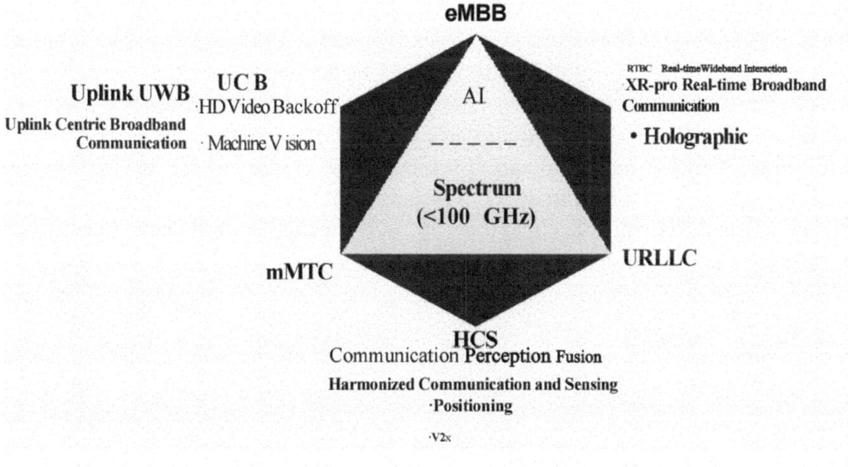

Fig. 8.2 Hexagon in a 5.5G scenario

8.2 Sustainable Prosperity of Industrial Ecology

"The 14th Five-Year Plan" is a critical period for the large-scale application of 5G in China. With the principle of "complementary advantages, win–win cooperation and common development", the company and its partners will jointly build a 5G industry service platform. Through industrial policy and environment research, training and empowering, testing and certification, project co-creation, vertical industry solution integration, and verification can be reached. At the same time, efforts will be exerted to explore 5G scenario-based innovative application solutions, promote the incubation of related technologies, facilitate the development and promotion of 5G standards in various industries, and promote the development of the industrial chain and related applications. Ultimately, the achievement is expected in mutual benefits and strategic win–win operations with partners, and solid support is anticipated for digital transformation, intelligent upgrading, and convergence innovation in various fields of economy and society.

8.2.1 Large-Scale Application of 5G Entering a Critical Period

5G is not only a representative and leading technology for the new round of technological revolution and industrial transformation but also a key infrastructure for empowering the digital transformation of economy and society. With its extremely strong spillover effects, 5G can promote high-quality economic and social development by stabilizing investment, promoting consumption, increasing employment, and adjusting structure. In the past two years, since the official commercialization of 5G in China, under the decision and deployment of CPC and the State Council, 5G has achieved leading global achievements in network development, industrial capacity, and application innovation under the joint efforts of the Ministry of Industry and Information Technology and all parties in the industry. In terms of application innovation, 5GtoB commercialization set sail,, and a large number of commercial practices were implemented in 2020. 5G+ industrial Internet, 5G+ smart education, 5G+ smart medical care, 5G+ smart tourism, and other applications have been landed one after another, among which, the application mode of 5G slicing+ virtual private network+ edge computing is the most striking one. The deep integration of 5G and the construction of new smart cities has accelerated the emergence of new business formats and models. In more than ten fields, such as factories, mines, ports, medical care, power grids, transportation, security, education, cultural tourism, and smart cities, 5G application scenarios have been initially formed that are expected to be commercially available on a large scale.

The breadth, depth, and technical innovation of China's 5G application are increasing, but due to immature application standards, business models, and industrial ecology, pilot demonstrations by head enterprises are still the main focus at this stage, and industry-wide scale applications have yet to be realized. The domestic 5G has accelerated from network construction into a new phase of application innovation. The key is to focus on developing key industry applications, leveraging the synergy of various parties, and forming a favorable situation to accelerate the development of digital transformation and upgrading in various industries.

5G Application Action Plan (2021–2023) (hereinafter referred to as *Action Plan*), jointly issued by the Ministry of Industry and Information Technology and other ten departments recently, clarifies the objectives of China's 5G development in three years, creating a new situation of China's 5G application innovation and development.

Action Plan defines the development path of 5G applications in three years and proposes eight special actions with 32 specific tasks, which are of great significance in promoting 5G applications, cultivating new momentum for economic and social development, and shaping new advantages for high-quality development. The release of *Action Plan* will accelerate the implementation of special applications in key areas, promote the fast development of key areas with solid foundation, clear patterns, and broad prospects, and create a good policy environment for further expanding 5G industry applications and promoting 5G empowerment in various industries.

8.2.2 5G Industry Service Platform: Ensure Connection and Transformation Between Policies and Market

China's 5G convergence application has achieved a breakthrough from nothing to something, from quantity to quality. However, it faces many challenges regarding supply, demand, talent, certification, standards, and ecology. 5G integration with the industry is still difficult in driving the industry to achieve digital transformation, but it will be an innovative process of overcoming difficulties. In the future, there is an urgent need to further play the leading role of *Action Plan*. Efforts should be put into driving the development of applications with technical and market advantages, stimulating enterprise investment, solving the key problems in the development of 5G convergence applications, and ensuring the connection and transformation between policies and the market.

Up to now, industry partners have done a lot of exploration in various industries. Over 2000 projects have been commercially deployed, involving more than 20 vertical industries. Among them, industrie like steel, coal mining, ports, manufacturing, and power have entered the 1-to-N replication phase. Based on its accumulation in connectivity, computing, cloud, and industry digitization, the industry has launched a one-stop platform solution for 5GtoB "sales, operation, and service". The industry has proposed a "1+N" 5G target network for the key base—5GtoB infrastructure network, as shown in Fig. 8.3. It is based on a wide pipeline infrastructure network with universal coverage to create the ultimate user experience while matching the needs of various industries and achieving N-dimensional capability overlay on demand to serve the digital upgrade of the industry.

The digital transformation of various industries through 5GtoB scale commercialization is not only a technical issue but also an urgent need to establish unified industry standards, cultivate sufficient high-quality talents, and create end-to-end industrial ecology and business models, etc.

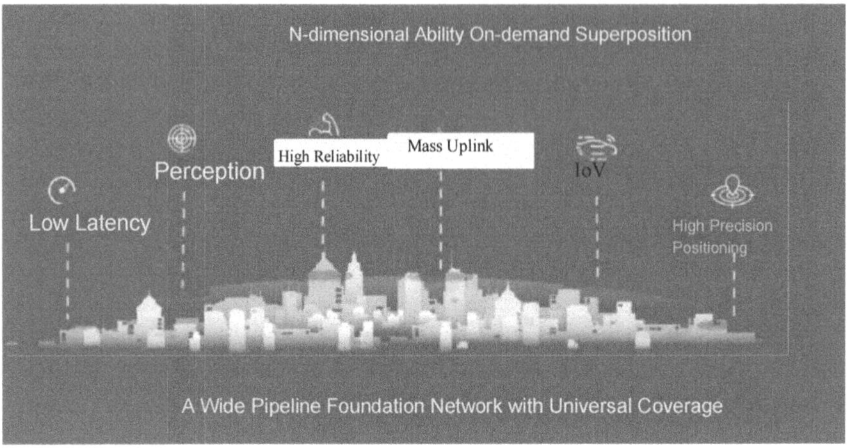

Fig. 8.3 5G target network: 1+N

The technical characteristics, demand scenarios, and professional thresholds of 5G applications in various fields vary greatly, and the strength of industries and the information and communication industry must be effectively brought into play. Thus, a new ecology of deep integration of IT (information technology), CT (communication technology), and OT (operation technology) is jointly promoted. The 5G industry service platform has been born to build a multi-win industrial ecology with industrial partners by giving full play to the platform's effectiveness and making concerted efforts in four major areas.

1. **Policy Standards**

In terms of policy, the deep integration of 5G with the social economy and various industries requires continuous optimization of the policy environment. It also requires authorities at all levels to jointly promote policy innovation. *Action Plan* proposes to strengthen departmental coordination and provincial linkage and to do a good job of organic convergence of standards, industry, applications, policies, and other aspects. The final objective is to work together with a joint effort across departments, fields, and industries to improve the policy system and promotion measures. At the same time, local governments are encouraged to actively implement policies about 5G application landing, ecological construction, industry cultivation, network construction, etc. They should be supported to create various leading areas of 5G+ application innovation by combining regional characteristics and industry advantages. In addition, to achieve 5GtoB scale commercialization, it is necessary to enable representatives and experts from different industries and fields to work together to define scenario requirements and explore solutions. In industries and scenarios with a high degree of standardization, it is easier to receive more achievements with less effort in promoting 5G technology under the guarantee of standards. It is expected that the research and release of industry IoT standards on 5G industry service platforms can strengthen the development and implementation of 5G application standards, open up the channel of 5G technology application in relevant industries, and lay the foundation for 5G product development and ecological application.

2. **Talent Cultivation**

Under the impetus of "New Infrastructure" and other policies, 5G is accelerating its integration with various industries. In combination with the development of industrial Internet, smart manufacturing, smart cities, and other construction, the market is calling for a large number of 5G talents. Therefore, it is necessary to strengthen the training and guidance of 5G integration applications for practitioners in various industries, as well as build a talent base to accelerate digitalization and intelligent transformation to provide the foundation for 5G integration applications. In this regard, the 5G industry service platform is working in two major directions: curriculum development and training certification. The development of emerging technologies is often accompanied by higher technical thresholds and the emergence of new application fields, which put forward higher requirements for the professional and technical skills and comprehensive capabilities of talents. Towards this end, it is necessary to strengthen the cooperation between schools and enterprises further and improve the training system for application-oriented talents. This mode can take advantage of universities' educational background and industry companies' leading technical advantages. A solid guarantee can be built for the output of 5G talents through the strong combination of industry-academia in-depth cooperation mode. Traditionally, textbooks have covered theoretical knowledge frameworks, but what is urgently needed are excellent application-oriented talents who have a solid theoretical foundation and can engage in engineering practice. So, it is crucial to establish a knowledge bridge from theory to engineering practice.

3. **Testing and Certification**

The end-to-end industrial ecology and business model are critical to the commercial success of 5GtoB. In the face of the overall rapid development of the 5G industry, terminal product testing and certification are of great practical importance to control the quality of listed products, such as performance, interconnection, and security. For example, the current 5GtoB project has been rapidly launched, but the insufficient terminal capacity of 5GtoB is affecting the development of the industry. For 5G industry application development, the terminal is an issue that cannot be ignored and should not become a bottleneck. The problem-oriented principle should be adhered to make a breakthrough in key technologies and products that restrict the scale application of 5G. It is also necessary to enhance the supply capacity of key industrial links such as modules, improve the performance-to-price ratio of industry terminal modules, and lay the foundation for developing 5G industry applications on a large scale. To address this key issue, a testing and certification platform will be established by relying on Open Labs, which industry partners jointly build. It also needs to unify test specifications, test cases, test processes, test methods and tools, carry out chip, module, and terminal IoT tests, and issue 5G certificates to certified terminal modules to improve end-side capabilities and end-network synergy. Thus, industrial ecology becomes prosperous.

4. **Project Co-creation**

The key to empowering various industries is to continuously improve the adaptability of 5G technology solutions to industry applications and expand the depth and breadth of industry applications. 5G application scenarios are complex with obvious personalized and diversified needs characteristics. Take the national economy industry as an example. According to the national standard *Industrial Classification for National Economic Activities* in China, there are mainly about 20 categories, including agriculture, forestry, animal husbandry and fishery, mining, manufacturing, electricity, heat, gas and water production and supply, transportation, storage, and postal services. Each major category can be subdivided into multiple industries, and these numerous industries will become potential directions for 5G applications.

As the status of industry informatization, production management operation mode and future development trend are different. It is necessary to focus on the specific needs of enterprises in terms of security, flexibility, and network construction cost based on the different capabilities of 5G in terms of latency, speed, reliability, and mobility. Different solutions and technology routes need to be designed to meet the differentiated needs of vertical industry applications according to local conditions. The practice has shown that in industries with mature industry integration and application development capabilities, 5G, as a key connectivity capability, can be easily and quickly embedded, which is more conducive to the rapid promotion of 5G technology. The 5G industry service platform has carried out project co-creation activities, including research and verification of 5G+ industry integration solutions and joint innovation of industrial incubation projects. This can open up the intermediate link between technology and application. It will also facilitate the exploration of business models for 5G applications, thus promoting the large-scale transformation of 5G technology and solutions.

8.2.3 More Refined Industrial Standards, More Promising Future

Promoting the large-scale development of 5G applications is an important step to promoting high-quality economic development, empower the manufacturing industry's digital transformation, and enhance the resilience of the industrial chain supply chain. This is important for focusing on the new development stage, implementing the new development concept, and constructing a new development pattern. The 5G+ application development is periodic, innovative, and complex. 5G industry service platform is a model of multi-party collaboration, tackling difficulties and fighting hard battles.

In 5G Rel-17, 3GPP will release a new technology standard, RedCap (Reduced Capability), for medium-speed connectivity. The application scenarios include wearable devices, industrial sensors, and video monitoring, and the number of connections is expected to reach tens of billions by 2030.

RedCap has two main advantages. First, it takes full advantage of the large bandwidth and high traffic volume of 5G to obtain very cost-effective traffic costs, which are 5–10 times lower than 4G traffic costs. Second, the cost of 5G terminal chips and modules will be significantly reduced, with the overall cost of RedCap expected to be only one-tenth of that of eMBB. It can be said that RedCap will be the best wireless connection for medium-speed connectivity, helping build a smart world of IoT.

8.3 Work Together for Commercial Success

Looking ahead, 5GtoB is on its way to large-scale commercial use. The whole industry should work together to realize the scale replication of 5G industry applications and eventually obtain commercial success.

5G has expanded the boundaries of previous generations of mobile communication technologies from human-to-human connection to IoT, i.e., the main battlefield is in the industry. With the three main features of eMMB, URLLC, and mMTC, the process of full penetration of mobile communication technology from toC to toB has started, with its influence far beyond the ICT industry. By 2035, 5G is expected to drive global output growth of 13.2 trillion US dollars in various industries, creating 22.3 million job opportunities.

5G private network is an important means for operators to open the 5GtoB market quickly. SLAs can ensure that the value of 5G connections is significantly increased and 5G traffic is quickly realized. 5G ecosystem is becoming increasingly mature. With the commercialization of various 5G CPE, 5G enterprise routers, 5G cameras, and other related products, 5G private networks can effectively use 5G networks for various scenarios, including SME Internet access, enterprise interconnection, live media, and video backhaul.

There are two sides to everything. While the application of 5G in the industry presents historic opportunities, it also places higher demands on the capabilities of 5G operators.

5G expansion requires stringent SLA guarantees and capabilities, such as edge computing, slicing, cloud, and AI, to meet scenario-based network and application requirements. Therefore, operators must build planning, construction, maintenance, optimization, and operation capabilities for toB services. For example, products and solutions are provided to meet industry requirements, such as dust prevention, explosion prevention, and high-temperature prevention, based on the ability to plan private networks in conjunction with the production environment. At the same time, constructions are needed to build business and ecological enabling platforms, provide standardized products and services, and build a healthy business model. Only with these capabilities can the development of 5GtoB be brought to a true commercial scale. All industries will be able to enjoy the dividends of 5G.

In the leap from the digital era to the smart era, the industry is also on the eve of a major explosion of 5G applications. By investing heavily in research and development annually in 5G, cloud, computing, and AI research and development, a fully connected, fully aware, and fully intelligent digital platform has been born.

At an altitude of 30 m in Zhoushan Port, Ningbo, Zhejiang, the gantry crane cab is empty while remote control operations are carried out in a comfortable operating room. Industry partners join hands to cut into the pain points of the port and achieve cost reduction and efficiency with a 5G network to realize remote control, automatic control, automatic driving, intelligent monitoring, and other technologies. Take the tire crane operation as an example, machinemachines assist 90% of the operations assist 90% of the operations, and the driver only needs to intervene in the two operations of "catching/unloading" containers. Previously, one person can only operate one unit of tire crane, but now he or she can operate 4–6 units.

In the Xinyuan Coal Mine of Shanxi Yangmei Group, in the operation pit 534 m below the surface, the inspection images transmitted to the ground workshop through the 5G private network are visible. The industry partners use SA networking mode and MEC technology and choose a split BOOK base station suitable for underground coverage. Meanwhile, for a large amount of upload data in the coal industry, it is the first one in the country to innovatively adopt a 3:1 time slot ratio, with an upload rate of over 800 Mbps and transmission delay of less than 20 ms, which can fully meet the demand of network for remote ultra-low delay control, ultra-high definition video simultaneous transmission, and industrial control scenarios.

From the beginning of scientific and systematic research on electrical engineering to the mid-nineteenth century, it took nearly a century for humans to enter the full-scale electrical era. The transformation to 5G intelligent era, supported by various technical theories and industry-wide practice, will not be as slow as before, but will not happen overnight.

In the electrical age, the "synergy of infrastructure", factories, and machines converts electrical energy into light, heat, kinetic energy, chemical energy, sound waves, electromagnetic waves, etc.; in the digital age, the "new infrastructure" of the digital era—the integration of 5G, cloud, computing, and AI—combines industry applications and gradually transforms and "refines" more application value from data. At the same time, a variety of "slow variables" will accumulate into qualitative change. Eventually, it will accelerate the improvement of productivity at different levels of enterprises, industries, and supply chains through digital spillover and become an engine to drive economic growth.

5G investment is driven not only by operators themselves but also by the real demand of verticalization in various industries.

The development of 5GtoB requires not only the efforts of operators and 5G equipment vendors but also the joint efforts of industry partners.

The dividends of 5G advanced technology will benefit enterprises, the ultimate billpayers. In addition, it is also necessary to consider the interests of the original industry eco-partners, such as industry equipment partners and industry integration partners who have been working in the industry for many years and have a strong technical and commercial influence on the practical implementation of emerging technologies. It is gratifying to note that cross-industry collaboration is being actively carried out. For example, in the cooperation project among China Mobile, Huawei, Southern Power Grid, and power partners, the proposal for power demand was adopted by 3GPP, completing the docking of power and communication at the standard level. Mining, steel, and other industries have also seen positive results, with several industry eco-alliances established and cooperation at the standard level already underway. Rel-16 not only enhances 5G basic network capabilities but more importantly, expands vertical industry-oriented capabilities, enabling the wide use of 5G from "useable" to "user-friendly" and eventually to "willing to use", which will further accelerate the development of 5GtoB.

5G is not only a disconnected upgrade of mobile communication technology but also a solution to realize IoT when it enters the industry. This represents a fundamental change in the social role of communication technology. The impact of 5G is cross-industrial and cross-disciplinary. People's daily lives, such as clothing, food, housing, and transportation, will change and become increasingly exciting due to 5G technology.

8.4 Summary

This chapter provides an understanding of the evolutionary process of mobile communication network development and an outlook on the evolutionary direction and technology route of 5G to 5.5G. Finally, the current situation of the 5G industry and the development of industrial parties are introduced, and suggestions are made for the follow-up industrial service platforms from the four major directions of policy standards, testing and certification, and project co-creation.

Exercise 8

8.1. What are the three main features of 5G technology?
8.2. What are the three aspects of 5.5G technical features that are expanded based on 5G?

The manufacturer's authorised representative in the EU is Springer Nature Customer Service Centre GmbH, Europaplatz 3, 69115 Heidelberg, Germany. If you have any concerns regarding our products, please contact ProductSafety@springernature.com

Printed and bound by CPI Group (UK) Ltd, Croydon, CR0 4YY
01/10/2026
02225962-0015